# Water Environment Pollution and Control

# Water Environment Pollution and Control

Editors

**Weiying Feng**
**Fang Yang**
**Jing Liu**

Basel • Beijing • Wuhan • Barcelona • Belgrade • Novi Sad • Cluj • Manchester

*Editors*

Weiying Feng
School of Space and
Environment
Beihang University
Beijing
China

Fang Yang
State Key Laboratory of
Environmental Criteria and
Risk Assessment
Chinese Research Academy
of Environmental Sciences
Beijing
China

Jing Liu
Environment Research
Institute
Shandong University
Qingdao
China

*Editorial Office*
MDPI
St. Alban-Anlage 66
4052 Basel, Switzerland

This is a reprint of articles from the Special Issue published online in the open access journal *Water* (ISSN 2073-4441) (available at: www.mdpi.com/journal/water/special_issues/water_environment_pollution_control).

For citation purposes, cite each article independently as indicated on the article page online and as indicated below:

Lastname, A.A.; Lastname, B.B. Article Title. *Journal Name* **Year**, *Volume Number*, Page Range.

**ISBN 978-3-7258-1332-2 (Hbk)**
**ISBN 978-3-7258-1331-5 (PDF)**
doi.org/10.3390/books978-3-7258-1331-5

© 2024 by the authors. Articles in this book are Open Access and distributed under the Creative Commons Attribution (CC BY) license. The book as a whole is distributed by MDPI under the terms and conditions of the Creative Commons Attribution-NonCommercial-NoDerivs (CC BY-NC-ND) license.

# Contents

**Preface** . . . . . . . . . . . . . . . . . . . . . . . . . . . . . . . . . . . . . . . . . . . . . . . . . . . . . . . . . . . . . . . . . . . . . . . vii

**Weiying Feng, Fang Yang and Jing Liu**
Water Environment Pollution and Control in the Dual-Carbon Background
Reprinted from: *Water* 2023, 15, 3082, doi:10.3390/w15173082 . . . . . . . . . . . . . . . . . . . . . . . . . . . . 1

**Zhiqiang Tian, Sheng Zhang, Junping Lu, Xiaohong Shi, Shengnan Zhao, Biao Sun, et al.**
Differences of Nitrogen Transformation Pathways and Their Functional Microorganisms in Water and Sediment of a Seasonally Frozen Lake, China
Reprinted from: *Water* 2023, 15, 2332, doi:10.3390/w15132332 . . . . . . . . . . . . . . . . . . . . . . . . . . . . 5

**Agata Witczak, Kamila Pokorska-Niewiada, Agnieszka Tomza-Marciniak, Grzegorz Witczak, Jacek Cybulski and Aleksandra Aftyka**
The Problem of Selenium for Human Health—Removal of Selenium from Water and Wastewater
Reprinted from: *Water* 2023, 15, 2230, doi:10.3390/w15122230 . . . . . . . . . . . . . . . . . . . . . . . . . . . 20

**Talal Alharbi and Abdelbaset S. El-Sorogy**
Health Risk Assessment of Nitrate and Fluoride in the Groundwater of Central Saudi Arabia
Reprinted from: *Water* 2023, 15, 2220, doi:10.3390/w15122220 . . . . . . . . . . . . . . . . . . . . . . . . . . . 34

**Hongbin Gao, Yanru Fan, Gang Wang, Lin Li, Rui Zhang, Songya Li, et al.**
The Sources of Sedimentary Organic Matter Traced by Carbon and Nitrogen Isotopes and Environmental Effects during the Past 60 Years in a Shallow Steppe Lake in Northern China
Reprinted from: *Water* 2023, 15, 2224, doi:10.3390/w15122224 . . . . . . . . . . . . . . . . . . . . . . . . . . . 46

**Cuicui Li and Wenliang Wu**
Analysis of the Driving Mechanism of Water Environment Evolution and Algal Bloom Warning Signals in Tai Lake
Reprinted from: *Water* 2023, 15, 1245, doi:10.3390/w15061245 . . . . . . . . . . . . . . . . . . . . . . . . . . . 62

**H. R. Robles-Jimarez, N. Jornet-Martínez and P. Campíns-Falcó**
New Green and Sustainable Tool for Assessing Nitrite and Nitrate Amounts in a Variety of Environmental Waters
Reprinted from: *Water* 2023, 15, 945, doi:10.3390/w15050945 . . . . . . . . . . . . . . . . . . . . . . . . . . . . 79

**Zakhar Slukovskii**
Geochemical Indicators for Paleolimnological Studies of the Anthropogenic Influence on the Environment of the Russian Federation: A Review
Reprinted from: *Water* 2023, 15, 420, doi:10.3390/w15030420 . . . . . . . . . . . . . . . . . . . . . . . . . . . . 90

**Jiaoxia Sun, Yao Zhou, Xueting Jiang and Jianxin Fan**
Different Adsorption Behaviors and Mechanisms of Anionic Azo Dyes on Polydopamine–Polyethyleneimine Modified Thermoplastic Polyurethane Nanofiber Membranes
Reprinted from: *Water* 2022, 14, 3865, doi:10.3390/w14233865 . . . . . . . . . . . . . . . . . . . . . . . . . . . 117

**Qi Zhang, Yafang Li, Qingfeng Miao, Guoxia Pei, Yanxia Nan, Shuyu Yu, et al.**
Distribution, Sources, and Risk of Polychlorinated Biphenyls in the Largest Irrigation Area in the Yellow River Basin
Reprinted from: *Water* 2022, 14, 3472, doi:10.3390/w14213472 . . . . . . . . . . . . . . . . . . . . . . . . . . . 135

**Ashraf Zohud and Lubna Alam**
A Review of Groundwater Contamination in West Bank, Palestine: Quality, Sources, Risks, and Management
Reprinted from: *Water* **2022**, *14*, 3417, doi:10.3390/w14213417 . . . . . . . . . . . . . . . . . . . . **151**

**Yimeng Zhang, Fang Yang, Haiqing Liao, Shugang Hu, Huibin Yu, Peng Yuan, et al.**
Variation in Spectral Characteristics of Dissolved Organic Matter and Its Relationship with Phytoplankton of Eutrophic Shallow Lakes in Spring and Summer
Reprinted from: *Water* **2022**, *14*, 2999, doi:10.3390/w14192999 . . . . . . . . . . . . . . . . . . . . **171**

**Yingming Guo, Ben Ma, Shengchen Yuan, Yuhong Zhang, Jing Yang, Ruifeng Zhang and Longlong Liu**
Simultaneous Removal of $COD_{Mn}$ and Ammonium from Water by Potassium Ferrate-Enhanced Iron-Manganese Co-Oxide Film
Reprinted from: *Water* **2022**, *14*, 2651, doi:10.3390/w14172651 . . . . . . . . . . . . . . . . . . . . **184**

# Preface

The water environment serves as the foundation for numerous habitats and ecosystems. The discharge and accumulation of pollutants can disrupt the balance of aquatic ecosystems, leading to species extinction, ecological disruptions, and direct or indirect impacts on human health. Against the backdrop of the global carbon peak and carbon neutrality, controlling water pollution has become increasingly important. Within the broad framework of "water environment pollution and control," this Special Issue aims to provide essential knowledge and establish a solid scientific foundation for the control and management of water pollution by studying the environmental behavior and bioavailability of various pollutants. It can serve as a scientific basis for developing effective pollution prevention and control strategies, and can provide guidance for establishing a clean, healthy, and sustainable environment. We believe the reprint of "Water Environment Pollution and Control", will offer readers comprehensive and in-depth knowledge, thereby promoting further research in the field of water environment pollution and control.

**Weiying Feng, Fang Yang, and Jing Liu**
*Editors*

*Editorial*

# Water Environment Pollution and Control in the Dual-Carbon Background

Weiying Feng [1], Fang Yang [2,*] and Jing Liu [3,*]

[1] School of Materials Science and Engineering, Beihang University, Beijing 100191, China; fengweiying@buaa.edu.cn
[2] State Key Laboratory of Environmental Criteria and Risk Assessment, Chinese Research Academy of Environmental Sciences, Beijing 100012, China
[3] Environment Research Institute, Shandong University, Qingdao 266237, China
* Correspondence: yang_fang@craes.org.cn (F.Y.); liu_jing@email.sdu.edu.cn (J.L.)

## 1. Introduction to the Special Issue

Water pollution and control are becoming increasingly important in the global context of carbon peaking and carbon neutrality. Water environment safety is important to keep humans healthy. In this wide "Water Environment Pollution and Control" framework, this Special Issue aimed to provide important knowledge and lay a sound scientific foundation for the control and management of water environment pollution by studying the environmental behaviors and bioavailabilities of various pollutants. The thirteen articles in this Special Issue focusing on water environmental pollution and control are mainly divided into four categories: (1) the composition characteristics and environmental behaviors of the pollutants in the surface water (i.e., the Chinese lakes Ulansuhai [1], Hulun Lake [2], Tai Lake [3], Shahu Lake [4], and Russian Ancient lakes [5]); (2) the assessment of groundwater and aquifer interaction [6] and the sources, risks, and management of groundwater pollution [7]; (3) the impact of polluted water on human health [8–10]; and (4) the removal of pollutants in water [11–13].

Since the call for papers was announced in 2022, and after a rigorous peer-review process, thirteen papers have been accepted for publication in the Special Issue [1–13], which include eleven research papers [1–4,6,8–13] and two reviews [5,7]. We offer brief highlights of the published papers below.

## 2. Overview of the Contribution of the Special Issue

The paper "Differences of Nitrogen Transformation Pathways and Their Functional Microorganisms in Water and Sediment of Seasonally Frozen Lake, China" [1] improves the understanding of the nitrogen cycle in seasonally frozen lakes. Shotgun metagenomic sequencing of subglacial water and sediment from Lake Ulansuhai was performed to identify and compare nitrogen metabolism pathways and microbes involved in these pathways. The study found that ammonia assimilation was the most prominent nitrogen transformation pathway, and bacteria and proteobacteria were the most abundant portion of microorganisms in nitrogen metabolism. Gene sequences devoted to nitrogen fixation, nitrification, denitrification, dissimilatory nitrate reduction to ammonium, and ammonia assimilation were significantly higher in sediment than in surface and subsurface water.

The paper "The Sources of Sedimentary Organic Matter Traced by Carbon and Nitrogen Isotopes and Environmental Effects during the Past 60 Years in a Shallow Steppe Lake in Northern China" [2] quantified the contribution of organic matter sources in the lake sediment via multiple mixing models based on the stoichiometric ratios and stable isotopic compositions. The results showed that the organic matter in the sediments from Hulun Lake mainly came from terrestrial organic matter: the proportion of terrestrial organic matter was more than 80%. The results of the SIAR mixing model further revealed that

Citation: Feng, W.; Yang, F.; Liu, J. Water Environment Pollution and Control in the Dual-Carbon Background. *Water* **2023**, *15*, 3082. https://doi.org/10.3390/w15173082

Received: 22 August 2023
Accepted: 24 August 2023
Published: 28 August 2023

**Copyright:** © 2023 by the authors. Licensee MDPI, Basel, Switzerland. This article is an open access article distributed under the terms and conditions of the Creative Commons Attribution (CC BY) license (https://creativecommons.org/licenses/by/4.0/).

the proportions of terrestrial C3 plant-derived organic matter, soil organic matter, and lake plankton-derived organic matter were 76.0%, 13.9%, and 10.1%, respectively.

The paper "Analysis of the Driving Mechanism of Water Environment Evolution and Algal Bloom Warning Signals in Tai Lake" [3] collected the long-term water quality indicators, ecological indexes, natural meteorological factors, and socio-economic indexes in Tai Lake and studied the environmental evolution of the lake ecosystem. The key time nodes and early warning signals of the steady-state transformation of Tai Lake were also identified, which could provide a theoretical basis for early indication of the transformation of lake ecosystems. Furthermore, the characteristics and driving mechanisms of the lake's ecosystem evolution were analyzed based on the physical and chemical indexes of its sediments and its long-term water quality indexes. These results have important theoretical and practical significance for pollution control and the management of eutrophic lakes.

The paper "Variation in Spectral Characteristics of Dissolved Organic Matter and Its Relationship with Phytoplankton of Eutrophic Shallow Lakes in Spring and Summer" [4] characterized the seasonal changes of dissolved organic matter as well as phytoplankton abundance and composition in Shahu Lake via three-dimensional fluorescence spectroscopy combined with parallel factor analysis. The relationship between the response of DOM and phytoplankton abundance was explored via Pearson correlation and redundancy analysis in the overlying water. Seasonal phytoplankton growth had an important influence on the composition of the DOM.

The paper "Geochemical Indicators for Paleolimnological Studies of the Anthropogenic Influence on the Environment of the Russian Federation: A Review" [5] reviewed the most significant studies of sequential accumulation of pollutants, including heavy metals in the lake sediments in Russia, where there are about 2 million lakes. It was found that sedimentation rates were significantly lower in pristine areas, especially in the Frigid zone, compared to urbanized areas and industrial territories. In addition, the excess concentrations of heavy metals in the sediments of lakes were directly affected by the source of pollution. Further prospects of developing paleolimnological studies in Russia were discussed in the context of the continuing anthropogenic impact on the environment.

The paper "New Green and Sustainable Tool for Assessing Nitrite and Nitrate Amounts in a Variety of Environmental Waters" [6] improved the selectivity and sensitivity of the quantitation of nitrite and nitrate in waters by liquid chromatography with a short analysis time of about 10 min and using low residues. Ion pair formation and ion exchange retention mechanisms were considered. The experimental scheme was optimized and a new research method was established.

The paper "A Review of Groundwater Contamination in West Bank, Palestine: Quality, Sources, Risks, and Management" [7] reviewed the four levels of domains used to evaluate the groundwater condition in the West Bank for the past 27 years, including (i) assessing the groundwater quality in the West Bank, (ii) identifying the sources of groundwater pollution, (iii) determining the degree of health risks associated with groundwater pollution, and (iv) determining the role of groundwater management in maintaining the quality and sustainability of these groundwater sources. A review matrix was developed based on these four core domains. The results showed that the contamination and shortages of drinking water in the West Bank were among the most important challenges facing the Palestinian National Authority (PA) and the population residing in all sectors.

The paper "Health Risk Assessment of Nitrate and Fluoride in the Groundwater of Central Saudi Arabia" [8] assessed the non-carcinogenic health risks posed by nitrate and fluoride to infants, children, and adults using the daily water intake (CDI), hazard quotient (HQ), and non-carcinogenic hazard index (HI). Groundwater samples were collected from 36 wells and boreholes in three central Saudi Arabian study areas for nitrate and fluoride analysis using ionic chromatography and fluoride selective electrodes, respectively. Fluoride in 30.55% of the samples exceeded the WHO recommendations for acceptable drinking water (1.5 mg/L). The average hazard index (HI) values for adults, children, and infants were 0.99, 2.59, and 2.77, respectively. Accordingly, water samples from Jubailah and Wadi Nisah may expose infants, children, and adults to non-cancer health concerns. Immediate attention and remedial measures must be implemented to protect residents from the adverse effects of $F^-$ in the study area.

The paper "The Problem of Selenium for Human Health—Removal of Selenium from Water and Wastewater" [9] analyzed the change in the content of selenium (Se) in drinking water, raw water, as well as treated and raw wastewater in an annual cycle in the city of Szczecin. Selenium content in raw water was the highest in the summer. The removal of Se from raw water and wastewater was difficult because Se is often present in many complex forms and can form various compounds with other elements. Treated wastewater could be a source of Se in the environment, and the discharge of treated wastewater can become a secondary source of Se in the surface water. Treating wastewater resulted in lowering the Se content in the wastewater by as much as 47%.

The paper "Distribution, Sources, and Risk of Polychlorinated Biphenyls in the Largest Irrigation Area in the Yellow River Basin" [10] studied samples in the Yellow River irrigation area in Inner Mongolia, China to determine the polychlorinated biphenyl (PCB) content and to investigate the contamination of PCBs in agricultural soils irrigated chronically with polluted water. The distribution and migration of PCBs under long-term irrigation were also studied with 100 farmland soil profile samples. Cluster analysis was used to identify possible sources of PCBs, and the USEPA Health Risk Evaluation Model was used to assess the health risks posed by PCBs to humans.

The paper "Different Adsorption Behaviors and Mechanisms of Anionic Azo Dyes on Polydopamine–Polyethyleneimine Modified Thermoplastic Polyurethane Nanofiber Membranes" [11] successfully developed a method for removal of anionic azo dyes using the polydopamine–polyethyleneimine (PEI)-modified TPU nanofiber membranes (PDA/PEI-TPU NFMs). After six iterations of adsorption–desorption, the adsorption performance of the PDA/PEI-TPU NFMs did not decrease significantly, which indicated that the PDA/PEI-TPU NFMs had a potential application for the removal of Cr molecules by adsorption from wastewater.

The paper "Treatment of Wastewater Effluent with Heavy Metal Pollution Using a Nano Ecological Recycled Concrete" [12] synthesized a new material (Nano ecological recycled concrete, Nano-ERC) for removing heavy metals from wastewater. The results showed that nano-ERC simultaneously reduced the treatment cost of the simulated wastewater effluents and the environmental burden of solid waste. The adsorption capacity of nano-ERC was presumed to be significantly enhanced by adding nano CuO. Nano-ERC can serve as a cost-effective approach for the further treatment of wastewater effluent and may be applied more widely in wastewater treatment to help relieve water stress.

The paper "Simultaneous Removal of $COD_{Mn}$ and Ammonium from Water by Potassium Ferrate-Enhanced Iron-Manganese Co-Oxide Film" [13] developed a stable and efficient method for removing water pollutants. The catalytic oxidation ability of iron–manganese co-oxide film (MeOx) was enhanced by dosage with potassium ferrate ($K_2FeO_4$) to achieve the simultaneous removal of $COD_{Mn}$ and $NH_4+$ from water in a pilot-scale experimental system. By adding $K_2FeO_4$ to enhance the activity of MeOx, the removal efficiencies of $COD_{Mn}$ and $NH_4^+$ were increased to 92% and 61%, respectively, and the pollutants were consistently and efficiently removed for more than 90 days. The mechanism of $K_2FeO_4$-enhanced MeOx for $COD_{Mn}$ removal was proposed by the analysis of the oxidation process.

## 3. Conclusions

The guest editors envision that the papers in this Special Issue will be of interest to researchers and practitioners and help identify further research directions. We also hope that the results and methods presented in these studies will shed light on the efficient removal of pollutants from water bodies, the pollution control of water ecosystems, the risk assessment of water quality to human health, the impact evaluation of climate change on the availability of surface and groundwater resources, and the interpretation of future policies for sustainable water environmental management.

**Author Contributions:** W.F.: writing—original draft preparation; F.Y. and J.L.: writing—review and editing. All authors have read and agreed to the published version of the manuscript.

**Funding:** This research received no external funding.

**Conflicts of Interest:** The authors declare no conflict of interest.

## References

1. Tian, Z.; Zhang, S.; Lu, J.; Shi, X.; Zhao, S.; Sun, B.; Wang, Y.; Li, G.; Cui, Z.; Pan, X.; et al. Differences of Nitrogen Transformation Pathways and Their Functional Microorganisms in Water and Sediment of a Seasonally Frozen Lake, China. *Water* **2023**, *15*, 2332. [CrossRef]
2. Gao, H.; Fan, Y.; Wang, G.; Li, L.; Zhang, R.; Li, S.; Wang, L.; Jiang, Z.; Zhang, Z.; Wu, J.; et al. The Sources of Sedimentary Organic Matter Traced by Carbon and Nitrogen Isotopes and Environmental Effects during the Past 60 Years in a Shallow Steppe Lake in Northern China. *Water* **2023**, *15*, 2224. [CrossRef]
3. Li, C.; Wu, W. Analysis of the Driving Mechanism of Water Environment Evolution and Algal Bloom Warning Signals in Tai Lake. *Water* **2023**, *15*, 1245. [CrossRef]
4. Zhang, Y.; Yang, F.; Liao, H.; Hu, S.; Yu, H.; Yuan, P.; Li, B.; Cui, B. Variation in Spectral Characteristics of Dissolved Organic Matter and Its Relationship with Phytoplankton of Eutrophic Shallow Lakes in Spring and Summer. *Water* **2022**, *14*, 2999. [CrossRef]
5. Slukovskii, Z. Geochemical Indicators for Paleolimnological Studies of the Anthropogenic Influence on the Environment of the Russian Federation: A Review. *Water* **2023**, *15*, 420. [CrossRef]
6. Robles-Jimarez, H.R.; Jornet-Martínez, N.; Campíns-Falcó, P. New Green and Sustainable Tool for Assessing Nitrite and Nitrate Amounts in a Variety of Environmental Waters. *Water* **2023**, *15*, 945. [CrossRef]
7. Zohud, A.; Alam, L. A Review of Groundwater Contamination in West Bank, Palestine: Quality, Sources, Risks, and Management. *Water* **2022**, *14*, 3417. [CrossRef]
8. Alharbi, T.; El-Sorogy, A.S. Health Risk Assessment of Nitrate and Fluoride in the Groundwater of Central Saudi Arabia. *Water* **2023**, *15*, 2220. [CrossRef]
9. Witczak, A.; Pokorska-Niewiada, K.; Tomza-Marciniak, A.; Witczak, G.; Cybulski, J.; Aftyka, A. The Problem of Selenium for Human Health—Removal of Selenium from Water and Wastewater. *Water* **2023**, *15*, 2230. [CrossRef]
10. Zhang, Q.; Li, Y.; Miao, Q.; Pei, G.; Nan, Y.; Yu, S.; Mei, X.; Feng, W. Distribution, Sources, and Risk of Polychlorinated Biphenyls in the Largest Irrigation Area in the Yellow River Basin. *Water* **2022**, *14*, 3472. [CrossRef]
11. Sun, J.; Zhou, Y.; Jiang, X.; Fan, J. Different Adsorption Behaviors and Mechanisms of Anionic Azo Dyes on Polydopamine–Polyethyleneimine Modified Thermoplastic Polyurethane Nanofiber Membranes. *Water* **2022**, *14*, 3865. [CrossRef]
12. Liu, J.; Su, J.; Zhao, Z.; Feng, W.; Song, S. Treatment of Wastewater Effluent with Heavy Metal Pollution Using a Nano Ecological Recycled Concrete. *Water* **2022**, *14*, 2334. [CrossRef]
13. Guo, Y.; Ma, B.; Yuan, S.; Zhang, Y.; Yang, J.; Zhang, R.; Liu, L. Simultaneous Removal of $COD_{Mn}$ and Ammonium from Water by Potassium Ferrate-Enhanced Iron-Manganese Co-Oxide Film. *Water* **2022**, *14*, 2651. [CrossRef]

**Disclaimer/Publisher's Note:** The statements, opinions and data contained in all publications are solely those of the individual author(s) and contributor(s) and not of MDPI and/or the editor(s). MDPI and/or the editor(s) disclaim responsibility for any injury to people or property resulting from any ideas, methods, instructions or products referred to in the content.

*Article*

# Differences of Nitrogen Transformation Pathways and Their Functional Microorganisms in Water and Sediment of a Seasonally Frozen Lake, China

Zhiqiang Tian [1,2], Sheng Zhang [1,*], Junping Lu [1,*], Xiaohong Shi [1], Shengnan Zhao [1], Biao Sun [1], Yanjun Wang [1,3], Guohua Li [1], Zhimou Cui [1], Xueru Pan [1], Guoguang Li [4] and Zixuan Zhang [1]

1. College of Water Conservancy and Civil Engineering, Inner Mongolia Agricultural University, Hohhot 010018, China; zhiqiang1709@126.com (Z.T.); wangyanjun@imau.edu.cn (Y.W.)
2. Department of Water Conservancy and Civil Engineering, Hetao College, Bayannur 015000, China
3. Vocational and Technical College of Inner Mongolia Agricultural University, Baotou 014109, China
4. College of Life Sciences, Inner Mongolia Agricultural University, Hohhot 010011, China
* Correspondence: shengzhang@imau.edu.cn (S.Z.); lujunping2008@iamu.edu.cn (J.L.)

**Abstract:** Nitrogen is one of the most important elements involved in ecosystem biogeochemical cycling. However, little is known about the characteristics of nitrogen cycling during the ice-covered period in seasonally frozen lakes. In this study, shotgun metagenomic sequencing of subglacial water and sediment from Lake Ulansuhai was performed to identify and compare nitrogen metabolism pathways and microbes involved in these pathways. In total, ammonia assimilation was the most prominent nitrogen transformation pathway, and Bacteria and Proteobacteria (at the domain and phylum levels, respectively) were the most abundant portion of microorganisms involved in nitrogen metabolism. Gene sequences devoted to nitrogen fixation, nitrification, denitrification, dissimilatory nitrate reduction to ammonium, and ammonia assimilation were significantly higher in sediment than in surface and subsurface water. In addition, 15 biomarkers of nitrogen-converting microorganisms, such as Ciliophora and Synergistetes, showed significant variation between sampling levels. The findings of the present study improve our understanding of the nitrogen cycle in seasonally frozen lakes.

**Keywords:** seasonally frozen lakes; ice-covered period; Lake Ulansuhai; nitrogen cycle; microorganisms; heterogeneity habitats

## 1. Introduction

Nitrogen is an essential element of life [1,2]. Its role as a limiting nutrient means that nitrogen runoff can exacerbate eutrophication in lakes [3,4]. The nitrogen cycle and its environmental effects are, therefore, a focus of research around the world. Most processes that can alter the chemical form of nitrogen depend on microorganisms [5,6]. For a long time, research on the nitrogen cycle and nitrogen-processing microorganisms focused on marine and terrestrial environments [7,8]. However, more recent studies focus on the nitrogen cycle of lake ecosystems, partially in response to the increasing need for eutrophication prevention and control methods.

Nitrogen can exist in nine known chemical forms with valence charges ranging from $-3$ to $+5$ and can be exchanged through 14 known redox reactions. The main nitrogen cycle includes two oxidation pathways (anaerobic ammonium oxidation and nitrification) and four reduction pathways (denitrification, nitrogen fixation, assimilatory nitrate reduction to ammonium [ANRA], and dissimilatory nitrate reduction to ammonium [DNRA]). Ammonification can also occur without a change in charge [9,10]. Ammonium is generally the most preferred nitrogen source for nitrogen assimilation; however, polyamines and monoamines have recently been found as alternative nitrogen sources for bacterial nitrogen assimilation [11]. New nitrogen cycle pathways have been discovered in recent years such as anaerobic ammonium oxidation (anammox) [12], complete ammonia oxidation (comammox) [13], ANRA [14], DNRA [15], and aerobic denitrification [16]. These nitrogen transformation pathways are mainly driven by Bacteria, but novel microorganisms were identified as part of these processes, including symbiotic heterotrophic nitrogen-fixing Cyanobacteria [17], ammonia-oxidizing Archaea [18], Streptomyces [19], and Eukaryota [20–22].

Although our understanding of microbial nitrogen cycling in lake ecosystems has increased, studies have primarily been performed in open water lakes, largely ignoring ice-covered lakes. In fact, half of the world's lakes, especially those located at high altitudes and latitudes in temperate and boreal climates, are covered with ice for more than 40% of the year [23–25]. Consequently, little information is known about microbial life and nitrogen cycling in these lakes. It is traditionally believed that lake ecosystems under the ice subjected to low light and low water temperature are "on hold" in winter [26,27]. Increased interest in declines in ice cover dynamics caused by global warming led to the discovery of an unexpectedly dynamic subglacial microbiome. For example, large-scale cyanobacterial blooms broke out under the ice of Lake Stechlin in Germany during the winter of 2009–2010 and triggered the active growth of heterotrophic bacteria [28]. Similarly, large-scale algal blooms have also occurred under the ice of Lake Michigan and Lake Erie [29,30]. Ice sheets have a significant role in shaping subglacial hydrodynamics, nutrient concentration, salinity, photosynthetically active radiation (PAR), and water temperature. These changes alter the subglacial microbial community structure, and the way nitrogen is processed [20,31].

Lake Ulansuhai, as the eighth largest freshwater lake in China, is an ideal study location. The lake has a mid-temperate continental climate with a current approximate ice cover duration of four months each year. Although this lake's ecological functions and eutrophication have been studied in detail [32–34], like other seasonal frozen lakes, the subglacial microecology of Ulansuhai in winter has not been given enough attention. Metagenomics has emerged as a major research tool in microbial ecology around the world as one of the most comprehensive approaches to characterizing microbial communities, revealing the functional diversity of microorganisms and the interactions between microorganisms. In the present study, we used a metagenomic approach to study the microbial processing of nitrogen throughout the water column. The objectives of this present study are to identify and compare the nitrogen transformation pathways and their functional microbes observed in subglacial water and sediments.

## 2. Materials and Methods

### 2.1. Case Study Lake

Lake Ulansuhai ($40°36'–41°03'$ N, $108°41'–108°57'$ E) is located in Bayannur City in the Inner Mongolia Autonomous Region, China (Figure 1). As the largest freshwater lake in the Yellow River Basin and the eighth largest freshwater lake in China, its entire area is 325.31 km$^2$, of which 123.11 km$^2$ is open water and the remaining area is inhabited by littoral *Phragmites* sp. [35]. It has a north–south length of 35–40 km and an east–west width of 5–10 km. As an important part of the Hetao Irrigation Area, one of the three largest irrigation areas in China, Lake Ulansuhai provides a reservoir capacity of 250–300 million m$^3$. The mean water depth is approximately 1.5 m. More than 90% of the farmland drainage from the Hetao Irrigation Area flows into the lake, with 81% inflow through the Main Drainage Channel and only 10% outflow into the Yellow River through the Retreating Channel. As a typical shallow lake in cold and arid areas, it has a mid-temperate continental climate with mean annual precipitation, annual average evaporation, and annual average air temperature of 224 mm, 1502 mm, and 7.2 °C, respectively. The multi-year average temperature during the ice-covered period from November to March is −10.24 °C, with an ice thickness of 0.3–0.6 m and snowfall of less than 10 cm [36].

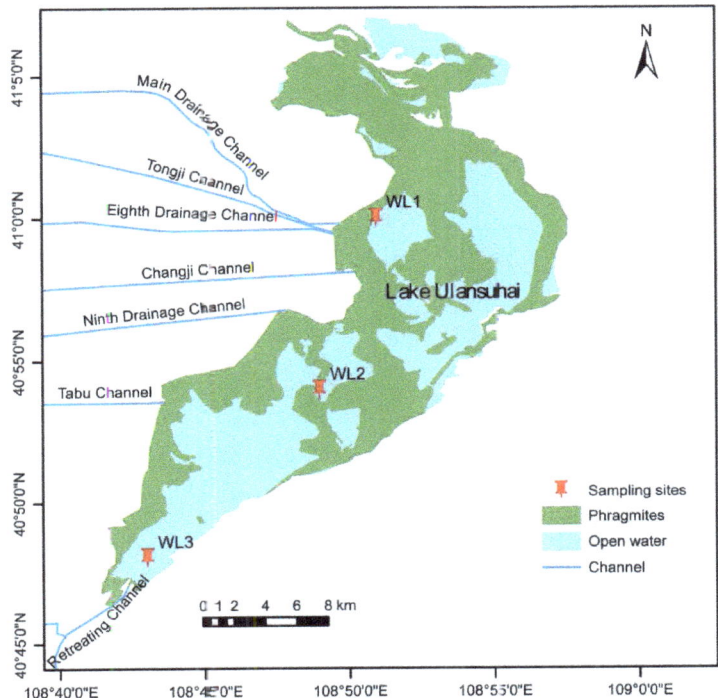

**Figure 1.** Locations of Lake Ulansuhai and sampling sites.

*2.2. Sample Collection and Treatment*

Three sampling sites (numbered WL1–WL3) were selected at the inlet, middle, and outlet according to observed water flow (Figure 1). During the ice-covered period in January 2021, holes were drilled at each sampling site using an ice auger. Surface water samples, bottom water samples, and sediment samples (numbered W1, W2, and S1, respectively) were collected from each sampling site. Surface water samples were collected at a depth of 0.5 m and bottom water samples were collected at 0.5 m above the water bottom. Duplicates of each sample were filled into 1 L sterile polyethylene sampling bottles. Water temperature (WT), oxidation-reduction potential (ORP), dissolved oxygen (DO), and pH were measured with a multi-parameter YSI Professional Plus handheld water quality monitor (YSI Inc., Yellow Spring, OH, USA) in situ. After water collection, surface sediments (0–10 cm) were collected using a Petersen grab sampler, and then were divided into two parts with a 5–10 g into sterilized tubes for DNA extractions using sterile spoons and the remaining portion in sterile plastic bags for physical and chemical analyses. Physical and chemical samples of water and sediment were kept on ice in the field and during transport, and at 4 degrees Celsius in the lab. All sediment and water samples collected for DNA extractions were frozen on dry ice and brought back to the lab for long-term storage at 80 °C.

*2.3. Physical and Chemical Analyses of Samples*

In addition to measuring the WT, pH, ORP, and DO of water samples by YSI, ammonium ($NH_4^+$-N), nitrate ($NO_3^-$-N), nitrite ($NO_2^-$-N), total nitrogen (TN), and total organic carbon (TOC) in water samples were analyzed in the laboratory according to methods reported previously [37,38]. The indophenol blue colorimetric method was utilized specifically to quantify the $NH_4^+$-N concentration. The amount of $NO_2^-$-N was determined calorimetrically using N-(1-naphthyl)-1,2-diaminoethane dihydrochloride, whereas the amount of $NO_3^-$-N was measured by the difference in UV absorbance at 220 and 275 nm.

Ultraviolet spectrophotometry was used to measure TN levels after alkaline potassium persulfate digestion. After taking the TC (total carbon) value and subtracting the TIC (total inorganic carbon) value, the TOC value was obtained. Meanwhile, the chemical characteristics of sediment, including TN, $NH_4^+$-N, $NO_3^-$-N, $NO_2^-$-N, dissolved organic nitrogen (DON), and TOC, were determined according to methods reported previously [39]. In detail, the concentrations of $NH_4^+$-N, $NO_3^-$-N, $NO_2^-$-N, and DON were measured by extracting 5 g of fresh sediment with 25 mL of KCl (2 mol/L) and then placing it in an oscillator at room temperature for 2 h (180 r/min). The aforementioned extract was filtered via a 0.45 m ANPEL membrane filter prior to analysis. The filtrate concentration was determined following standard protocol using a UV-VIS spectrophotometer (SHIMADZU UV-1700, Kyoto, Japan). To measure TN and TOC, we filtered some dried sediment over a 100-mesh screen using an elemental analyzer (ELEMEN-TAR, Frankfurt, Germany). Dissolved inorganic nitrogen (DIN) was calculated by the addition of $NH_4^+$-N, $NO_3^-$-N, and $NO_2^-$-N; the total organic nitrogen (TON) in water samples was obtained by the difference between the concentration of TN and DIN (particulate inorganic nitrogen did not represent a significant fraction of any samples). Here, C/N equaled TOC/TN.

### 2.4. DNA Extraction, Library Construction and Metagenomics Sequencing

All water and sediment samples were processed in accordance with the manufacturer's instructions for the FastDNA™ SPIN Kit for Soil (MP Biomedicals, Santa Ana, CA, USA) to extract genomic DNA. The concentration, purity, and integrity of the extracted DNA were assessed with a Quantus™ Fluorometer (Promega, Madison, WI, USA), NanoDrop™ 2000 spectrophotometer (NanoDrop Technologies, Wilmington, DE, USA), and 1% agarose gel electrophoresis, respectively. After the genomic DNA was qualified, it was fragmented to an average size of 400 bp using Covaris M220 (Gene Company Limited, Hong Kong, China) for paired-end library construction using NEXTflex™ Rapid DNA-Seq Kit (Bioo Scientific, Austin, TX, USA). The paired-end sequencing was performed on Illumina NovaSeq platform (Illumina Inc., San Diego, CA, USA) by Majorbio Bio-Pharm Technology Co., Ltd. (Shanghai, China) according to the manufacturer's instructions.

### 2.5. Sequence Processing and Bioinformatics Analysis

To generate the clean reads from metagenome sequencing, we utilized the fastp software (version 0.20.0) [40] to remove adaptor sequences, trim, and eliminate low-quality reads. These included reads with unknown nucleotide "N" bases, a minimum length threshold of 50 bp, and a minimum quality threshold of 20. This resulted in 743,197,588 high-quality clean reads from the initial 765,555,050 raw reads with 13,228,319,062 total bases. We then assembled these clean reads into contigs using MEGAHIT (version 1.1.2) [41] which employs succinct de Bruijn graphs. The 8,604,784 contigs with a minimum length of 300 bp were selected as the final assembling result. Open reading frames (ORFs) in contigs were identified using MetaGene (http://metagene.cb.k.u-tokyo.ac.jp/ (accessed on 1 May 2022)) [42]. A total of 10,673,571 genes were predicted from ORFs with a minimum length of 100 bp, retrieved, and translated into amino acid sequences. A comprehensive gene catalog with 6,445,882 microbial genes (spanning 3,114,468,057 bp) was created using CD-HIT (http://www.bioinformatics.org/cd-hit/ (accessed on 1 May 2022), version 4.6.1) [43]. The catalog was built with 90% sequence identity and 90% coverage. To determine gene abundances, SOAPaligner (http://soap.genomics.org.cn/ (accessed on 4 May 2022), version 2.21) [44] was used to map the clean reads of each sample to the catalog with 95% identity. All metagenomic sequencing statistics are displayed in Table S1 (Online Resource).

To annotate nitrogen-related genes and abundances, the non-redundant gene catalog of nitrogen metabolism was constructed from the PATHWAY subdatabase of the Kyoto Encyclopedia of Genes and Genomes (KEGG) database. The non-redundant gene catalog of nitrogen metabolism was translated into amino acid sequences. These translated amino acid sequences were annotated based on the integrated non-redundant (NR) database of the NCBI (https://ftp.ncbi.nlm.nih.gov/blast/db/FASTA/ (accessed on 11 May 2022), version 20200604) using blastp as implemented in the Diamond software (http://www.diamondsearch.org/index.php (accessed on 11 May 2022), version 0.8.35) with an e-value cutoff of $1 \times 10^{-5}$ for N taxonomic annotations [45]. The KEGG annotation was conducted using Diamond against the KEGG Orthology (KO) database (http://www.genome.jp/keeg/ (accessed on 11 May 2022), version 94.2) (e-value = $1 \times 10^{-5}$) for N functional analyses.

*2.6. Statistical Analysis*

Samples were grouped according to spatial location: surface water ($n$ = 3), bottom water ($n$ = 3), and lakebed sediment ($n$ = 3). Physicochemical data of samples were expressed as mean ± standard error (SE). Relative abundances of nitrogen metabolic genes were measured in terms of instances per million sequencing reads, simplified as "parts per million" (ppm). Relative abundances of the taxonomic and functional profiles of nitrogen-related genes were also defined in ppm. Differences in the physicochemical properties of water samples and nitrogen transformation pathways were performed by a non-parametric Kruskal–Wallis test of independent samples in IBM SPSS Statistics 19.0 (IBM Corporation, Armonk, NY, USA). $p \leq 0.05$ was defined as a statistically significant difference. The Bray–Curtis similarity between functional and taxonomic profiles consisting of relative abundances was ordinated using principal coordinate analysis (PCoA) with the significance of groupings assessed using analysis of similarities (ANOSIM) with 999 random permutations. Significantly different taxonomic profiles in multiple groups were identified using a linear discriminant analysis (LDA) effect size (LEfSe) method. Functional contributions of observed microbial taxa (phylum level) to the functional pathways were explored using customized R scripts.

## 3. Results

*3.1. Physical and Chemical Properties of Samples*

The physicochemical properties of the water and sediment samples are shown in Tables S2 and S3 (Online Resource). The concentrations of TN, $NH_4^+$-N, $NO_3^-$-N, $NO_2^-$-N, DIN, and TON were higher in the bottom water than in the surface water, but the differences between them were not significant ($p$ > 0.05). The concentration of DIN was higher than that of TON in the water column, and its proportions in surface water and bottom water were 77.09% and 78.17%, respectively. The concentrations of $NH_4^+$-N and $NO_3^-$-N were much higher than that of $NO_2^-$-N in the water column. Overall, the concentration of nitrogen in Lake Ulansuhai was relatively high during the ice-covered period, and DIN (dominated by $NH_4^+$-N and $NO_3^-$-N) was the main form of nitrogen. The bottom water also displayed a higher TOC concentration than the surface water. There were significant differences ($p \leq 0.05$) between the physical properties of the surface water and bottom water in WT and DO; however, the values of pH and ORP were comparable ($p$ > 0.05). The concentration of DON was higher than that of DIN in the sediment, and $NH_4^+$-N was the main form of DIN. C/N was also present in higher concentrations in the sediment than in either water depth sampled.

*3.2. Detection Frequency of Nitrogen-Related Genes*

In this study, we analyzed the frequency of detection of nitrogen-related genes in the samples. After aligning the high-quality clean reads to the non-redundant gene catalog of nitrogen metabolism from the KEGG Pathway database, we identified a total of 588,894 nitrogen-related gene sequences. The abundance of nitrogen-related genes in each sample ranged from 1278 to 6247 ppm. On average, the surface water samples had 1751 ppm, bottom water samples had 2198 ppm, and sediment samples had 4655 ppm of nitrogen-related genes. Composition and comparison of nitrogen transformation pathways

KEGG Orthology (KO) is a collection of genes with the same or similar function in the KEGG database that can directly characterize specific metabolic pathways and identify the number of functionally equivalent genes (KEGG orthologs, or KOs). All nitrogen-related genes identified in this survey were classified by aligning them to the KO database. According to the annotation results, 53, 54, and 54 N functional genes (KOs) were identified in the surface water, bottom water, and sediment, respectively, involving seven, seven, and eight nitrogen transformation pathways. In the surface water and bottom water, ammonia assimilation and ammonification were the major nitrogen transformation pathways, and their abundances ranged from 1097.16 to 1402.57 ppm and from 287.47 to 321.04 ppm, respectively. In the sediment, the predominant pathways were ammonia assimilation, DNRA, and denitrification, and their abundances were 2632.20 ppm, 831.16 ppm, and 431.21 ppm, respectively (Figure 2). The functional gene abundance of ammonia assimilation was higher than that of other nitrogen transformation pathways, indicating that ammonia assimilation was a major nitrogen transformation pathway in the water and sediment of Lake Ulansuhai during the ice-covered period. In comparison, the functional gene abundance of anammox was the lowest (0.39 ppm), and this pathway did not exist in detectable quantities in surface water or bottom water.

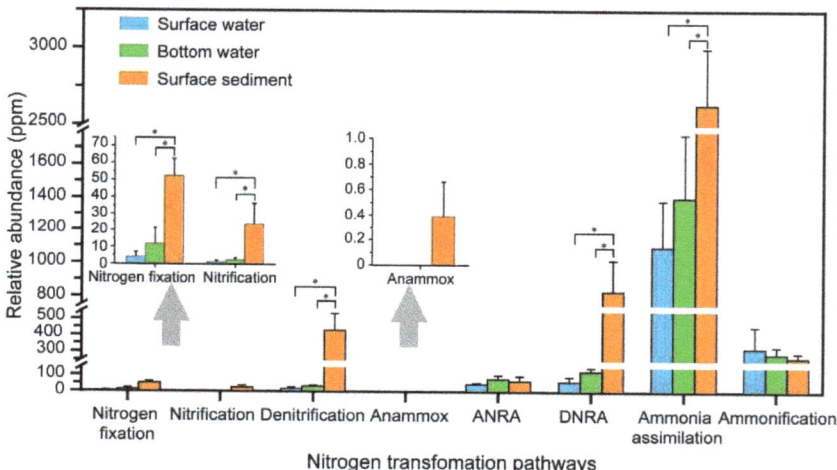

**Figure 2.** The relative abundances and differences of nitrogen transformation pathways in water and sediment during the ice-covered period. The error bars represent the standard error of the mean ($n$ = 3). * represents a statistically significant difference at the 0.05 level. Abbreviation: ANRA, assimilatory nitrate reduction to ammonium; DNRA, dissimilatory nitrate reduction to ammonium.

Ordination of functional profiles based on the relative abundance of KOs demonstrated that genetic content variation with respect to water depth was significantly smaller than the variation between water and sediment samples (Figure 3). The ANOSIM analysis strongly supported this clustering (R = 0.56, $p$ = 0.04), suggesting notable distinctions in nitrogen transformation pathways between subglacial water and sediment. A non-parametric Kruskal–Wallis test showed that the KO abundances of nitrogen fixation, nitrification, denitrification, DNRA, and ammonia assimilation in sediment were significantly ($p \leq 0.05$) higher than those in subglacial water. No such significant difference was observed ($p > 0.05$) between surface water and bottom water. Anammox, ANRA, and ammonification showed no significant genetic signature difference ($p > 0.05$) between surface water, bottom water, and sediment samples (Figure 2).

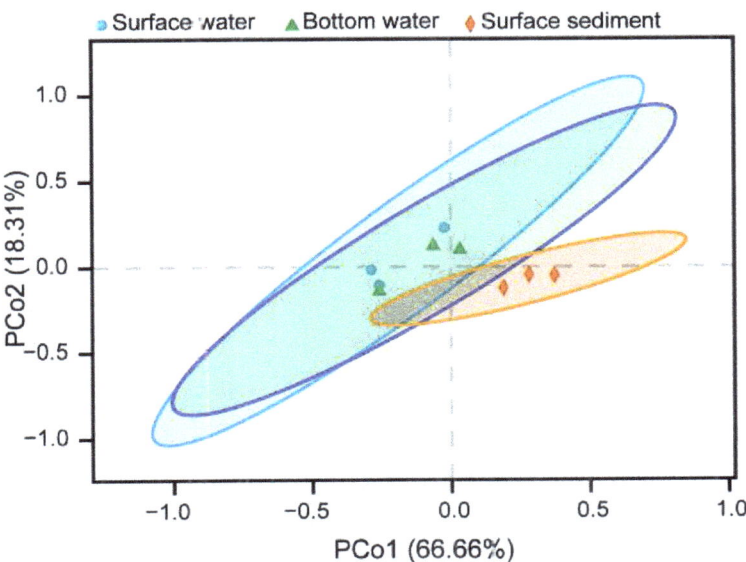

**Figure 3.** PCoA plots using Bray–Curtis distance of KOs in different samples from the ice-covered period.

*3.3. Composition and Comparison of Nitrogen-Processing Microbiome*

To determine the makeup of the microbiome at each level, all discovered N functional genes were cross-referenced with the NR database. At the domain level, Bacteria (82.08–99.46%) were the most common among all samples, with an average relative abundance of 86.22% in surface water, 91.92% in bottom water, and 96.76% in sediment. Intriguingly, the relative abundances of Eukaryota and Archaea varied greatly between water samples and sediment samples. A high percentage of Eukaryota (13.62% and 7.96%) were observed in surface water and bottom water, respectively, while the predicted abundance of Eukaryota was only 0.07% in sediment. Contrarily, the relative percentage of Archaea in sediment samples (3.07%) was greater than the measured percentages in surface water and bottom water (0.10% and 0.06%) (Figure 4A).

At the phylum level, a total of 121 phyla of nitrogen-transforming microorganisms were identified in all samples including 67 phyla in surface water, 77 phyla in bottom water, and 89 phyla in sediment. Proteobacteria was the dominant phylum in all samples, with a relative abundance ranging from 34.21% to 69.04%. In surface water, the average relative abundance of Proteobacteria was 42.99%, followed by Actinobacteria (23.20%) and Bacteroidetes (13.22%). Similar abundances of Proteobacteria (48.59%), Actinobacteria (20.83%), and Bacteroidetes (8.69%) were observed in the bottom water. Notably, compared with Bacteroidetes, the abundances of Cyanobacteria (7.32%) and Bacillariophyta (6.80%) (Eukaryota domain) were relatively high in WL2_W1, and Cyanobacteria (6.59%), Bacillariophyta (6.71%), and Verrucomicrobiota (6.59%) were more abundant in WL2_W2. Proteobacteria made up the majority of the phyla in the sediment, accounting for 54.88% of it, followed by Chloroflexota (14.67%) and Actinobacteria (6.61%) (Figure 4B).

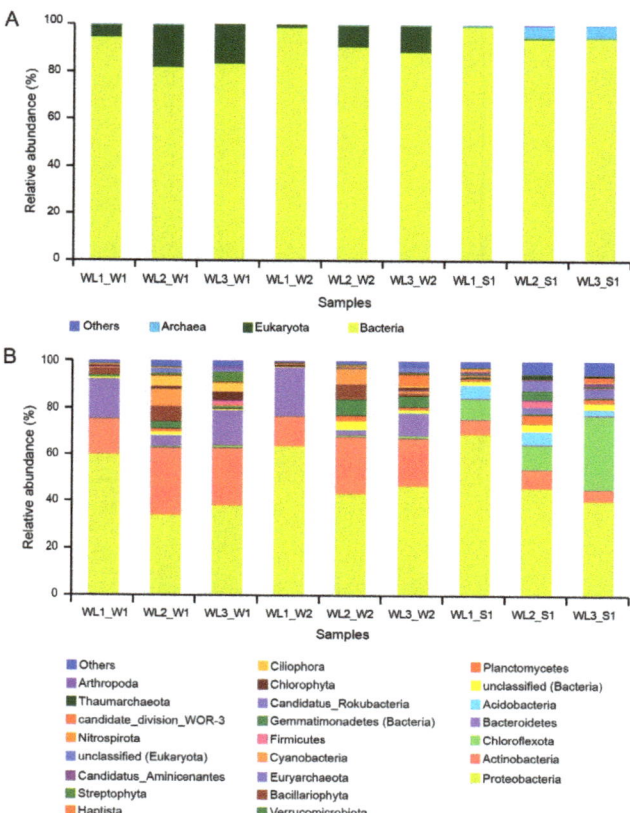

**Figure 4.** The relative abundances of the microbial community at (**A**) domain and (**B**) phylum levels in different samples during the ice-covered period. The unclassified, unidentified, and sequences with a relative abundance <1% are in the "Others" group.

The sample grouping observed for functional profiles (Figure 3) was also strongly reflected in the taxonomic profiles (Figure 5a,b). Whether at the domain or phylum level, there were two distinct clusters: one consisting of water samples from the surface and bottom water, and a second cluster consisting of sediment samples, based on PCoA plots with Bray–Curtis dissimilarity distance. These clusters were strongly supported by ANOSIM analysis (R = 0.47, $p$ = 0.05; R = 0.55, $p$ = 0.04), indicating significant differences in the community composition of nitrogen-transforming microorganisms at the domain and phylum levels between subglacial water and detritus, with fewer differences observed between surface and bottom water (Figure 5a,b). To further identify the difference in taxa between sample clusters, we conducted LEfSe analysis to compare the average relative abundances of the microbial community at different taxonomic levels. Considering the LEfSe outcomes, 15 species with an LDA score of at least two were considered to be significantly different in composition between clusters, which are hereafter referred to as taxa biomarkers. Among these taxa biomarkers, 5 and 10 taxa were significantly enriched in the surface water and sediment, respectively. In the surface water, an unclassified microbe derived from the domain Archaea and Ciliophora derived from the domain Eukaryota earned a significantly higher LDA score. In the sediment, Chordata derived from domain Eukaryota and candidate_division_Zixibacteria, Synergistetes, Candidatus_Latescibacteria, and candidate_division_NC10 derived from domain Bacteria were substantially more prevalent than those in the water column (Figure 6a–f).

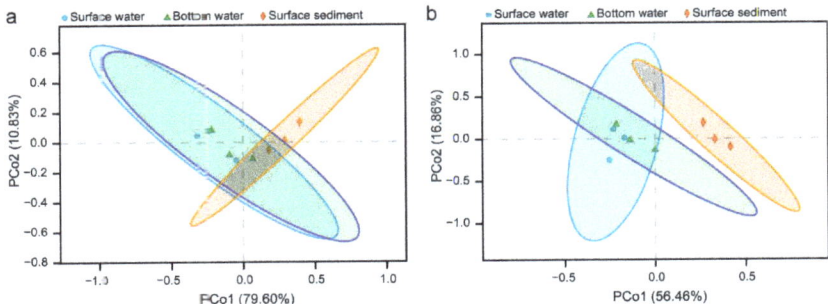

**Figure 5.** PCoA plots using Bray–Curtis distance of microbial community at (**a**) domain and (**b**) phylum levels in different samples during the ice-covered period.

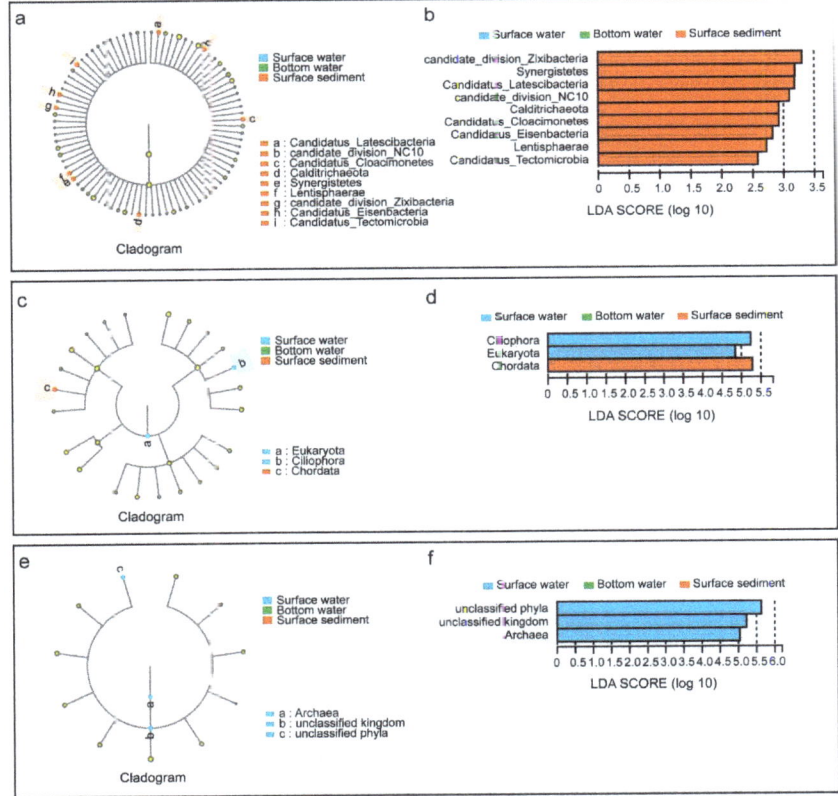

**Figure 6.** Comparison and difference of microbial community in surface water, bottom water, and surface sediment. (**a**,**b**) Cladogram and LDA score (log10) of LEfSe analysis involved in Bacteria (domain–kingdom–phylum). (**c**,**d**) Cladogram and LDA score (log10) of LEfSe analysis involved in Eukaryota (domain–kingdom–phylum). (**e**,**f**) Cladogram and LDA score (log10) of LEfSe analysis involved in Archaea (domain–kingdom–phylum).

## 4. Discussion

The nitrogen transformation pathways and the community structures of functional microorganisms in the water and sediment of Lake Ulansuhai during the ice-covered period require greater study. To the best of our knowledge, this is the only comprehensive analysis to date of the characteristics of the nitrogen cycle in subglacial water and sediment during the ice-covered period of Lake Ulansuhai.

### 4.1. Characterization of Nitrogen Transformation Pathways

Marker gene abundance in a pathway is usually used to determine nitrogen cycling capacity. A nitrogen transformation pathway often involves multiple enzyme-catalyzed reactions, so multiple functional genes are required to work together as marker genes to ensure the integrity of metabolic pathways [46–48]. Therefore, in our study, all the functional genes annotated to each nitrogen transformation pathway are considered to characterize the corresponding pathway, so that the results are more reliable. For example, we used the abundance of *nifDHK* gene clusters to jointly characterize the incidence of nitrogen fixation pathways.

### 4.2. Nitrogen Transformation Pathways and Their Differences

Among the eight nitrogen transformation pathways we analyzed, ammonia assimilation was the most frequently detected nitrogen transformation pathway during the ice-covered period. In low-temperature conditions, microorganisms may need more organic nitrogen to support cell synthesis and growth, such as amino acids and proteins [5]. Ammonia assimilation is a main nitrogen conversion pathway that can convert ammonia nitrogen into organic nitrogen. At the same time, ammonia nitrogen is a nitrogen source that is more easily used by microorganisms and can be used by almost all microorganisms through ammonia assimilation. As a result, the detection rate of genes with ammonia assimilation function is the highest [49,50]. For example, the only detected glutamine synthetase (GS) type-1 (GSI) with abundances ranging from 247.45 to 895.29 ppm. The actual ammonia assimilation pathway has long been known [51], and as the dominant pathway of nitrogen transformation, similar results have been observed not only in frozen lakes but also in a number of environments [52,53].

Nitrogen fixation, nitrification, and anammox pathways were detected less frequently in the two media, especially in anammox pathways. The previous study showed that there is a negative correlation between nitrogenase activity and ammonia nitrogen concentration [54], and the physical and chemical characteristics of our samples included high ammonia nitrogen concentration and low biological nitrogen fixation in subglacial water and sediments. These characteristics were also found in the permanently ice-covered Bonney Lake in Antarctica [55]. A weak potential for nitrification was found in this study, similar to the findings in other frozen lakes [56]. According to our investigation and a prior study [57], the level of DO in Ulansuhai's subglacial water was higher than anticipated, which ruled out the possibility of anammox. In contrast to water, denitrification and anaerobic ammonia oxidation occur naturally and are significant in sediments. However, related studies have revealed that anaerobic ammoxidation's standard free energy is lower than denitrification, making denitrification more thermodynamically feasible [58,59]. In addition, compared with anammox bacteria, denitrifying bacteria have a higher growth yield, so denitrification is often dominant [60,61] in an environment where two pathways exist at the same time. The seasonal death of aquatic plants in the lake's littoral zone releases large amounts of organic carbon into the water and sediment, which hinders the anaerobic ammoxidation process [62].

There are significant differences in this study's observed functional gene abundance of nitrogen fixation, nitrification, denitrification, DNRA, and ammonia assimilation in subglacial water and sediment. Generally speaking, compared with water bodies, sediments contain more nitrogen-metabolizing microbes, as seen in our results.

Active nitrogen is crucial to aquatic environments, and biological nitrogen fixation is a major contributor. Traditionally, nitrogen-fixing cyanobacteria are generally considered to be the main nitrogen-fixing bacteria in lakes. However, heteromorphic bacteria, chemotrophic bacteria, and archaea can also perform nitrogen fixation [63] in dark sediments. Lakebed sediment is generally rich in microorganisms carrying *nifDHK* gene clusters that demonstrate higher nitrogen fixation activity than microbes in the water column, which has been confirmed by related studies [64]. Iron and molybdenum are necessary components for synthesizing nitrogenase, and the content of these heavy metal elements in sediments is generally high [65,66]. These factors likely contribute to the disparity we observe in the distribution of nitrogen fixation genes. A large number of studies have shown that denitrification and nitrogen fixation occur simultaneously in sediments because the nitrogen loss caused by denitrification can be compensated by nitrogen fixation [50,67]. Temperature is also an important environmental factor affecting the nitrogen conversion pathway. When the temperature is below 5 °C, denitrification and anammox are negligible [62,68]. Our study shows that the temperature of the subglacial water body is only $5.0 \pm 1.84$ °C in the bottom water, while the sediment temperature is relatively high. Although we did not measure the sediment temperature, the heat released from the overlying water body during the ice-sealing period can be used as evidence of higher sediment temperature than that of the water column [69]. Therefore, denitrification should occur at different rates in the lake's subglacial water and sediment. Supplementary Tables S2 and S3 (Online Resource) show that sediment samples contained higher levels of C/N and $NH_4^+$-N, which may be the main reason for the difference between DNRA and ammonia assimilation in water and sediments. Although we have discussed a variety of factors that may have produced the differences we observed, a definitive model of nitrogen transformation pathways in Lake Ulansuhai would require more thorough surveying of the lake.

### 4.3. Nitrogen Transformation Functional Microorganisms and Their Differences

It is well known that there are a surprising number of microorganisms that can transform nitrogen. Bacteria, Eukaryota, and Archaea may all drive nitrogen transformation, but relatively speaking, Bacteria are the main participants in nitrogen transformation. This has also been well verified in our study (Figure 4A). Proteobacteria was the most abundant phylum in the subglacial water and sediment of Ulansuhai during the ice-covered period, which corresponded with the outcomes of experiments conducted in Hulun Lake, Chaohu Lake, and Erhai Lake [70–72], indicating that Proteobacteria is the dominant species driving nitrogen transformation in lakes during both ice-covered and ice-free periods. In addition, Actinobacteria, and Bacteroidetes were detected at higher concentrations in subglacial water, while a greater proportion of Chloroflexota and Actinobacteria were detected in sediment, which was similar to the microbial community structure in aquaculture ponds and sediments [73]. Previous studies have shown that, regardless of nutritional status, Proteobacteria, Actinobacteria, Bacteroidetes, and Chloroflexota generally occupy a dominant position in freshwater lakes, and they actively participate in the nitrogen cycle process [74–76]. Our study also confirmed that the distribution of microbial communities is not uniform but shows a small number of dominant species and a large number of rare species, and a small number of common dominant flora can be observed in different environments [77].

For lake water, although there were significant differences in the environmental factors between surface water and water near the lakebed (Table S2 (Online Resource)), their microbial communities were similar. Fifteen species with statistically significant differences were identified in the two environmental media, with 5 and 10 taxa significantly enriched in surface water and sediment, respectively. Ciliophora was found in freshwater in a planktonic state for most times of the year [78], which may account for the greater abundance of Ciliophora in surface water in our study. The significant enrichment of Synergistetes in the sediment may be related to its environmental characteristics for anaerobic existence [79]. It is worth noting that these differential species are not the few common dominant species, indicating that they can adapt to the environment in surprisingly specific ways. These observations may reflect a wide range of environmental heterogeneity. Species differences caused by habitat heterogeneity are not very common at higher classification levels but occur at lower classification levels such as the genus or species level [80,81]. Our study showed that the habitat heterogeneity of different species in the water and sediment of Lake Ulansuhai during the ice-covered period also existed at the domain–kingdom–phylum level.

Although we generalized the functionality of microbial genes through KEGG Orthography, some functional genes will play a role in multiple nitrogen transformation pathways at the same time. For these multi-purpose functional genes, we only classified them as one nitrogen transformation pathway and did not distinguish their contribution to other pathways. Additionally, we discussed the characteristics and differences of nitrogen transformation in subglacial heterogeneous habitats based on transformation pathways and community structure; however, quantifying the relationship between them is a problem that needs to be solved in the future.

## 5. Conclusions

During the ice-covered period of Lake Ulansuhai, the characteristics of the nitrogen cycle were analyzed through the metagenomic approach of subglacial water and sediment. We found that ammonia assimilation was the main nitrogen transformation pathway used by water column and sediment microbes, and domain Bacteria and phylum Proteobacteria were the dominant taxa driving nitrogen transformation. Habitat heterogeneity had a significant impact on nitrogen transformation pathways and species taxa, and surface sediment was crucial to the nitrogen cycle in the lake during the time when the lake was covered in ice. A small number of common dominant taxa were similar in different habitats, while taxonomical differences were concentrated in a large number of rarer species.

**Supplementary Materials:** The following supporting information can be downloaded at: https://www.mdpi.com/article/10.3390/w15132332/s1, Table S1: Statistics of metagenomic sequencing data; Table S2: Physical and chemical properties of water samples; Table S3: Physical and chemical properties of sediment samples.

**Author Contributions:** Z.T.: conceptualization, investigation, formal analysis, methodology, visualization, and writing—original draft; S.Z. (Sheng Zhang): conceptualization, funding acquisition, supervision, and writing—review and editing; J.L., X.S. and S.Z. (Shengnan Zhao): data curation, funding acquisition, project administration, resources, and visualization; B.S. and Y.W.: formal analysis and supervision; G.L. (Guohua Li), Z.C., X.P., G.L. (Guoguang Li) and Z.Z.: data curation and investigation. All authors have read and agreed to the published version of the manuscript.

**Funding:** This work was funded by the Research Program of Science and Technology at Universities of Inner Mongolia Autonomous Region (NJZY21180 and NJZY21503), the National Natural Science Foundation of China (52260029, 52060022, 52260028, and 52160021), the Inner Mongolia Autonomous Region Science and Technology Plan (2021GG0089), and the National Key Research and Development Program of China (2017YFE0114800 and 2019YFC0409204).

**Institutional Review Board Statement:** The research complies with ethical standards.

**Informed Consent Statement:** Not applicable.

**Data Availability Statement:** The National Center for Biotechnology Information (NCBI) Short Read Archive database has received the sequencing raw data related to this project (Accession Numbers: SRR18899872-SRR18899880). The other data generated or analyzed during this study are not publicly available due but are available from the corresponding author on reasonable request.

**Acknowledgments:** We acknowledge Shanghai Majorbio Bio-Pharm Technology Co., Ltd., for their technical assistance with sequencing. We would also like to thank LetPub (www.letpub.com, accessed on 20 June 2023) for its linguistic assistance during the preparation of this manuscript.

**Conflicts of Interest:** The authors declare no conflict of interest.

## References

1. Conley, D.J.; Paerl, H.W.; Howarth, R.W.; Boesch, D.F.; Seitzinger, S.P.; Havens, K.E.; Lancelot, C.; Likens, G.E. Ecology. Controlling eutrophication: Nitrogen and phosphorus. *Science* **2009**, *323*, 1014–1015. [CrossRef]
2. Parsons, C.; Stueken, E.E.; Rosen, C.J.; Mateos, K.; Anderson, R.E. Radiation of nitrogen-metabolizing enzymes across the tree of life tracks environmental transitions in Earth history. *Geobiology* **2021**, *19*, 18–34. [CrossRef]
3. Qin, B.; Yang, L.; Chen, F.; Zhu, G.; Zhang, L.; Chen, Y. Mechanism and control of lake eutrophication. *Chin. Sci. Bull.* **2006**, *51*, 2401–2412. [CrossRef]
4. Lin, S.S.; Shen, S.L.; Zhou, A.; Lyu, H.M. Assessment and management of lake eutrophication: A case study in Lake Erhai, China. *Sci. Total. Environ.* **2021**, *751*, 141618. [CrossRef] [PubMed]
5. Stein, L.Y.; Klotz, M.G. The nitrogen cycle. *Curr. Biol.* **2016**, *26*, R94–R98. [CrossRef]
6. Ollivier, J.; Towe, S.; Bannert, A.; Hai, B.; Kastl, E.M.; Meyer, A.; Su, M.X.; Kleineidam, K.; Schloter, M. Nitrogen turnover in soil and global change. *FEMS Microbiol. Ecol.* **2011**, *78*, 3–16. [CrossRef]
7. Karl, D.M.; Michaels, A.F. Nitrogen Cycle. In *Encyclopedia of Ocean Sciences*, 2nd ed.; Steele, J.H., Ed.; Academic Press: Oxford, UK, 2001; pp. 32–39. ISBN 978-0-12-374473-9.
8. Batlle-Aguilar, J.; Brovelli, A.; Porporato, A.; Barry, D.A. Modelling soil carbon and nitrogen cycles during land use change. *Sustain. Agric.* **2011**, *2*, 499–527. [CrossRef]
9. Masclaux-Daubresse, C.; Daniel-Vedele, F.; Dechorgnat, J.; Chardon, F.; Gaufichon, L.; Suzuki, A. Nitrogen uptake, assimilation and remobilization in plants: Challenges for sustainable and productive agriculture. *Ann. Bot.* **2010**, *105*, 1141–1157. [CrossRef]
10. Kuypers, M.M.M.; Marchant, H.K.; Kartal, B. The microbial nitrogen-cycling network. *Nat. Rev. Microbiol.* **2018**, *16*, 263–276. [CrossRef] [PubMed]
11. Krysenko, S.; Wohlleben, W. Polyamine and Ethanolamine Metabolism in Bacteria as an Important Component of Nitrogen Assimilation for Survival and Pathogenicity. *Med. Sci.* **2022**, *10*, 40. [CrossRef]
12. Zhang, Q.; Fan, N.-S.; Fu, J.-J.; Huang, B.-C.; Jin, R.-C. Role and application of quorum sensing in anaerobic ammonium oxidation (anammox) process: A review. *Crit. Rev. Environ. Sci. Technol.* **2020**, *51*, 626–648. [CrossRef]
13. van Kessel, M.A.; Speth, D.R.; Albertsen, M.; Nielsen, P.H.; Op den Camp, H.J.; Kartal, B.; Jetten, M.S.; Lucker, S. Complete nitrification by a single microorganism. *Nature* **2015**, *528*, 555–559. [CrossRef]
14. Hochman, A.; Nissany, A.; Amizur, M. Nitrate reduction and assimilation by a moderately halophilic, halotolerant bacterium Ba1. *Biochim. Biophys. Acta (BBA) Gen. Subj.* **1988**, *965*, 82–89. [CrossRef]
15. Li, Q.; Bu, C.; Ahmad, H.A.; Guimbaud, C.; Gao, B.; Qiao, Z.; Ding, S.; Ni, S.Q. The distribution of dissimilatory nitrate reduction to ammonium bacteria in multistage constructed wetland of Jining, Shandong, China. *Environ. Sci. Pollut. Res. Int.* **2021**, *28*, 4749–4761. [CrossRef] [PubMed]
16. Song, T.; Zhang, X.; Li, J.; Wu, X.; Feng, H.; Dong, W. A review of research progress of heterotrophic nitrification and aerobic denitrification microorganisms (HNADMs). *Sci. Total. Environ.* **2021**, *801*, 149319. [CrossRef]
17. Thompson, A.W.; Foster, R.A.; Krupke, A.; Carter, B.J; Musat, N.; Vaulot, D.; Kuypers, M.M.; Zehr, J.P. Unicellular cyanobacterium symbiotic with a single-celled eukaryotic alga. *Science* **2012**, *337*, 1546–1550. [CrossRef] [PubMed]
18. Jin, T.; Zhang, T.; Yan, Q. Characterization and quantification of ammonia-oxidizing archaea (AOA) and bacteria (AOB) in a nitrogen-removing reactor using T-RFLP and qPCR. *Appl. Microbiol. Biotechnol.* **2010**, *37*, 1167–1176. [CrossRef]
19. Romero-Rodríguez, A.; Maldonado-Carmona, N.; Ruiz-Villafán, B.; Koirala, N.; Rocha, D.; Sánchez, S. Interplay between carbon, nitrogen and phosphate utilization in the control of secondary metabolite production in Streptomyces. *Antonie Van Leeuwenhoek* **2018**, *111*, 761–781. [CrossRef]
20. Hayatsu, M.; Tago, K.; Saito, M. Various players in the nitrogen cycle: Diversity and functions of the microorganisms involved in nitrification and denitrification. *Soil Sci. Plant Nutr.* **2008**, *54*, 33–45. [CrossRef]
21. Kamp, A.; Hogslund, S.; Risgaard-Petersen, N.; Stief, P. Nitrate Storage and Dissimilatory Nitrate Reduction by Eukaryotic Microbes. *Front. Microbiol.* **2015**, *6*, 1492. [CrossRef]
22. Woehle, C.; Roy, A.S.; Glock, N.; Wein, T.; Weissenbach, J.; Rosenstiel, P.; Hiebenthal, C.; Michels, J.; Schonfeld, J.; Dagan, T. A Novel Eukaryotic Denitrification Pathway in Foraminifera. *Curr. Biol.* **2018**, *28*, 2536–2543.e2535. [CrossRef]
23. Walsh, S.E.; Vavrus, S.J.; Foley, J.A.; Fisher, V.A.; Wynne, R.H.; Lenters, J.D. Global patterns of lake ice phenology and climate: Model simulations and observations. *J. Geophys. Res. Atmos.* **1998**, *103*, 28825–28837. [CrossRef]
24. Weyhenmeyer, G.A.; Livingstone, D.M.; Meili, M.; Jensen, O.; Benson, B.; Magnuson, J.J. Large geographical differences in the sensitivity of ice-covered lakes and rivers in the Northern Hemisphere to temperature changes. *Glob. Chang. Biol.* **2011**, *17*, 268–275. [CrossRef]
25. Verpoorter, C.; Kutser, T.; Seekell, D.A.; Tranvik, L.J. A global inventory of lakes based on high-resolution satellite imagery. *Geophys. Res. Lett.* **2014**, *41*, 6396–6402. [CrossRef]
26. Salonen, K.; Leppäranta, M.; Viljanen, M.; Gulati, R.D. Perspectives in winter limnology: Closing the annual cycle of freezing lakes. *Aquat. Ecol.* **2009**, *43*, 609–616. [CrossRef]
27. Powers, S.M.; Baulch, H.M.; Hampton, S.E.; Labou, S.G.; Lottig, N.R.; Stanley, E.H. Nitrification contributes to winter oxygen depletion in seasonally frozen forested lakes. *Biogeochemistry* **2017**, *136*, 119–129. [CrossRef]

28. Üveges, V.; Tapolczai, K.; Krienitz, L.; Padisák, J. Photosynthetic characteristics and physiological plasticity of an *Aphanizomenon flos-aquae* (Cyanobacteria, Nostocaceae) winter bloom in a deep oligo-mesotrophic lake (Lake Stechlin, Germany). *Hydrobiologia* **2012**, *698*, 263–272. [CrossRef]
29. Vanderploeg, H.A.; Bolsenga, S.J.; Fahnenstiel, G.L.; Liebig, J.R.; Gardner, W.S. Plankton ecology in an ice-covered bay of Lake Michigan: Utilization of a winter phytoplankton bloom by reproducing copepods. *Hydrobiologia* **1992**, *243*, 175–183. [CrossRef]
30. Twiss, M.; McKay, R.; Bourbonniere, R.; Bullerjahn, G.; Carrick, H.; Smith, R.; Winter, J.; D'souza, N.; Furey, P.; Lashaway, A. Diatoms abound in ice-covered Lake Erie: An investigation of offshore winter limnology in Lake Erie over the period 2007 to 2010. *J. Great Lakes Res.* **2012**, *38*, 18–30. [CrossRef]
31. Bertilsson, S.B.A.; Carey, C.C.; Fey, S.B.; Grossart, H.P.; Grubisic, L.M.; Jones, I.D.; Kirillin, G.; Lennon, J.T.; Shade, A.; Smyth, R.L. The under-ice microbiome of seasonally frozen lakes. *Limnol. Oceanogr.* **2013**, *58*, 1998–2012. [CrossRef]
32. Quan, D.; Shi, X.; Zhao, S.; Zhang, S.; Liu, J. Eutrophication of Lake Ulansuhai in 2006–2017 and its main impact factors. *J. Lake Sci.* **2019**, *31*, 1259–1267. [CrossRef]
33. He, Y.; Sun, D.; Wu, J.; Sun, Y. Factors controlling the past ~150-year ecological dynamics of Lake Wuliangsu in the upper reaches of the Yellow River, China. *Holocene* **2015**, *25*, 1394–1401. [CrossRef]
34. Sun, H.; Lu, X.; Yu, R.; Yang, J.; Liu, X.; Cao, Z.; Zhang, Z.; Li, M.; Geng, Y. Eutrophication decreased $CO_2$ but increased $CH_4$ emissions from lake: A case study of a shallow Lake Ulansuhai. *Water Res.* **2021**, *201*, 117363. [CrossRef]
35. Li, G.; Zhang, S.; Shi, X.; Zhan, L.; Zhao, S.; Sun, B.; Liu, Y.; Tian, Z.; Li, Z.; Arvola, L.; et al. Spatiotemporal variability and diffusive emissions of greenhouse gas in a shallow eutrophic lake in Inner Mongolia, China. *Ecol. Indic.* **2022**, *145*, 109578. [CrossRef]
36. Yang, T.; Hei, P.; Song, J.; Zhang, J.; Zhu, Z.; Zhang, Y.; Yang, J.; Liu, C.; Jin, J.; Quan, J. Nitrogen variations during the ice-on season in the eutrophic lakes. *Environ. Pollut.* **2019**, *247*, 1089–1099. [CrossRef]
37. Chen, X.; Yang, L.; Xiao, L.; Miao, A.; Xi, B. Nitrogen removal by denitrification during cyanobacterial bloom in Lake Taihu. *J. Freshw. Ecol.* **2012**, *27*, 243–258. [CrossRef]
38. Shi, Y.; Xu, L.; Gong, D.; Lu, J. Effects of sterilization treatments on the analysis of TOC in water samples. *J. Environ. Sci.* **2010**, *22*, 789–795. [CrossRef]
39. Hu, L.; Shi, X.; Yu, Z.; Lin, T.; Wang, H.; Ma, D.; Guo, Z.; Yang, Z. Distribution of sedimentary organic matter in estuarine–inner shelf regions of the East China Sea: Implications for hydrodynamic forces and anthropogenic impact. *Mar. Chem.* **2012**, *142–144*, 29–40. [CrossRef]
40. Chen, S.; Zhou, Y.; Chen, Y.; Gu, J. fastp: An ultra-fast all-in-one FASTQ preprocessor. *Bioinformatics* **2018**, *34*, i884–i890. [CrossRef]
41. Li, D.; Liu, C.M.; Luo, R.; Sadakane, K.; Lam, T.W. MEGAHIT: An ultra-fast single-node solution for large and complex metagenomics assembly via succinct de Bruijn graph. *Bioinformatics* **2015**, *31*, 1674–1676. [CrossRef] [PubMed]
42. Noguchi, H.; Park, J.; Takagi, T. MetaGene: Prokaryotic gene finding from environmental genome shotgun sequences. *Nucleic Acids Res.* **2006**, *34*, 5623–5630. [CrossRef]
43. Fu, L.; Niu, B.; Zhu, Z.; Wu, S.; Li, W. CD-HIT: Accelerated for clustering the next-generation sequencing data. *Bioinformatics* **2012**, *28*, 3150–3152. [CrossRef] [PubMed]
44. Li, R.; Li, Y.; Kristiansen, K.; Wang, J. SOAP: Short oligonucleotide alignment program. *Bioinformatics* **2008**, *24*, 713–714. [CrossRef] [PubMed]
45. Buchfink, B.; Xie, C.; Huson, D.H. Fast and sensitive protein alignment using DIAMOND. *Nat. Methods* **2015**, *12*, 59–60. [CrossRef]
46. Morales, S.E.; Cosart, T.; Holben, W.E. Bacterial gene abundances as indicators of greenhouse gas emission in soils. *ISME J.* **2010**, *4*, 799–808. [CrossRef]
47. Dos Santos, P.C.; Fang, Z.; Mason, S.W.; Setubal, J.C.; Dixon, R. Distribution of nitrogen fixation and nitrogenase-like sequences amongst microbial genomes. *BMC Genom.* **2012**, *13*, 162. [CrossRef] [PubMed]
48. Albright, M.B.; Timalsina, B.; Martiny, J.B.; Dunbar, J. Comparative genomics of nitrogen cycling pathways in bacteria and archaea. *Microb. Ecol.* **2019**, *77*, 597–606. [CrossRef] [PubMed]
49. Jackson, L.E.; Schimel, J.P.; Firestone, M.K. Short-term partitioning of ammonium and nitrate between plants and microbes in an annual grassland. *Soil Biol. Biochem.* **1989**, *21*, 409–415. [CrossRef]
50. Newell, S.E.; Pritchard, K.R.; Foster, S.Q.; Fulweiler, R.W. Molecular evidence for sediment nitrogen fixation in a temperate New England estuary. *PeerJ* **2016**, *4*, e1615. [CrossRef]
51. Tempest, D.W.; Meers, J.L.; Brown, C.M. Synthesis of glutamate in *Aerobacter aerogenes* by a hitherto unknown route. *Biochem. J.* **1970**, *117*, 405–407. [CrossRef]
52. Souza, R.C.; Hungria, M.; Cantão, M.E.; Vasconcelos, A.T.R.; Nogueira, M.A.; Vicente, V.A. Metagenomic analysis reveals microbial functional redundancies and specificities in a soil under different tillage and crop-management regimes. *Appl. Soil Ecol.* **2015**, *86*, 106–112. [CrossRef]
53. Qu, X.; Zhang, M.; Yang, Y.; Xie, Y.; Ren, Z.; Peng, W.; Du, X. Taxonomic structure and potential nitrogen metabolism of microbial assemblage in a large hypereutrophic steppe lake. *Environ. Sci. Pollut. Res. Int.* **2019**, *26*, 21151–21160. [CrossRef]
54. Fernandez-Valiente, E.; Quesada, A.; Howard-Williams, C.; Hawes, I. $N_2$-Fixation in Cyanobacterial Mats from Ponds on the McMurdo Ice Shelf, Antarctica. *Microb. Ecol.* **2001**, *42*, 338–349. [CrossRef] [PubMed]
55. Paerl, H.W.; Priscu, J.C. Microbial Phototrophic, Heterotrophic, and Diazotrophic Activities Associated with Aggregates in the Permanent Ice Cover of Lake Bonney, Antarctica. *Microb. Ecol.* **1998**, *36*, 221–230. [CrossRef] [PubMed]

56. Cavaliere, E.; Baulch, H.M. Winter nitrification in ice-covered lakes. *PLoS ONE* **2019**, *14*, e0224864. [CrossRef]
57. Song, S.; Li, C.; Shi, X.; Zhao, S.; Tian, W.; Li, Z.; Bai, Y.; Cao, X.; Wang, Q.; Huotari, J.; et al. Under-ice metabolism in a shallow lake in a cold and arid climate. *Freshw. Biol.* **2019**, *64*, 1710–1720. [CrossRef]
58. Jetten, M.S.; Strous, M.; van de Pas-Schoonen, K.T.; Schalk, J.; van Dongen, U.G.; van de Graaf, A.A.; Logemann, S.; Muyzer, G.; van Loosdrecht, M.C.; Kuenen, J.G. The anaerobic oxidation of ammonium. *FEMS Microbiol. Rev.* **1998**, *22*, 421–437. [CrossRef] [PubMed]
59. Ahn, Y.-H. Sustainable nitrogen elimination biotechnologies: A review. *Process. Biochem.* **2006**, *41*, 1709–1721. [CrossRef]
60. El Hassan, G.A.; Zablotowicz, R.M.; Focht, D.D. Kinetics of denitrifying growth by fast-growing cowpea rhizobia. *Appl. Environ. Microbiol.* **1985**, *49*, 517–521. [CrossRef]
61. Murray, R.E.; Parsons, L.L.; Smith, M.S. Aerobic and anaerobic growth of rifampin-resistant denitrifying bacteria in soil. *Appl. Environ. Microbiol.* **1990**, *56*, 323–328. [CrossRef]
62. Kumar, M.; Lin, J.G. Co-existence of anammox and denitrification for simultaneous nitrogen and carbon removal—Strategies and issues. *J. Hazard. Mater.* **2010**, *178*, 1–9. [CrossRef]
63. Severin, I.; Confurius-Guns, V.; Stal, L.J. Effect of salinity on nitrogenase activity and composition of the active diazotrophic community in intertidal microbial mats. *Arch. Microbiol.* **2012**, *194*, 483–491. [CrossRef]
64. Zhang, B.; Zhang, L. Spatial-seasonal variations of nitrogen fixation of water column in Taihu Lake. *Acta Scentiae Circunstantiae* **2016**, *36*, 1129–1136. [CrossRef]
65. Bellenger, J.P.; Darnajoux, R.; Zhang, X.; Kraepiel, A.M.L. Biological nitrogen fixation by alternative nitrogenases in terrestrial ecosystems: A review. *Biogeochemistry* **2020**, *149*, 53–73. [CrossRef]
66. Tian, L.; Yan, Z.; Wang, C.; Xu, S.; Jiang, H. Habitat heterogeneity induces regional differences in sediment nitrogen fixation in eutrophic freshwater lake. *Sci. Total Environ.* **2021**, *772*, 145594. [CrossRef]
67. Halm, H.; Musat, N.; Lam, P.; Langlois, R.; Musat, F.; Peduzzi, S.; Lavik, G.; Schubert, C.J.; Sinha, B.; LaRoche, J.; et al. Co-occurrence of denitrification and nitrogen fixation in a meromictic lake, Lake Cadagno (Switzerland). *Environ. Microbiol.* **2009**, *11*, 1945–1958. [CrossRef]
68. Pang, Y.; Zhang, Y.; Yan, X.; Ji, G. Cold Temperature Effects on Long-Term Nitrogen Transformation Pathway in a Tidal Flow Constructed Wetland. *Environ. Sci. Technol.* **2015**, *49*, 13550–13557. [CrossRef]
69. Fang, X.; Stefan, H.G. Dynamics of heat exchange between sediment and water in a lake. *Water Resour. Res.* **1996**, *32*, 1719–1727. [CrossRef]
70. Fan, Y.Y.; Li, B.B.; Yang, Z.C.; Cheng, Y.Y.; Liu, D.F.; Yu, H.Q. Mediation of functional gene and bacterial community profiles in the sediments of eutrophic Chaohu Lake by total nitrogen and season. *Environ. Pollut.* **2019**, *250*, 233–240. [CrossRef] [PubMed]
71. Pan, X.; Lin, L.; Huang, Z.; Chen, J. Differentiation of Nitrogen and Microbial Community in the Sediments from Lake Erhai, Yunnan–Kweichow Plateau, China. *Geomicrobiol. J.* **2020**, *37*, 818–825. [CrossRef]
72. Shang, Y.; Wu, X.; Wang, X.; Wei, Q.; Ma, S.; Sun, G.; Zhang, H.; Wang, L.; Dou, H.; Zhang, H. Factors affecting seasonal variation of microbial community structure in Hulun Lake, China. *Sci. Total Environ.* **2022**, *805*, 150294. [CrossRef]
73. Deng, M.; Hou, J.; Song, K.; Chen, J.; Gou, J.; Li, D.; He, X. Community metagenomic assembly reveals microbes that contribute to the vertical stratification of nitrogen cycling in an aquaculture pond. *Aquaculture* **2020**, *520*, 734911. [CrossRef]
74. Lew, S.; Koblížek, M.; Lew, M.; Medová, H.; Glińska-Lewczuk, K.; Owsianny, P.M. Seasonal changes of microbial communities in two shallow peat bog lakes. *Folia Microbiol.* **2015**, *60*, 165–175. [CrossRef] [PubMed]
75. Dai, Y.; Yang, Y.; Wu, Z.; Feng, Q.; Xie, S.; Liu, Y. Spatiotemporal variation of planktonic and sediment bacterial assemblages in two plateau freshwater lakes at different trophic status. *Appl. Microbiol. Biotechnol.* **2016**, *100*, 4161–4175. [CrossRef] [PubMed]
76. Tang, X.; Chao, J.; Gong, Y.; Wang, Y.; Wilhelm, S.W.; Gao, G. Spatiotemporal dynamics of bacterial community composition in large shallow eutrophic Lake Taihu: High overlap between free-living and particle-attached assemblages. *Limnol. Oceanogr.* **2017**, *62*, 1366–1382. [CrossRef]
77. McGill, B.J.; Etienne, R.S.; Gray, J.S.; Alonso, D.; Anderson, M.J.; Benecha, H.K.; Dornelas, M.; Enquist, B.J.; Green, J.L.; He, F.; et al. Species abundance distributions: Moving beyond single prediction theories to integration within an ecological framework. *Ecol. Lett.* **2007**, *10*, 995–1015. [CrossRef]
78. Lynn, D.H. Ciliophora. In *Handbook of the Protists*; Archibald, J.M., Simpson, A.G.B., Slamovits, C.H., Eds.; Springer International Publishing: Cham, Switzerland, 2017; pp. 679–730. ISBN 978-3-319-28149-0.
79. Jumas-Bilak, E.; Roudiere, L.; Marchandin, H. Description of 'Synergistetes' phyl. nov. and emended description of the phylum 'Deferribacteres' and of the family Syntrophomonadaceae, phylum 'Firmicutes'. *Int. J. Syst. Evol. Microbiol.* **2009**, *59*, 1028–1035. [CrossRef]
80. Tamames, J.; Abellán, J.J.; Pignatelli, M.; Camacho, A.; Moya, A. Environmental distribution of prokaryotic taxa. *BMC Microbiol.* **2010**, *10*, 85. [CrossRef]
81. Segata, N.; Izard, J.; Waldron, L.; Gevers, D.; Miropolsky, L.; Garrett, W.S.; Huttenhower, C. Metagenomic biomarker discovery and explanation. *Genome Biol.* **2011**, *12*, R60. [CrossRef]

**Disclaimer/Publisher's Note:** The statements, opinions and data contained in all publications are solely those of the individual author(s) and contributor(s) and not of MDPI and/or the editor(s). MDPI and/or the editor(s) disclaim responsibility for any injury to people or property resulting from any ideas, methods, instructions or products referred to in the content.

Article

# The Problem of Selenium for Human Health—Removal of Selenium from Water and Wastewater

Agata Witczak [1,*], Kamila Pokorska-Niewiada [1], Agnieszka Tomza-Marciniak [2], Grzegorz Witczak [3], Jacek Cybulski [1] and Aleksandra Aftyka [4]

[1] Department of Toxicology, Dairy Technology and Food Storage, Faculty of Food Sciences and Fisheries, West Pomeranian University of Technology, 70-310 Szczecin, Poland; kamila.pokorska@zut.edu.pl (K.P.-N.); jacek.cybulski@zut.edu.pl (J.C.)
[2] Department of Animal Reproduction Biotechnology and Environmental Hygiene, Faculty of Biotechnology and Animal Husbandry, West Pomeranian University of Technology, 70-310 Szczecin, Poland; agnieszka.tomza-marciniak@zut.edu.pl
[3] Department of Gynecological Surgery and Gynecological Oncology of Adults and Adolescents, Pomeranian Medical University, 70-204 Szczecin, Poland; grzegorzwitczak9@gmail.com
[4] Provincial Veterinary Inspectorate, 71-337 Szczecin, Poland; ola.aftyka@wp.pl
* Correspondence: agata.witczak@zut.edu.pl

**Abstract:** Selenium is a trace element that can be poisonous in small quantities. The aim of this study was to analyze the change in the content of selenium in drinking water, raw water, as well as treated and raw wastewater in an annual cycle in the city of Szczecin. The concentration of Se in samples was determined using the spectrofluorometric method at a 518 nm emission wavelength and a 378 nm excitation wavelength. The amount of selenium in drinking water ranged from <LOD to 0.007 µg/mL, in raw water, from 0.001 to 0.006 µg/mL, in raw wastewater, from 0.001 to 0.008 µg/mL, and in treated wastewater, from 0.001 to 0.009 µg/mL. The selenium content did not exceed the maximum allowable concentration (MAC), 0.010 µg/mL, in any of the water samples tested.

**Keywords:** selenium; drinking water; water treatment; wastewater treatment

## 1. Introduction

Selenium (Se) is a trace element that is widespread in the environment, but it is particularly found in igneous and sedimentary rocks, waters, soils, plants, and living organisms. Selenium is released from both natural and anthropogenic sources (metallurgical, glass, and pigment-producing industries [1,2]. It occurs most frequently in one of two forms—organic, including seleno-amino acids, selenopeptides, and selenoproteins—and inorganic. Inorganic forms are more toxic and less bioavailable than organic forms [3]. Selenium in waters occurs in the forms of seleno-amino acids, selenides, selenates, and dimethyl and trimethyl derivatives. The form of occurrence depends on water pH and redox potential (Eh).

Surface waters and groundwaters contain highly variable concentrations of this element. Selenium concentrations in natural marine and fresh waters ranges from 0.01 to 0.1 µg/L [4]. In some areas selenium concentrations in groundwaters can reach levels as high as 6.000 µg/L; however, in water supply networks in developed countries it occurs in quantities that do not exceed 10 µg/L [5]. Industrial wastewater can contain much higher concentrations ranging from 0.1 to 20 mg/L [6,7].

Selenium is recognized as an essential but also potentially toxic element, and the margin between selenium deficiency and toxicity is very narrow [7]. Nutritional deficiencies occur at an intake of less than 40 µg/day, and can lead to the development of many disorders, such as Keshan disease [8,9]. It is recommended to take a prophylactic dose of Se of approximately 200 µg/day to reduce the risk of cancer in humans. However, an Se dose of >400 µg/day is considered to be potentially toxic [10,11]. Excess Se can lead to the development of cancers, type II diabetes, and diseases of the circulatory and endocrine systems [7,12]. As much as 80% of this element can be absorbed in the gastrointestinal tract [13]. Concentrations of selenium in humans are determined mainly by their diet, but the intake of it can be increased by drinking water that has high concentrations of selenium [14].

Selenium requirements differ depending on age, and range from 17 to 50 µg/day, but requirements increase with excess stress, overconsumption of alcohol and nicotine, and increased physical exertion [15]. Among other things, selenium is a component of selenoproteins that are involved in antioxidant defense, DNA synthesis, and thyroid hormone production, and are essential to reproduction. Se inhibits the synthesis of osteopontin, an important protein in cancer metastasis [16]. Links have been confirmed between selenium deficiency and the occurrence of anxiety, depression, and affective disorders [13]. The recommended daily allowance (RDA) of selenium for adults is 55 µg [17].

Considering the health benefits and toxic effects of selenium in humans, the current study was undertaken with the aim of assessing changes in the selenium content of drinking water throughout the year in the city of Szczecin. The study also aimed to estimate the impact of water treatment on the selenium content of drinking water. Given that treated wastewater is discharged into surface waters and can become a secondary source of selenium in treated water, the extent to which wastewater treatment affects the final selenium content was also analyzed.

## 2. Materials and Methods

### 2.1. Materials

The study material was raw water, drinking water, and treated and raw wastewater. Raw water was collected from Lake Miedwie (Figure 1), which is the main source of water (80%) for the city of Szczecin.

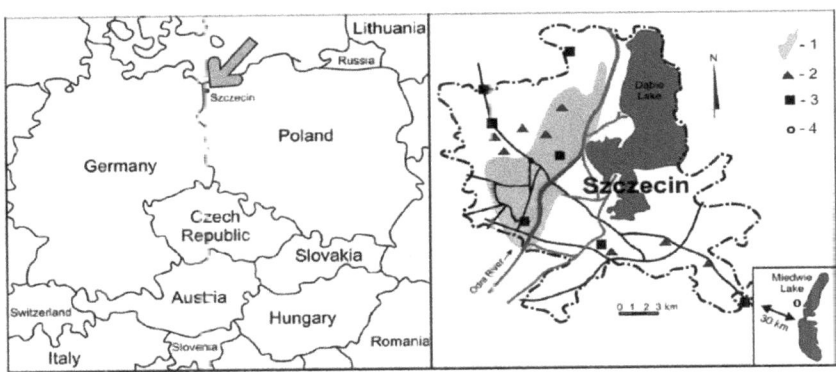

**Figure 1.** Location of the study area: 1: area supplied with water from Lake Miedwie—left-bank part of Szczecin (north, west and downtown districts); 2: pumping stations; 3: water treatment plant; 4: water sampling sites at the Żelewo Water Treatment Plant [18].

Water samples were taken from an intake located 6 m above the bottom and 16–18 m below the water surface of Lake Miedwie. The wastewater tested was from the Pomorzany Wastewater Treatment Plant (Szczecin, Poland), from which raw wastewater was sampled at the grating station, and treated wastewater was collected at the outflow of the canal. The study began in March 2018 and ran until March, 2019.

#### 2.1.1. Drinking Water Treatment

Raw water from Lake Miedwie is treated at the Żelewo Water Treatment Plant (Stare Czarnowo, Poland) (Figure 2).

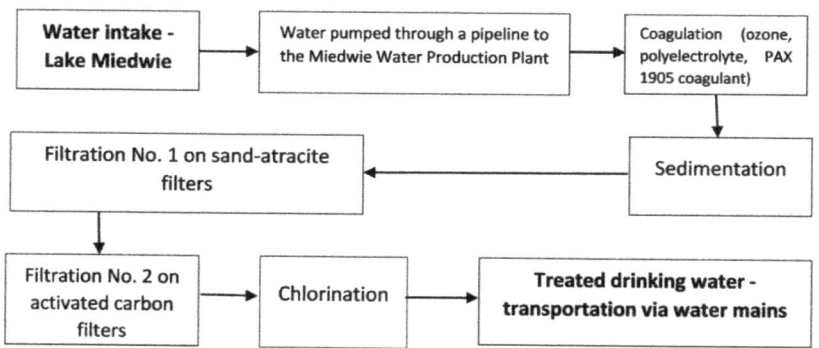

Figure 2. Model of the water treatment process in the Water Treatment Plant, Żelewo.

The first stage of treatment is pre-oxidation with an ozone dose of 1–2 g/m$^3$. The highly alkaline coagulant PAX XL 1905 (Kemipol, Police, Poland, with the following properties: pH—3.6 ± 0.4; alkalinity—85 ± 5%; density—1150 kg/m$^3$; aluminum content—6.0 ± 0.5%; chlorides—5.0 ± 1.0%) is added to the water with the ozone. Next, the water reacts with coagulants for 3 to 6 min. Then, the water flows into labyrinth chambers (20–40 min.), where a polyelectrolyte is added at a dose that depends on the current water quality and the amount of coagulant used. Sludge sedimentation can happen in the chambers, and this is washed out with an additional stream of water, after which, the water flows into horizontal settling tanks with a capacity for sedimentation of 1600 m$^3$ each. Sludge is removed continuously to sludge funnels and discharged into a sludge canal and then into the wastewater system. The water from the settling tanks is purified on carbon filter beds and then on anthracite-sand filters. The rapid anthracite-sand filters comprise 12 filtration chambers with a surface area of 46.17 m$^2$. The filter beds are 0.6 m layers of quartz sand and 40 cm layers of anthracite. The maximum speed of filtration is 10 m/h. Water from the filtration chamber flows into two indirect ozonation chambers where viruses and bacteria are eliminated and most pesticides are oxidized. Water is retained in the ozonation chambers for about 15 min at a flow rate of 3000 m$^3$/h. The water is filtered through 8 carbon filters arranged in two rows between which there is an ozonation line. In the last stage, the water is disinfected with chlorine dioxide and chlorine gas. When the water is appropriately pure, it is collected in two tanks with a combined capacity of 10,000 m$^3$, and from there, it is distributed through the water supply network to the city of Szczecin (materials from ZWiK, Department of Waterworks, Szczecin, Poland) [19].

2.1.2. Wastewater Treatment

Szczecin's wastewater is treated at the Pomorzany Wastewater Treatment Plant (Figure 3).

Wastewater flows into the expansion chamber, then through a canal in which there is a sampling station, and then, to the grating station fitted with two 40 mm mesh screens and six 6 mm mesh screens. The flow rate in the grating station is 5.4 m$^3$/s. The largest solids are removed with a screw compacting press, while the particle sizes of the solids retained on the fine screens are further reduced.

Once the largest solids are removed, the wastewater flows to the aerated grit chambers constructed of reinforced concrete with a degreasing chamber on the side. A bottom scraper collects fat and grease into the grease chamber, and then it is pumped into fermentation chambers. Minerals that sediment to the tank bottom are funneled into sand separators. From here, the wastewater flows to pre-settling tanks. The sludge that accumulates is collected with chain scrapers and is pumped into the sludge chamber. Fat and grease that rise to the surface are collected in a gutter and pumped to the fermentation chamber along with the fat and grease from the grit chambers. Mechanical treatment is followed by biological treatment. This part of the plant is equipped with devices that take measurements and steer processes such as how much air is released into the nitrification or denitrification chambers or the quantities required of iron coagulants PIX 113 (Kemipol, Poland, with the following properties: total iron $11.8 \pm 0.4\%$; density in kg/m$^3$ (20 °C) 1500–1570; PAX 16 (Kemipol, Poland, with the following properties: Al$_2$O$_3$ content $15.5 \pm 0.4\%$; chlorides (Cl$^-$) $19.0 \pm 2.0\%$; alkalinity $37.0 \pm 5.0\%$; density in kg/m$^3$ (20 °C) $1330 \pm 20$; pH $1.0 \pm 0.2$).

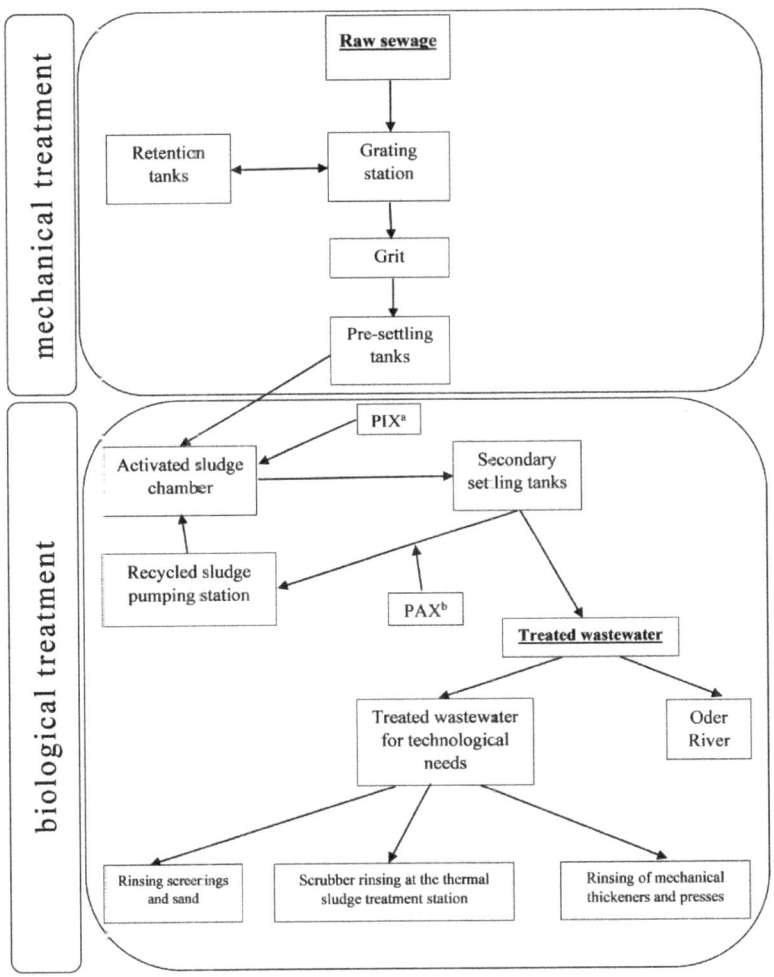

$^a$PIX- iron coagulant $^b$PAX- aluminum coagulant

**Figure 3.** Model of mechanical and biological treatment at the Pomorzany wastewater treatment plant.

## 2.2. Methods

High-purity analytical reagents from Merck (Darmstadt, Germany) were used in the study. Selenium concentrations in samples were determined with the spectrofluorimetric method. Water samples were digested in HNO$_3$ for 180 min at a temperature of 230 °C and in HClO$_4$ for 20 min at a temperature of 310 °C. After mineralization, a solution of 9% HCl was added to the samples to reduce selenate to selenite. Then, the Se was derivatized in an acidic environment (pH 1–2), which resulted in the formation of a selenodiazole complex, which was then extracted with cyclohexane. Selenium concentrations were determined with an RF-5001 PC fluorescence spectrophotometer (Shimadzu, Tokyo, Japan) at a emission wavelength of 518 nm and an excitation wavelength of 378 nm.

The accuracy of the analytical method was tested with Certified Reference Materials: Trace Metals in Drinking Water Solution A (CRM-TMDW-A; High-Purity Standards, North Charleston, SC, USA) (for water and Certified Reference Material ERM®-CA713 (IRMM, Geel, Belgium) for sewage. The recoveries of these reference materials were, respectively, 96.3 ± 3.2% and −98.8 ± 2.9%. In addition our own standard material, selenium content was used, whose recoveries ranged from 91 to 95%. The analytical procedure was verified with blank samples (20 replicates). The relative standard deviation (RSD%) of the determinations was 2.96%, while the limit of detection (LOD, x + 3σ) was 0.001 µg/mL.

Results for the remaining parameters, including, among others, alkalinity and nitrates, were provided by the laboratories at ZWIK Szczecin (Szczecin, Poland). The results of the study were analyzed with Statistica 13.0 (StatSoft, Kraków, Poland), and they are presented as arithmetic means with standard deviations (SD) and minimum and maximum values. The significance of the differences was tested with Tukey's post hoc test ($p < 0.05$).

## 3. Results

### 3.1. Analysis of Selenium in Water

The selenium content in raw water during the study period fluctuated from 0.0020 to 0.0068 µg/mL (average 0.0037 ± 0.0016 µg/mL), while in drinking water, it ranged from <LOD to 0.0052 µg/mL (average 0.0024 ± 0.0013 µg/mL) (Figure 4).

A positive significant correlation ($p < 0.05$) was noted in the content of Se in raw water and alkalinity, quantity of nitrates, COD, and UV absorbance $m^{-1}$ (Table 1). High water alkalinity can prevent soil components from absorbing Se in aquatic ecosystems and prevent high Se concentrations in groundwater. Under these conditions, Se is predominantly present in the form of labile selenate [20]. High nitrate concentrations are known to create weak oxidation conditions that inhibit microbial Se fixation, which can also cause increased Se concentrations in water [21].

Table 1. Correlations between selenium content and selected parameters and chemical compounds in drinking water and raw water.

| Drinking Water | | Raw Water | |
| --- | --- | --- | --- |
| Parameter | Pearson's Correlation Coefficient $p < 0.05$ | Parameter | Pearson's Correlation Coefficient $p < 0.05$ |
| pH | −0.440 | Absorbance in UV $m^{-1}$ | 0.314 |
| alkalinity (mmol/L) | 0.353 | alkalinity (mmol/L) | 0.762 |
| nitrates ($NO_3$/L) | 0.362 | nitrates ($NO_3$/L) | 0.350 |
| chlorine dioxide (mg $ClO_2$/L) | 0.502 | COD with the permanganate method mg $O_2$/L | 0.287 |

The differences in selenium content between raw and drinking water in individual months were statistically insignificant ($p < 0.05$) except in June, July, August, and October. The significance of differences ($p < 0.05$) in selenium content in individual months of the period analyzed is presented in Table A1 (Appendix A). Raw water treatment was shown to significantly reduce Se in drinking water by 4 to 70% (Figure 5).

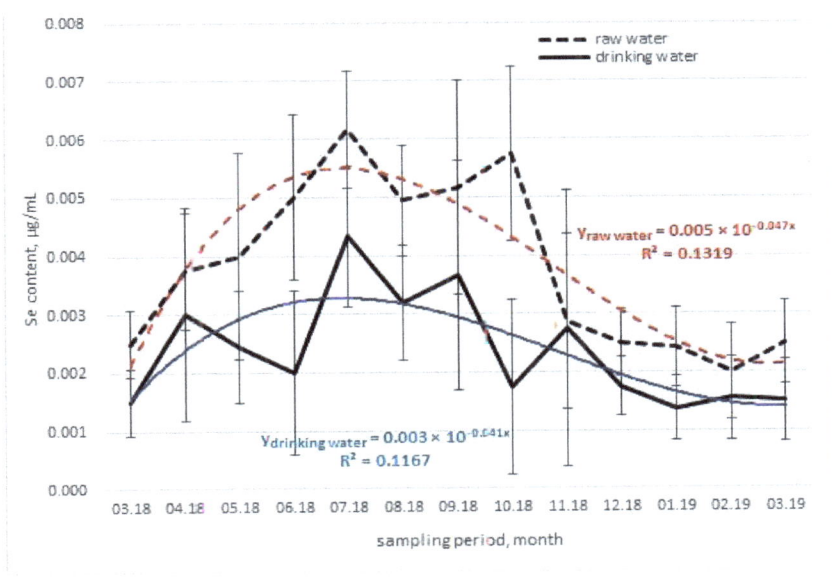

**Figure 4.** Changes in selenium content in drinking water and raw water in an annual cycle. MAC—the maximum allowable concentration [22].

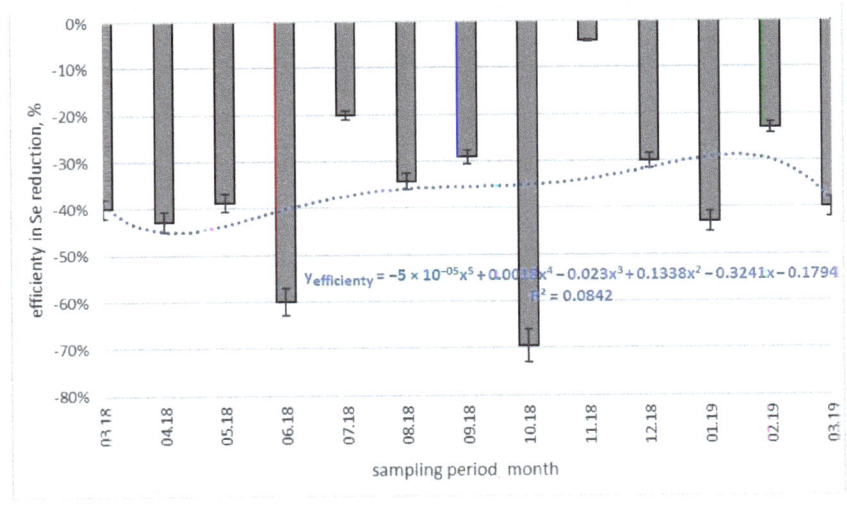

**Figure 5.** Changes in selenium content in water after treatment.

### 3.2. Analyses of Selenium Content in Treated and Raw Wastewater

The average selenium contents in raw and treated wastewater during the study period were $0.007 \pm 0.002$ µg/mL and $0.005 \pm 0.002$ µg/mL, respectively (Figure 6). The statistical analysis (Tukey's test) did not reveal significant ($p < 0.05$) differences between the selenium contents in treated and raw wastewater.

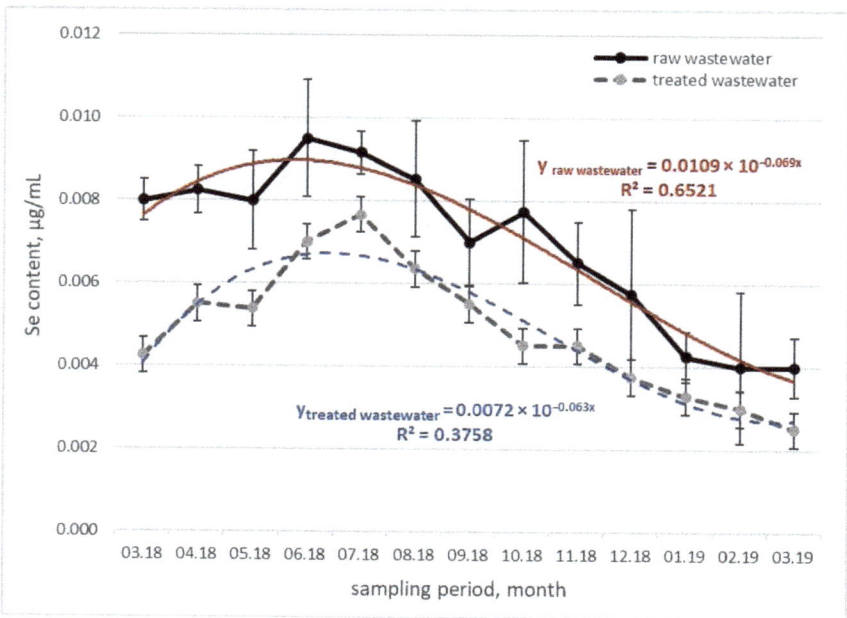

Figure 6. Selenium content in raw and treated wastewater in an annual cycle.

Significantly more selenium was noted in wastewater treated in June and July, 2018 than in other months. Conversely, significantly lower Se content was noted in wastewater treated in the winter months. The significance of the differences (Tukey's test, $p < 0.05$) between the selenium content in treated water and raw sewage in the different months of the period analyzed are presented in Table A2 (Appendix A). Wastewater treatment reduced selenium content significantly (Figure 7).

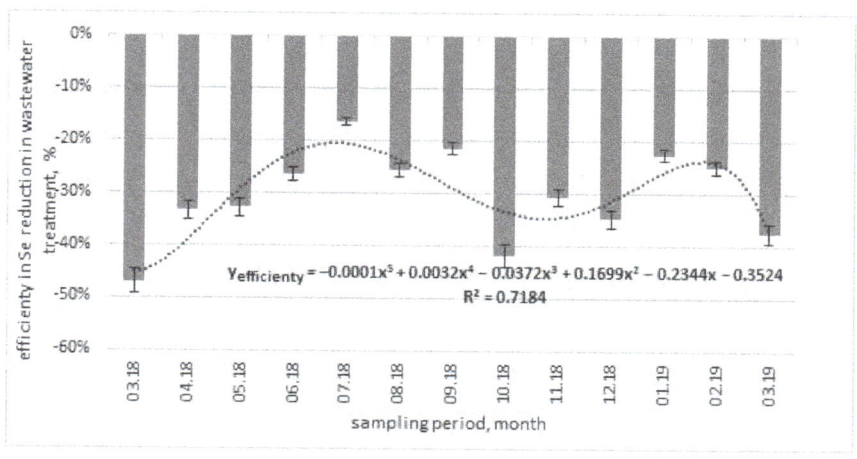

Figure 7. Changes in wastewater selenium content after treatment.

## 3.3. Selenium Consumption with Drinking Water and Consumer Safety

The permissible content of selenium in drinking water in Poland should not exceed 0.010 μg/mL [22]. Based on the results obtained, the amount of Se in drinking water was at a low level (Figure 8). In July 2018, however, the content of this element was over 40% of the maximum residue level (MAC; Figure 8).

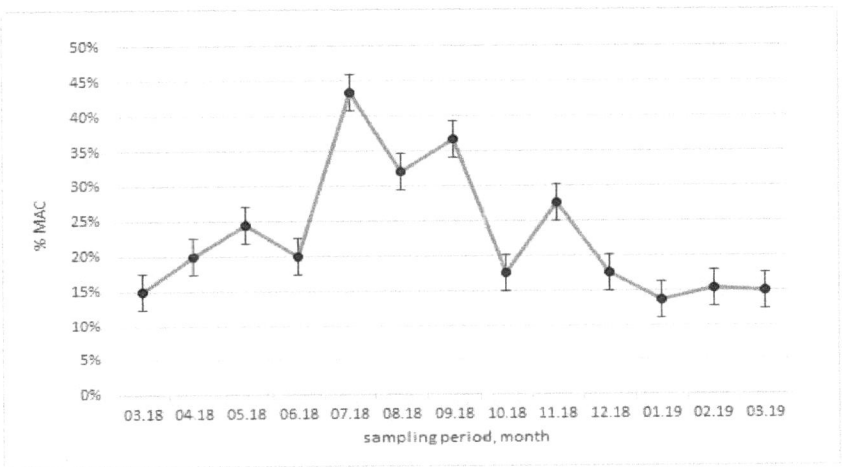

**Figure 8.** Selenium content in drinking water compared to MAC values.

## 4. Discussion

The content of selenium in surface water is influenced primarily by the geological environment of water reservoirs, e.g., selenium-rich rocks that are washed out over time. Another factor influencing Se content in water is contact with selenium-rich wastes from metallurgical, chemical, and photoelectric industries. The US EPA [23] established limits for lentic and lotic waters of 1.5 μg/L and 3.1 μg/L, respectively. According to the Regulation of the Minister of Maritime Economy and Inland Navigation [24], the permissible selenium concentration in surface waters in Poland is 10 μg/L. Given the risk of ingesting large amounts of selenium with drinking water, the process of removing this element at individual stages of water treatment was examined [6,25,26].

Selenium in drinking water, groundwater, and wastewater is a global problem. Although selenium health benefits and toxicity are well known, to date, no safe, optimal content of this element in drinking water has been established [27,28]. European standards are not uniform. The World Health Organization (WHO) standard for drinking water is 40 μg/L [29], while the US EPA [30] upper limit is 50 μg/L. The highest, permissible selenium concentration in drinking water set forth in Directive (EU) [31] is 20 μg/L. In Poland, the maximum allowable concentration (MAC) for selenium in drinking water was set forth in a regulation of the Minister of Health at 0.010 μg/mL [22].

Similar problems regarding a lack of standards are found in the selenium content of wastewater [3,28,32,33]. In most countries, Se content in drinking water is less than 10 μg/L [21,29]. Selenium content in drinking water samples from the USA, Canada, and Australia rarely exceed this value [14,34,35], but values as high as 160 μg/L have been reported in China [36]. Selenium content varies in drinking water in Europe; in Germany, values of 0.02–0.03 μg/L [37,38] have been noted, while in Slovenia, the value is 0.2 μg/L [39]. In the current study, the average content of this element in drinking water fluctuated around 2.4 μg/L, but it did not exceed the safe value for consumers. According to the WHO, selenium content in groundwater and surface water globally range from 0.06 ng/mL to 0.4 μg/mL, while in specific cases, it can reach 6 μg/mL [3,29]. The minimum and maximum values cited above do not differ from the selenium content in raw water in Szczecin.

In the current study, the slightly higher selenium content in raw water than in drinking water could have come from field and pasture fertilisers, the main component of which is selenium. This element can also reach water reservoirs with precipitation. Changes in selenium content in raw water can also be affected by the presence of other elements with which it forms compounds. Changes in selenium content in treated water could also be caused by the use of PAX aluminum coagulants that can alter water pH, among other things.

Water that is treated for drinking water usually contains less than 0.1 mg Se/L, while values in industrial water are above 1 mg/L. Wastewater that contains selenium is often associated with other substances and high salinity [40,41]. Consequently, the choice and effectiveness of treatment processes depend not only on the degree of selenium oxidation but also on the presence and concentrations of other contaminants and on other factors, including existing treatment plants and processes, treatment objectives, and concerns regarding waste treatment and costs [28]. There is no single method that ensures adequate water treatment, and, in practice, various chemical, biological, and physical methods are applied to attain the desired water quality [28,42]. For drinking water, certain local management solutions are implemented to maintain selenium below threshold levels without increasing costs to consumers. For example, groundwater can be cleaned through sand filtration combined with ion exchange resins and membrane treatments, e.g., microfiltration and nanofiltration, which can remove up to 95% of selenium [42]. However, such methods are often poorly adapted, not highly selective, and expensive. Thus, innovation is required to develop water treatment methods that are efficient, inexpensive, technologically feasible, and environmentally friendly. In France, these solutions generally aim to either request that the responsible authorities grant operational exemptions, or, most frequently, to seek other water sources [28]. When removing selenium on an industrial scale, the first possible method is to use iron co-precipitation and adsorption, and, if necessary, to combine this with coagulation–flocculation [27,28]. The principal technologies used to remove Se are the following: a nanofiltration membrane-based process [43], coagulation [44], phytoremediation [45], precipitation [46], and adsorption [6,47]. The water treatment solutions described in the present study fulfilled their roles.

Water tests were supplemented with wastewater analyses from the same study period. The highest allowable selenium concentration in all types of water is 1 mg/L [48]. At no time during the period analyzed was this limit exceeded in the current study. The causes of fluctuations in selenium content in treated wastewater could have been the PIX iron coagulant used, the components of which could have formed compounds with selenium. Coagulants are added to wastewater during treatment in various quantities depending on, among other things, oxygen and nitrogen concentrations. PAX coagulants, which can also influence Se contents in wastewater, are added when filamentous bacteria occur, which is also linked with fluctuations. The results of this study indicated that the likelihood of treated wastewater potentially increasing selenium concentrations in sources used to produce drinking water was low.

Despite the wastewater treatment, it is impossible to completely eliminate selenium, which returns to water reservoirs together with the treated wastewater. Therefore, treated wastewater may be a source of this element in the environment. Consumers and industrial plants can influence Se content. As a result of using drinking water, the level of selenium can be decreased or increased by discharging substances rich in this element.

Human requirements for water vary depending primarily on age and sex. The Institute of Food and Nutrition in Warsaw set recommended standards for $H_2O$ consumption, taking into consideration the liquid form and water consumed with foods (Table 2). The recommended daily allowance (RDA) of selenium for adults is 55 μg/day (Table 2), and the tolerable upper intake is 400 μg/day, as set by the US Institute of Medicine of the National Academy of Sciences [49], while selenium intake in excess of 5 mg/day can be fatal [50,51].

Table 2. Selenium requirements in different age categories [52].

| Age | RDA—Recommended Daily Allowance for Selenium (µg/Day) * | Recommended Water Consumption mL/Day | Daily Se Requirement Met by Drinking the Water Tested (%) |
|---|---|---|---|
| 1–3 | 20 | 1250 | 15 |
| 4–9 | 30 | 1600 (4 YOA)–1750 (9 YOA) | 13 (4 YOA)–14 (9 YOA) |
| 10–12 | 40 | 1900 ♀; 2100 ♂ | 11 ♀; 13 ♂ |
| 13–15 | 55 | 1950 ♀; 2350 ♂ | 9 ♀; 10 ♂ |
| ≥16 | 55 | 2000 ♀; 2500 ♂ | 9 ♀; 11 ♂ |
| Pregnant women | 60 | 2300 | 9 |
| Breastfeeding women | 70 | 2700 | 9 |

Note(s): * nutrition standards for the Polish population according to Jarosz et al. [52].

## 5. Conclusions

The selenium content in surface waters is influenced, among others, by the type of geochemical environment, leaching processes from rocks, and environmental pollution. Lake Miedwie, which was the source of the tested water, is characterized by a high content of organic matter, which affects the amount of nitrates in the water. In addition, the catchment area of Lake Miedwie is mainly agricultural land, where, among others, nitrogen fertilizers are also used. These compounds, as well as other polluting chemicals, may enter the waters of Lake Miedwie as a result of soil leachate and surface runoff of rainwater.

The amount of precipitation during the year also affects the presence of selenium in the analyzed water reservoir.

Selenium content in raw water was the highest in summer, which correlated with higher contents of nitrates (III). Usually in the summer period (June–August), there is also the largest leachate from the soil to the lake caused by heavier rainfall, which is short and very intense.

The removal of selenium from raw water and wastewater is difficult because the metalloid is often present in various complex forms and can also form various compounds with other elements.

The higher content of Se in raw water may be caused by fertilizing fields and pastures with fertilizers whose main component is selenium.

The water treatment process lowered selenium content in drinking water by a range of 4 to 70%.

The processes used during water treatment affect the reduction in the selenium content in the water. The changes in the selenium content in water after treatment could also be caused by the use of PAX aluminum coagulants, causing, among others, changes in water pH.

As a results of treatment process, the selenium content in drinking water was low and ranged from 14 to 43% of the MAC value.

Treated wastewater can be a source of selenium in the environment and, discharged into surface waters, can become a secondary source of selenium in treated water. In our study, treating wastewater resulted in lowering selenium content in treated wastewater by as much as 47%.

**Author Contributions:** Conceptualization, A.W., K.P.-N. and G.W.; methodology, A.T.-M.; validation, A.W.; formal analysis, A.W. and K.P.-N.; investigation, A.W., A.T.-M. and J.C.; writing—A.W., K.P.-N. and A.A.; visualization, A.W. and K.P.-N. All authors have read and agreed to the published version of the manuscript.

**Funding:** This research received no external funding.

**Institutional Review Board Statement:** Not applicable.

**Informed Consent Statement:** Not applicable.

**Data Availability Statement:** Data available on request.

**Conflicts of Interest:** The authors declare no conflict of interest.

## Appendix A

Table A1. Significant differences (in bold type) in selenium content in raw and drinking water among sampling periods ($p < 0.05$)—Tukey's test.

| Sampling Period | Raw Water | | | | | | | | | | | |
|---|---|---|---|---|---|---|---|---|---|---|---|---|
| | 03.18 | 04.18 | 05.18 | 06.18 | 07.18 | 08.18 | 09.18 | 10.18 | 11.18 | 12.18 | 01.19 | 02.19 |
| 03.18 | | | | | | | | | | | | |
| 04.18 | 0.9947 | | | | | | | | | | | |
| 05.18 | 0.8871 | 1.0000 | | | | | | | | | | |
| 06.18 | 0.7302 | 0.9911 | 0.9998 | | | | | | | | | |
| 07.18 | 0.0826 | 0.6210 | 0.7527 | 1.0000 | | | | | | | | |
| 08.18 | 0.3020 | 0.9383 | 0.9912 | 1.0000 | 0.9999 | | | | | | | |
| 09.18 | 0.1560 | 0.7971 | 0.9166 | 1.0000 | 1.0000 | 1.0000 | | | | | | |
| 10.18 | **0.0320** | 0.3750 | 0.7425 | 1.0000 | 1.0000 | 0.9983 | 1.0000 | | | | | |
| 11.18 | 1.0000 | 0.9999 | 0.9855 | 0.8859 | 0.2082 | 0.5650 | 0.3475 | 0.0923 | | | | |
| 12.18 | 1.0000 | 0.9947 | 0.8871 | 0.7302 | 0.0826 | 0.3020 | 0.1560 | **0.0320** | 1.0000 | | | |
| 01.19 | 1.0000 | 0.9894 | 0.8456 | 0.6881 | 0.0658 | 0.2548 | 0.1272 | **0.0249** | 1.0000 | 1.0000 | | |
| 02.19 | 1.0000 | 0.8871 | 0.5574 | 0.4661 | **0.0192** | 0.0947 | **0.0409** | **0.0066** | 0.9985 | 1.0000 | 1.0000 | |

| Sampling Period | Drinking Water | | | | | | | | | | | |
|---|---|---|---|---|---|---|---|---|---|---|---|---|
| | 03.18 | 04.18 | 05.18 | 06.18 | 07.18 | 08.18 | 09.18 | 10.18 | 11.18 | 12.18 | 01.19 | 02.19 |
| 03.18 | | | | | | | | | | | | |
| 04.18 | 1.0000 | | | | | | | | | | | |
| 05.18 | 0.9965 | 1.0000 | | | | | | | | | | |
| 06.18 | 1.0000 | 1.0000 | 1.0000 | | | | | | | | | |
| 07.18 | 0.0999 | 0.3155 | 0.3321 | 0.8034 | | | | | | | | |
| 08.18 | 0.7738 | 0.9756 | 0.9976 | 0.9989 | 0.9276 | | | | | | | |
| 09.18 | 0.4269 | 0.7929 | 0.8889 | 0.9783 | 0.9992 | 1.0000 | | | | | | |
| 10.18 | 1.0000 | 1.0000 | 0.9998 | 1.0000 | 0.1851 | 0.9075 | 0.6151 | | | | | |
| 11.18 | 0.9660 | 0.9997 | 1.0000 | 1.0000 | 0.8422 | 1.0000 | 0.9975 | 0.9945 | | | | |
| 12.18 | 1.0000 | 1.0000 | 0.9998 | 1.0000 | 0.1851 | 0.9075 | 0.6151 | 1.0000 | 0.9945 | | | |
| 01.19 | 1.0000 | 0.9999 | 0.9897 | 1.0000 | 0.0715 | 0.6866 | 0.3416 | 1.0000 | 0.9334 | 1.0000 | | |
| 02.19 | 1.0000 | 1.0000 | 0.9977 | 1.0000 | 0.1113 | 0.8003 | 0.4571 | 1.0000 | 0.9737 | 1.0000 | 1.0000 | |

Table A2. Significant differences (in bold type) in selenium content in treated and raw wastewater among sampling periods ($p < 0.05$).

| Sampling Period | Treated Wastewater | | | | | | | | | | | |
|---|---|---|---|---|---|---|---|---|---|---|---|---|
| | 03.18 | 04.18 | 05.18 | 06.18 | 07.18 | 08.18 | 09.18 | 10.18 | 11.18 | 12.18 | 01.19 | 02.19 |
| 03.18 | | | | | | | | | | | | |
| 04.18 | 0.9592 | | | | | | | | | | | |
| 05.18 | 0.9817 | 1.0000 | | | | | | | | | | |
| 06.18 | 0.5562 | 0.9887 | 0.9784 | | | | | | | | | |
| 07.18 | **0.0140** | 0.3884 | 0.0889 | 1.0000 | | | | | | | | |
| 08.18 | 0.4342 | 0.9984 | 0.9701 | 1.0000 | 0.8040 | | | | | | | |
| 09.18 | 0.9592 | 1.0000 | 1.0000 | 0.9887 | 0.1332 | 0.9900 | | | | | | |
| 10.18 | 1.0000 | 0.9931 | 0.9980 | 0.6932 | **0.0311** | 0.6276 | 0.9931 | | | | | |
| 11.18 | 1.0000 | 0.9931 | 0.9980 | 0.6932 | **0.0311** | 0.6276 | 0.9931 | 1.0000 | | | | |
| 12.18 | 1.0000 | 0.7064 | 0.7934 | 0.3017 | **0.0026** | 0.1511 | 0.7064 | 0.9996 | 0.9996 | | | |
| 01.19 | 0.9953 | 0.3596 | 0.4492 | 0.1442 | **0.0006** | **0.0428** | 0.3596 | 0.9682 | 0.9682 | 1.0000 | | |
| 02.19 | 0.9592 | 0.1939 | 0.2567 | 0.0838 | **0.0003** | **0.0172** | 0.1939 | 0.8661 | 0.8661 | 0.9996 | 1.0000 | |
| 03.19 | 0.9621 | 0.4210 | 0.4874 | **0.0297** | **0.0064** | 0.1107 | 0.4210 | 0.9061 | 0.9061 | 0.9978 | 1.0000 | 1.0000 |

Table A2. Cont.

| Sampling Period | Raw Wastewater | | | | | | | | | | | |
|---|---|---|---|---|---|---|---|---|---|---|---|---|
| | 03.18 | 04.18 | 05.18 | 06.18 | 07.18 | 08.18 | 09.18 | 10.18 | 11.18 | 12.18 | 01.19 | 02.19 |
| 03.18 | | | | | | | | | | | | |
| 04.18 | 1.0000 | | | | | | | | | | | |
| 05.18 | 1.0000 | 1.0000 | | | | | | | | | | |
| 06.18 | 0.9995 | 0.9999 | 0.9995 | | | | | | | | | |
| 07.18 | 0.9987 | 0.9999 | 0.9919 | 1.0000 | | | | | | | | |
| 08.18 | 1.0000 | 1.0000 | 1.0000 | 1.0000 | 1.0000 | | | | | | | |
| 09.18 | 0.9997 | 0.9976 | 0.9980 | 0.9565 | 0.5964 | 0.9357 | | | | | | |
| 10.18 | 1.0000 | 1.0000 | 1.0000 | 0.9978 | 0.5925 | 1.0000 | 1.0000 | | | | | |
| 11.18 | 0.9878 | 0.9595 | 0.9878 | 0.8590 | 0.5891 | 0.8922 | 1.0000 | 0.9976 | | | | |
| 12.18 | 0.8058 | 0.6812 | 0.8058 | 0.5975 | 0.2236 | 0.5274 | 0.9976 | 0.9007 | 1.0000 | | | |
| 01.19 | 0.1259 | 0.0780 | 0.1259 | 0.1348 | 0.0104 | 0.0441 | 0.5424 | 0.1951 | 0.8058 | 0.9878 | | |
| 02.19 | 0.0780 | 0.0467 | 0.0780 | 0.0969 | 0.0057 | 0.0255 | 0.4070 | 0.1259 | 0.6812 | 0.9595 | 1.0000 | |

## References

1. Tan, L.C.; Nancharaiah, Y.V.; van Hullebusch, E.D.; Lens, P.N. Selenium: Environmental significance, pollution, and biological treatment technologies. *Biotechnol. Adv.* **2016**, *34*, 886–907. [CrossRef] [PubMed]
2. Naga Jyothi, M.S.V.; Ramaiah, B.J.; Maliyekkal, S.M. Occurrence, contamination, speciation and analysis of selenium in the environment. In *Energy, Environment, and Sustainability*; Springer: Singapore, 2020; pp. 245–269. [CrossRef]
3. Devi, P.; Singh, P.; Malakar, A.; Snow, D. *Selenium Contamination in Water*; John Wiley & Sons Ltd.: Hoboken, NJ, USA, 2021. [CrossRef]
4. Matulová, M.; Bujdoš, M.; Miglierini, M.B.; Cesnek, M.; Duborská, E.; Mosnáčková, K.; Vojtková, H.; Kmječ, T.; Dekan, J.; Matúš, P.; et al. The effect of high selenite and selenate concentrations on ferric oxyhydroxides transformation under alkaline conditions. *Int. J. Mol. Sci.* **2021**, *22*, 9955. [CrossRef] [PubMed]
5. Nordberg, G.F.; Fowler, B.A.; Nordberg, M.; Friberg, L. *Handbook on the Toxicology of Metals*, 3rd ed.; Academic Press: Burlington, NJ, USA, 2007; p. 733.
6. Benis, K.Z.; McPhedran, K.N.; Soltan, J. Selenium removal from water using adsorbents: A critical review. *J. Hazard. Mater.* **2022**, *424*, 127603. [CrossRef] [PubMed]
7. Stefaniak, J.; Dutta, A.; Verbinnen, B.; Shakya, M.; Rene, E.R. Selenium removal from mining and process wastewater: A systematic review of available technologies. *J. Water Supply Res. Technol.–AQUA* **2018**, *67*, 903–918. [CrossRef]
8. WHO. *Trace Elements in Human Nutrition and Health*; WHO: Geneva, Switzerland, 1996.
9. Wu, J.; Huang, X.; Chen, H.; Gou, M.; Zhang, T. Fabrication of Cu-Al2O3/ceramic particles by using brick particles as supports for highly-efficient selenium adsorption. *J. Environ. Chem. Eng.* **2020**, *9*, 105008. [CrossRef]
10. Hammouh, F.; Zein, S.; Amr, R.; Ghazzawi, H.; Muharib, D.; Al Saad, D.; Subih, H. Assessment of dietary selenium intake of Jordanian adults in Madaba: A cross sectional study. *Nutr. Food Sci.* **2020**, *51*, 494–506. [CrossRef]
11. Rayman, M.P. The importance of selenium to human health. *Lancet* **2000**, *356*, 233–241. [CrossRef]
12. Chawla, R.; Filippini, T.; Loomba, R.; Ciloni, S.; Dhillon, K.S.; Vinceti, M. Exposure to a high selenium environment in Punjab, India: Biomarkers and health conditions. *Sci. Total Environ.* **2020**, *719*, 134541. [CrossRef]
13. Mehdi, Y.; Hornick, J.L.; Istasse, L.; Dufrasne, I. Selenium in the environment, metabolism and involvement in body functions. *Molecules* **2013**, *18*, 3292–3311. [CrossRef]
14. Ćurković, M.; Sipos, L.; Puntarić, D.; Dodig-Ćurković, K.; Pivac, N.; Kralik, K. Arsenic, copper, molybdenum, and selenium exposure through drinking water in rural Eastern Croatia. *Pol. J. Environ. Stud.* **2016**, *25*, 981–992. [CrossRef] [PubMed]
15. Ambroziak, A. Selen-niedoceniony antyoksydant. Selenium—An underestimated antioxidant. *Prz. Mlecz.* **2014**, *12*, 45–49. (In Polish)
16. Golonko, A.; Matejczyk, M. Dwa oblicza selenu. Selected aspects of biological activity of selenium. *Bud. Inż. Środ.* **2018**, *9*, 65–74. (In Polish)
17. Regulation (EU) No 1169/2011 of the European Parliament and of the Council of 25 October 2011 on the Provision of Food Information to Consumers, Amending Regulations (EC) No 1924/2006 and (EC) No 1925/2006 of the European Parliament and of the Council, and Repealing Commission Directive 87/250/EEC, Council Directive 90/496/EEC, Commission Directive 1999/10/EC, Directive 2000/13/EC of the European Parliament and of the Council, Commission Directives 2002/67/EC and 2008/5/EC and Commission Regulation (EC) No 608/2004 Text with EEA Relevance. Available online: https://eur-lex.europa.eu/legal-content/EN/ALL/?uri=CELEX:32011R1159 (accessed on 19 March 2023).
18. Górski, J.; Siepak, M. Assessment of metal concentrations in tapwater—From source to the tap: A case study from Szczecin, Poland. *Geologos* **2014**, *20*, 25–33. [CrossRef]
19. ZWIK. *Szczecin Water and Sewerage Department. Parameters of the Water Tested*; Unpublished work; ZWIK: Szczecin, Poland, 2019. (In Polish)

20. Korobova, E.M.; Ryzhenko, B.N.; Cherkasova, E.V.; Sedykh, E.M.; Korsakova, N.V.; Danilova, V.N.; Khushvakhtova, S.D.; Berezkin, V.Y. Iodine and selenium speciation in natural waters and their concentrating at landscape-geochemical barriers. *Geochem. Int.* **2014**, *52*, 500–514. [CrossRef]
21. Golubkina, N.; Erdenetsogt, E.; Tarmaeva, I.; Brown, O.; Tsegmed, S. Selenium and drinking water quality indicators in Mongolia. *Environ. Sci. Pollut. Res.* **2018**, *25*, 28619–28627. [CrossRef]
22. Journal of Laws, Item. 2294. Regulation of the Minister of Health of 7 December 2017 on the Quality of Water Intended for Human Consumption. 2017. Available online: https://isap.sejm.gov.pl/isap.nsf/DocDetails.xsp?id=WDU20170002294 (accessed on 13 April 2023). (In Polish)
23. U.S. EPA. Aquatic Life Ambient Water Quality Criterion for Selenium in Freshwater 2016. U.S. Environmental Protection Agency. June 2016, Washington, DC. EPA 822-R-16-006, 2016. Available online: https://www.epa.gov/sites/production/files/2016-07/documents/aquatic_life_awqc_for_selenium_-_freshwater_2016.pdf (accessed on 14 March 2023).
24. Journal of Laws, Item. 1747. Regulation of the Minister of Maritime Economy and Inland Navigation of 29 August 2019 on the Requirements for Surface Waters Used to Supply the Population with Water Intended for Human Consumption. 2019. Available online: https://isap.sejm.gov.pl/isap.nsf/DocDetails.xsp?id=WDU20190001747 (accessed on 13 April 2023). (In Polish)
25. Sharma, V.K.; Sohn, M.; McDonald, T.J. Remediation of selenium in water. A review. In *Advances in Water Purifcation Techniques*; Ahuja, S., Ed.; Chapter 8; Elsevier: Amsterdam, The Netherlands, 2019; pp. 203–218.
26. He, Y.; Xiang, Y.; Zhou, Y.; Yang, Y.; Zhang, J.; Huang, H.; Shang, C.; Luo, L.; Gao, J.; Tang, L. Selenium contamination, consequences and remediation techniques in waters and soils: A review. *Environ. Res.* **2018**, *164*, 288–301. [CrossRef]
27. Santos, M.D.; da Silva Júnior, F.M.R.; Zurdo, D.V.; Baisch, P.R.M.; Muccillo-Baisch, A.L.; Madrid, Y. Selenium and mercury concentration in drinking water and food samples from a coal mining area in Brazil. *Environ. Sci. Pollut. Res. Int.* **2019**, *26*, 15510–15517. [CrossRef]
28. Lichtfouse, E.; Morin-Crini, N.; Bradu, C.; Boussouga, Y.A.; Aliaskari, M.; Schäfer, A.I.; Das, S.; Wilson, L.D.; Ike, M.; Inoue, D.; et al. Technologies to remove selenium from water and wastewater. In *Environmental Chemistry for a Sustainable World*; Eric Lichtfouse, E., Schwarzbauer, J., Robert, D., Eds.; *Emerging Contaminants* 2(66); Springer Nature: Cham, Switzerland, 2021; pp. 207–304. ISBN 978-3-030-69089-2. [CrossRef]
29. WHO. *Guidelines for Drinking-Water Quality*; Organization of the United Nations: Geneva, Switzerland, 2011.
30. US EPA. *Ground and Drinking Water Fact Sheets: Selenium*; US EPA: Washington, DC, USA, 2015. Available online: https://safewater.zendesk.com/hc/en-us/sections/202346227-Selenium (accessed on 10 March 2023).
31. Directive (EU) 2020/2184 of the European Parliament and of the Council of 16 December 2020 on the Quality of Water Intended for Human Consumption. 2020. Available online: http://data.europa.eu/eli/dir/2020/2184/oj (accessed on 13 April 2023).
32. Kumrong, P.; LeBlanc, K.L.; Mercier, P.H.J.; Mester, Z. Selenium analysis in waters. Part 1: Regulations and standard methods. *Sci. Total Environ.* **2018**, *640*, 1611–1634. [CrossRef]
33. LeBlanc, K.L.; Kumkrong, P.; Mercier, P.H.J.; Mester, Z. Selenium analysis in waters. Part 2: Speciation methods. *Sci. Total Environ.* **2018**, *640*, 1635–1651. [CrossRef]
34. Paikaray, S. Origin, Mobilization and Distribution of Selenium in a Soil/Water/Air System: A Global Perspective with Special Reference to the Indian Scenario. *Clean-Soil Air Water* **2016**, *44*, 474–487. [CrossRef]
35. Abejón, R. A Bibliometric Analysis of Research on Selenium in Drinking Water during the 1990–2021 Period: Treatment Options for Selenium Removal. *Int. J. Environ. Res. Public Health* **2022**, *19*, 5834. [CrossRef] [PubMed]
36. World Health Organization. *Selenium in Drinking-Water*; World Health Organization: Geneva, Switzerland, 2003.
37. Veber, M.; Cujes, K.; Gomiscek, S. Determination of selenium and arsenic in mineral waters with hydride generation atomic absorption spectrometry. *J. Anal. At. Spectrom.* **1994**, *9*, 285–290. [CrossRef]
38. Tao, G.; Hansen, E.H. Determination of Ultra-trace Amounts of Selenium(IV) by Flow Injection Hydride Generation Atomic Absorption Spectrometry with On-line Preconcentration by Coprecipitation with Lanthanum Hydroxide. *Analyst* **1994**, *119*, 333–337. [CrossRef]
39. Niedzielski, P.; Siepak, M.; Siepak, J. Occurrence and content of arsenic, antimony and selenium in waters and other elements of the environment. *Rocz. Ochr. Srodowiska* **2020**, *2*, 317–340. (In Polish)
40. Kumar, B.S.; Priyadarsini, K.I. Selenium nutrition: How important is it? *Biomed. Prev. Nutr.* **2014**, *4*, 333–341. [CrossRef]
41. Kapoor, A.; Tanjore, S.; Viraraghavan, T. Removal of selenium from water and wastewater. *Int. J. Environ. Stud.* **1995**, *49*, 137–147. [CrossRef]
42. Crini, G.; Morin-Crini, N.; Fatin-Rouge, N.; Déon, S.; Fievet, P. Metal removal from aqueous media by polymer-assisted ultrafiltration with chitosan. *Arab. J. Chem.* **2017**, *10*, S3826–S3869. [CrossRef]
43. Malhotra, M.; Pal, M.; Pal, P. A response surface optimized nanofiltration-based system for efficient removal of selenium from drinking water. *J. Water Process. Eng.* **2020**, *33*, 101007. [CrossRef]
44. Kalaitzidou, K.; Bakouros, L.; Mitrakas, M. Techno-Economic Evaluation of Iron and Aluminum Coagulants on Se(IV) Removal. *Water* **2020**, *12*, 672. [CrossRef]
45. Monei, N.L.; Puthiya Veetil, S.K.; Gao, J.; Hitch, M. Selective removal of selenium by phytoremediation from post/mining coal wastes: Practicality and implications. *Int. J. Min. Reclam. Environ.* **2021**, *35*, 69–77. [CrossRef]
46. Geoffroy, N.; Demopoulos, G.P. The elimination of selenium(IV) from aqueous solution by precipitation with sodium sulfide. *J. Hazard. Mater.* **2011**, *185*, 148–154. [CrossRef]

47. Suazo-Hernández, J.; Sepúlveda, P.; Marquián-Cerda, K.; Ramírez-Tagle, R.; Rubio, M.A.; Bolan, N.; Sarkar, B.; Arancibia-Miranda, N. Synthesis and characterization of zeolite-based composites functionalized with nanoscale zero-valent iron for removing arsenic in the presence of selenium from water. *J. Hazard. Mater.* **2019**, *373*, 810–819. [CrossRef] [PubMed]
48. Journal of Laws, Item. 1311. Regulation of the Minister of Maritime Economy and Inland Navigation of 12 July 2019 on Substances Particularly Harmful to the Aquatic Environment and the Conditions to be Met When Discharging Sewage into Waters or Ground, as Well as When Discharging Rainwater or Meltwater into Waters or into Aquatic Devices. 2019. Available online https://isap.sejm.gov.pl/isap.nsf/DocDetails.xsp?id=WDU20190001311 (accessed on 13 April 2023). (In Polish)
49. Institute of Medicine (US) Panel on Dietary Antioxidants and Related Compounds. Vitamin C, vitamin E, selenium, and β-carotene and other carotenoids: Overview, antioxidant definition, and relationship to chronic disease. In *Dietary Reference Intakes for Vitamin C, Vitamin E, Selenium, and Carotenoids*; National Academies Press (US): Washington, DC, USA, 2000.
50. Mayer, M.A. Band structure engineering of $ZnO1-xSex$ alloys. *Appl. Phys. Lett.* **2010**, *97*, 022104. [CrossRef]
51. Sharma, S.; Sharma, A. Selenium Distribution and Chemistry in Water and Soil. In *Selenium Contamination in Water*; John Wiley & Sons Ltd.: Hoboken, NJ, USA, 2021. [CrossRef]
52. Jarosz, M.; Rychlik, E.; Stoś, K.; Charzewska, J. *Nutrition Standards for the Polish Population and Their Application*; National Institute of Public Health–National Institute of Hygiene: Warsaw, Poland, 2020; ISBN 978-83-65870-28-5. (In Polish)

**Disclaimer/Publisher's Note:** The statements, opinions and data contained in all publications are solely those of the individual author(s) and contributor(s) and not of MDPI and/or the editor(s). MDPI and/or the editor(s) disclaim responsibility for any injury to people or property resulting from any ideas, methods, instructions or products referred to in the content.

Article

# Health Risk Assessment of Nitrate and Fluoride in the Groundwater of Central Saudi Arabia

Talal Alharbi * and Abdelbaset S. El-Sorogy

Geology and Geophysics Department, College of Science, King Saud University, Riyadh 11451, Saudi Arabia; asmohamed@ksu.edu.sa
* Correspondence: tgalharbi@ksu.edu.sa

**Abstract:** High nitrate and fluoride contamination in groundwater cause a variety of disorders, including methemoglobinemia, teratogenesis, and dental and skeletal fluorosis. The present work assesses the non-carcinogenic health risks posed by nitrate and fluoride in infants, children, and adults using the daily water intake (CDI), hazard quotient (HQ), and non-carcinogenic hazard index (HI). Groundwater samples were collected from 36 wells and boreholes in three central Saudi Arabian study areas for nitrate and fluoride analysis using ionic chromatography and fluoride selective electrode, respectively. Nitrate concentrations varied from 0.70 to 47.00 mg/L. None of the 36 studied boreholes had nitrate levels that exceeded WHO guidelines (50.00 mg/L). Fluoride ranged from 0.63 to 2.00 mg/L, and 30.55% of the fluoride samples (11 out of 36) exceeded the WHO recommendations for acceptable drinking water (1.5 mg/L). The average hazard index (HI) values for adults, children, and infants were 0.99, 2.59, and 2.77, respectively. Water samples surpassed the safety level of 1 for adults, children, and infants at 44.44, 97.22, and 100%, respectively. Accordingly, water samples from Jubailah and a few from Wadi Nisah may expose infants, children, and adults to non-cancer health concerns. Infants and children are more vulnerable to non-carcinogenic health risks than adults, possibly due to their lower body weight. Immediate attention and remedial measures must be implemented to protect residents from the adverse effects of F- in the study area.

**Keywords:** nitrate; fluoride; hazard index; groundwater; Saudi Arabia

Citation: Alharbi, T.; El-Sorogy, A.S. Health Risk Assessment of Nitrate and Fluoride in the Groundwater of Central Saudi Arabia. *Water* **2023**, *15*, 2220. https://doi.org/10.3390/w15122220

Academic Editors: Weiying Feng, Fang Yang and Jing Liu

Received: 9 May 2023
Revised: 6 June 2023
Accepted: 12 June 2023
Published: 13 June 2023

**Copyright:** © 2023 by the authors. Licensee MDPI, Basel, Switzerland. This article is an open access article distributed under the terms and conditions of the Creative Commons Attribution (CC BY) license (https://creativecommons.org/licenses/by/4.0/).

## 1. Introduction

Groundwater is an important resource for drinking and irrigation, especially in dry and semi-arid regions [1,2]. However, shallow groundwater is vulnerable to contamination from various geogenic and anthropogenic sources, such as rock weathering, cation exchange during oxidation or reduction, rapid industrialization, urbanization, and excessive fertilizer use [3,4]. Recently, there has been a growing concern about groundwater quality and its influence on human health due to water consumption with high concentrations of nitrate and fluoride. Nitrate and fluoride concentrations in groundwater can negatively affect human health. These two toxic ions are listed as non-carcinogens by the US Environmental Protection Agency (USEPA) and have received worldwide attention for their devastating effects on human health [2].

Nitrate is an inorganic ion ($NO_3^-$) that naturally occurs in the nitrogen cycle and is widely found in nitrogen-containing fertilizers. Nitrate can enter groundwater through different natural and anthropogenic sources, e.g., rock-water interaction from the weathering of nitrite-bearing rocks, septic tanks, dairy lagoons, wastewater effluents, livestock waste, agricultural land, landfill leachate, fertilizers, pesticides, and manure application [5,6]. High levels of $NO_3$ in water bodies and drinking water create eutrophication, toxic algal blooms, and various diseases, including blue infant disorder (methemoglobinemia), thyroid disorders, teratogenesis, and mutagenesis [2,7,8]. Pregnant women, infants, and young children are most susceptible to the harmful effects of $NO_3$ [9].

The USEPA lists fluorine as a potentially harmful chemical pollutant [10,11]. Fluoride enters water bodies through natural and artificial sources, e.g., weathering of fluoride-bearing minerals, aluminum smelters, coal-based power stations, phosphatic fertilizer plants, brick manufacturing, steel production, coal combustion, sewerage, over-withdrawal of groundwater, and electroplating industries [12,13]. Drinking water with excessive fluoride can cause side effects including tooth decay (0.50 mg/L), fluorosis (1.50–5 mg/L), and skeletal fluorosis (5–40 mg/L) [14,15]. Arthritis, neurological problems, thyroid disease, cancer, infertility, hypertension, and a low fetus-to-sperm ratio are all linked to high fluoride concentrations (>10 mg/L). Moreover, fluoride alters DNA structure, which impacts teeth and bones [14,16,17].

The groundwater in central Saudi Arabia has been subjected to intense study in the last two decades regarding water resources, groundwater quality for drinking and agricultural usage, and general hydrochemical evaluations, e.g., [18–23]. Previous studies in the Northwest Riyadh area indicated that average concentrations of TDS, $Ca^{2+}$, $Na^+$, $K^+$, $Cl^-$, $SO_4^{2-}$, and $F^-$ exceeded the WHO's permissible limits for drinking water. The evaluation of groundwater quality from Wasia and Biyadh aquifers in Wasia Well Field (Northeast Riyadh) concluded the suitability of Wasia samples for drinking and agricultural purposes but not for industrial ones. Conversely, water samples from Biyadh are unsuitable for drinking, industrial, or agricultural purposes. The groundwater quality in Wadi Nisah, south of Riyadh, is unsuitable for drinking. Moreover, results of quality and groundwater contamination of Wadi Hanifa indicated concentrations of $SO_4^{2-}$, $Cl^-$, $Ca^{2+}$, $HCO_3^-$, $Na^+$, $Mg^{2+}$, and $K^+$ that are higher than the WHO standards for drinking water [24]. Previous studies concluded that extensive and repeated irrigation could increase ion levels in general. They attributed the higher $NO_3^-$ and $F^-$ values to the precipitation/dissolution of the carbonates and evaporites, as well as the widespread application of fertilizers and pesticides in the study area. In dry and semi-dry regions where groundwater is the main source of drinking water, nitrate and fluoride are two of the most common and hazardous toxins found in the water supply.

None of the previously mentioned studies on the groundwater of central Saudi Arabia addressed the health risk impacts of nitrate and fluoride on the people inhabiting the study area. Therefore, the main objectives of this study are to (1) determine the nitrate and fluoride contamination and distribution in groundwater of central Saudi Arabia, (2) document the potential sources of nitrate and fluoride in the collected groundwater, and (3) determine the human health risks of nitrate and fluoride via the ingestion pathway for adults, children, and infants using a methodology suggested by the USEPA. The results of this investigation will benefit risk management, the safeguarding of groundwater quality, and health professionals' decision-making.

## 2. Geological Setting

The exposed sedimentary succession in central Saudi Arabia is represented mainly by Triassic–Cretaceous rocks and is subdivided into the following formations from older to younger Minjur, Marrat, Dhruma, Tuwaiq Mountain Limestone, Hanifa, Jubaila, Arab, Hith, Wasia, and Aruma formations. The Jurassic and Cretaceous rocks in central Saudi Arabia have been described from many points of view, such as stratigraphy, paleontology, sedimentology, and depositional history, e.g., [25–39].

The major aquifer systems identified in the Wasia Well Field are Wasia (Middle Cretaceous) and Biyadh (Lower Cretaceous). Both Biyadh and Wasia have a primarily continental origin. The Biyadh aquifer has mainly cross-bedded quartzite and sandstone with some thin shale, marl, dolomite, and ironstone. The Wasia aquifer comprises medium- to coarse-grained, well-sorted, non-cemented, and poorly cemented rock. The Wadi Nisah area consists of sedimentary formations ranging from the Upper Triassic to the Quaternary period, with outcrops decreasing in age from west to east. Hydrogeologically, the study area consists of a multi-layered aquifer system with the Manjur, Biyadh, and Jurassic Limestone

Formations, Cretaceous Wasia, and Quaternary alluvial deposits forming the main water supply sources.

## 3. Material and Methods

Groundwater samples were collected from 36 groundwater wells mainly used for the agricultural water supply in central Saudi Arabia: 12 from Wadi Nisah, 14 from the Al Jubailah area, and 10 from the Wasia Well Field (Figure 1). The Al Jubailah area is located 40–55 km northwest of Riyadh, at 24°53′14.4″ to 24°54′59.4″ N and 46°20′41.5″ to 46°25′47.2″ E in Wadi Hanifa, which runs south through Riyadh. The Wasia Well Field is located some 110 km northeast of Riyadh between latitudes 25°09′ and 25°14′ N and longitudes 47°28′–47°33′ E. The samples were collected in pre-rinsed plastic bottles and filtered through 0.45-mm pore-size filters. Nitric acid was added to the samples for preservation, and the bottles were stored in cooling boxes at temperatures below 5 °C. $F^-$ and $NO_3^-$ levels were analyzed using fluoride selective electrode and ionic chromatography, respectively, in the laboratories of King Saud University. The accuracy of the results of the chemical analyses was checked by calculating the charge balance error of each sample, which showed a charge balance error of less than 5%, which is acceptable.

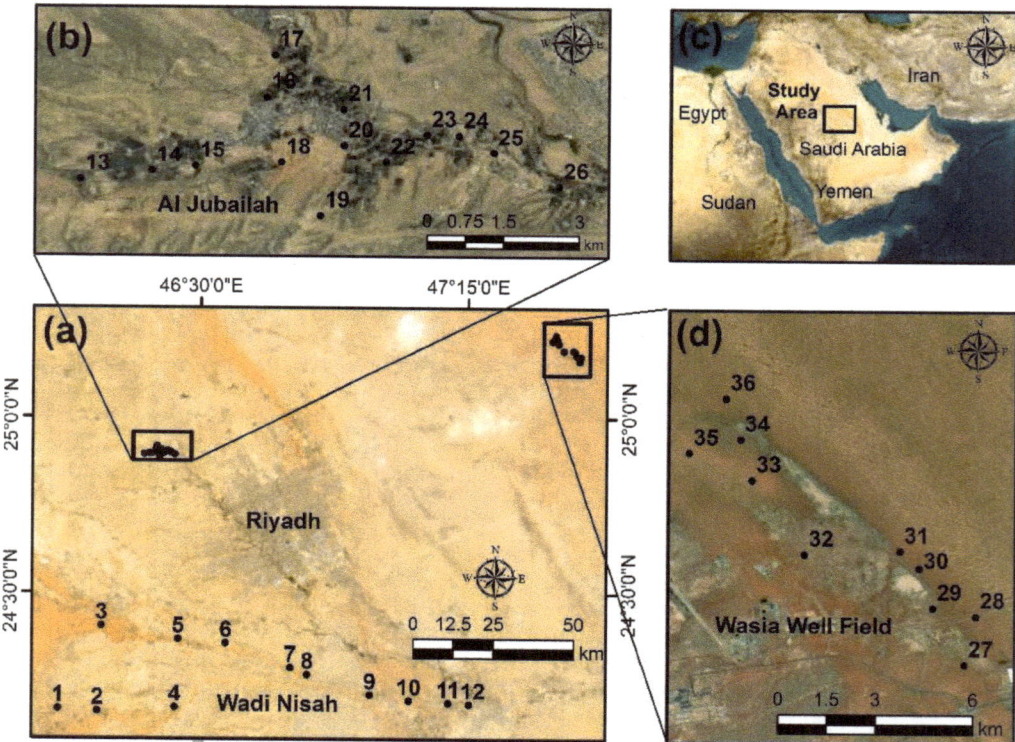

**Figure 1.** Locations of the groundwater samples in central Saudi Arabia. (**a**). Wadi Nisah, (**b**). Al Jubaillah, (**c**). Location of study in Saudi Arabia, (**d**). Wasia Well Field.

In this study, we assessed the danger that $NO_3^-$ and $F^-$ pose to human health for adults, children, and infants via the oral channel. The following equations calculate the daily water intake (CDI), hazard quotient (HQ), and non-carcinogenic hazard index (HI) associated with drinking water [40,41].

$$CDI = (C \times DI \times F \times ED)/(BW \times AT)$$

$$HQ = CDI/RfD$$

$$HI = \Sigma \, (HQfluoride + HQnitrate)$$

where C is the concentration of nitrate and fluoride in the water in milligrams per liter; DI is the amount of water consumed daily in liters; F is the number of days per year of exposure; ED is the number of years of exposure; BW is the weight of the age group under consideration in kilograms; AT is the average timing in days; and RfD is the reference dose ($NO_3^-$ = 1.6 and $F^-$ = 0.06 mg/kg/day) [42]. Table 1 describes the parameter values applied to the health exposure assessment.

Table 1. Parameters applied for health exposure assessment through drinking water and hazard index classification of in this work.

| Risk Exposure Factors | Unit | Adults | Children | Infants |
|---|---|---|---|---|
| DI | L/d | 2.0 | 1.5 | 0.8 |
| F | d/year | 365 | 365 | 365 |
| ED | years | 40 | 10 | 1.0 |
| BW | kg | 70 | 20 | 10 |
| AT | d | 14,600 | 3650 | 365 |
| HI $\leq$ 1 | no health risk to humans | | | |
| HI > 1 | higher level of hazard | | | |

## 4. Results

*4.1. Concentration and Distribution of Nitrate and Flouride*

The shallow groundwater levels provide a favorable condition for the leaching of nitrate into groundwater, resulting in the widespread nitrate contamination of groundwater in the study area [43]. It has been documented that the overused fertilizer cannot be fully utilized by plant roots and the maximum efficiency of N uptake from added inorganic fertilizers is around 50%, and consequently it can either be lost through denitrification, leaching, and volatilization, or be retained within the soil [44,45]. In groundwater, $F^-$ is naturally occurring or artificially affected [12]. The breakdown of fluoride-bearing minerals such as fluorite, hornblende, amphiboles, biotite, apatite, and muscovite are the natural source of $F^-$ [13]. Phosphatic fertilizer plants, over-withdrawal of groundwater, brick manufacturing, coal combustion, and sewerage are among the anthropogenic sources of $F^-$ in groundwater [46].

The spatial distribution of nitrate and fluoride in groundwater may vary depending on the local hydrogeological characteristics, such as aquifer type, depth, recharge rate, hydraulic conductivity, and porosity. The nitrate concentration in the study areas varied from 0.70 mg/L in borehole 33 (Wasia Well Field) to 47.0 mg/L in borehole 22 (Jubailah), with an average of 22.59 mg/L (Table 2). None of the 36 boreholes studied had nitrate levels exceeding WHO guidelines (50.0 mg/L). However, by comparing these three study areas, we noticed that the lowest nitrate levels were recorded in the Wasia Well Field, averaging 2.79 mg/L. By contrast, the highest concentrations were recorded in Jubailah, averaging 37.93 mg/L. The Wadi Nisah levels fell in between, averaging 21 mg/L (Figure 2). Similarly, fluoride ranged from 0.63 mg/L in borehole 31 from Wasia Well Field to 2.00 mg/L in borehole 25 from Jubailah, averaging 1.23 mg/L. Eleven out of thirty-six fluoride samples (30.55%) exceeded the WHO's acceptable limit for drinking water (1.5 mg/L). The lowest nitrate levels were recorded in Wasia Well Field, averaging 0.82 mg/L, whereas the highest concentrations were recorded in Jubailah, averaging 1.70 mg/L. Wadi Nisah fell in between, averaging 1.00 mg/L (Figure 3).

Table 2. Concentrations of nitrate and fluoride in the study areas.

| S.N. | Lat. | Long. | Nitrates | Fluoride | S.N. | Lat. | Long. | Nitrates | Fluoride |
|---|---|---|---|---|---|---|---|---|---|
| 1 | 24.16958 | 46.10469 | 18 | 0.92 | 21 | 24.90678 | 46.39139 | 36 | 1.3 |
| 2 | 24.16344 | 46.21425 | 16 | 0.86 | 22 | 24.89717 | 46.39886 | 47 | 1.8 |
| 3 | 24.40542 | 46.22508 | 26 | 0.86 | 23 | 24.90225 | 46.40614 | 40 | 1.5 |
| 4 | 24.17483 | 46.43194 | 16 | 1.1 | 24 | 24.90211 | 46.41181 | 43 | 1.7 |
| 5 | 24.36869 | 46.44094 | 32 | 1.32 | 25 | 24.899 | 46.41792 | 44 | 2 |
| 6 | 24.35672 | 46.57375 | 24 | 0.87 | 26 | 24.89272 | 46.42953 | 42 | 1.8 |
| 7 | 24.28703 | 46.75603 | 20 | 0.91 | 27 | 25.16055 | 47.5633 | 5.1 | 1.12 |
| 8 | 24.26814 | 46.80336 | 18 | 0.95 | 28 | 25.17417 | 47.56645 | 2.1 | 1.05 |
| 9 | 24.21053 | 46.97817 | 19 | 1.05 | 29 | 25.17642 | 47.55445 | 3.2 | 0.7 |
| 10 | 24.19528 | 47.08814 | 28 | 1.11 | 30 | 25.18752 | 47.55073 | 4.7 | 0.65 |
| 11 | 24.1885 | 47.19881 | 22 | 1.12 | 31 | 25.19228 | 47.54542 | 4.8 | 0.63 |
| 12 | 24.18442 | 47.25731 | 13 | 0.88 | 32 | 25.19123 | 47.51897 | 1.3 | 0.87 |
| 13 | 24.89375 | 46.34486 | 33 | 1.5 | 33 | 25.21192 | 47.50437 | 0.7 | 0.76 |
| 14 | 24.89539 | 46.3575 | 28 | 2 | 34 | 25.2234 | 47.50117 | 2.1 | 0.88 |
| 15 | 24.89628 | 46.36517 | 46 | 1.7 | 35 | 25.2195 | 47.487 | 1.1 | 0.78 |
| 16 | 24.90889 | 46.37775 | 18 | 2 | 36 | 25.23465 | 47.49712 | 2.8 | 0.79 |
| 17 | 24.9165 | 46.37911 | 44 | 1.3 | Min. | | | 0.7 | 0.63 |
| 18 | 24.89706 | 46.38039 | 44 | 1.9 | Max. | | | 47 | 2 |
| 19 | 24.88733 | 46.38731 | 26 | 1.8 | Aver. | | | 22.59 | 1.23 |
| 20 | 24.90019 | 46.39144 | 40 | 1.5 | | | | | |

Notes: 1–12 (Wadi Nisah); 13–26 (Jubailah); 27–36 (Wasia Well Field).

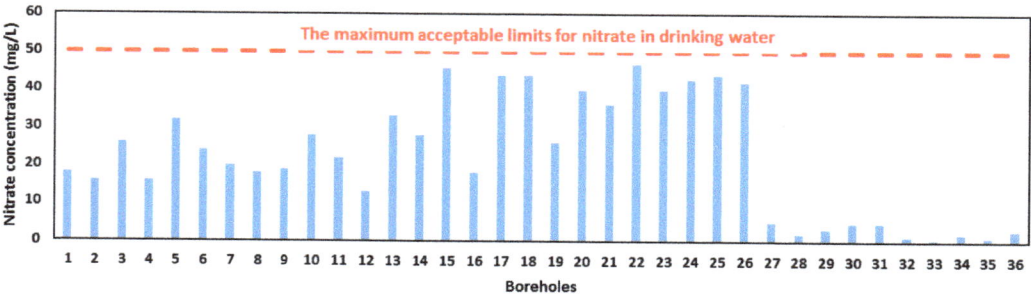

Figure 2. Distribution of nitrate concentrations (mg/L) in the groundwater of the study areas.

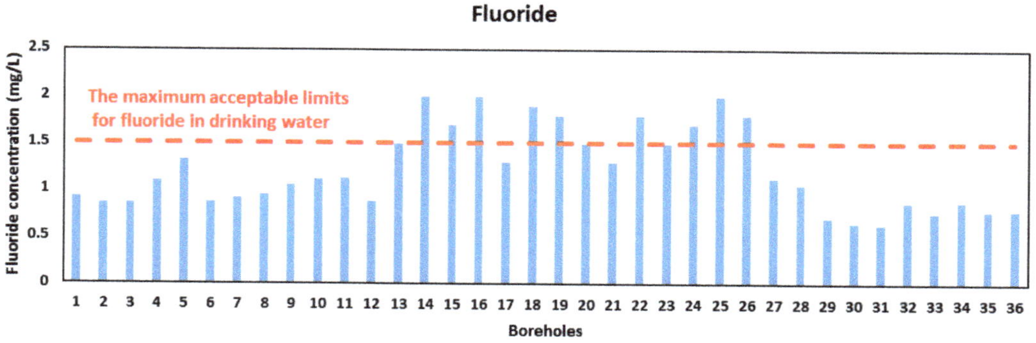

Figure 3. Distribution of fluoride concentrations (mg/L) in groundwater of the study areas.

## 4.2. Human Health Risk

Nitrate's chronic daily intake (CDI) values (mg/kg/d) for adults, children, and infants ranged from 0.020 to 1.343 (average of 0.646), 0.053 to 3.525 (average of 1.695), and 0.056 to 3.76 (average of 1.808), respectively. The CDI values for fluoride in adults, children, and infants varied from 0.018 to 0.057 (average of 0.035), 0.047 to 0.150 (average of 0.092), and 0.050 to 0.160 (average of 0.098), respectively (Table 3). The average HQ values of nitrate and fluoride for adults, children, and infants were 0.40 and 0.58, 1.06 and 1.53, and 1.13 and 1.64, respectively (Table 4).

**Table 3.** Chronic daily intake (CDI) (mg/kg/d) of nitrate and fluoride for adults, children, and infants.

| S.N. | $NO_3^-$ | CDI | | | $F^-$ | CDI | | |
|---|---|---|---|---|---|---|---|---|
| | | Infants | Children | Adults | | Infants | Children | Adults |
| 1 | 18 | 1.44 | 1.35 | 0.514 | 0.92 | 0.074 | 0.069 | 0.026 |
| 2 | 16 | 1.28 | 1.2 | 0.457 | 0.86 | 0.069 | 0.065 | 0.025 |
| 3 | 26 | 2.08 | 1.95 | 0.743 | 0.86 | 0.069 | 0.065 | 0.025 |
| 4 | 16 | 1.28 | 1.2 | 0.457 | 1.1 | 0.088 | 0.083 | 0.031 |
| 5 | 32 | 2.56 | 2.4 | 0.914 | 1.32 | 0.106 | 0.099 | 0.038 |
| 6 | 24 | 1.92 | 1.8 | 0.686 | 0.87 | 0.07 | 0.065 | 0.025 |
| 7 | 20 | 1.6 | 1.5 | 0.571 | 0.91 | 0.073 | 0.068 | 0.026 |
| 8 | 18 | 1.44 | 1.35 | 0.514 | 0.95 | 0.076 | 0.071 | 0.027 |
| 9 | 19 | 1.52 | 1.43 | 0.543 | 1.05 | 0.084 | 0.079 | 0.03 |
| 10 | 28 | 2.24 | 2.1 | 0.8 | 1.11 | 0.089 | 0.083 | 0.032 |
| 11 | 22 | 1.76 | 1.65 | 0.629 | 1.12 | 0.09 | 0.084 | 0.032 |
| 12 | 13 | 1.04 | 0.98 | 0.371 | 0.88 | 0.07 | 0.066 | 0.025 |
| 13 | 33 | 2.64 | 2.48 | 0.943 | 1.5 | 0.12 | 0.113 | 0.043 |
| 14 | 28 | 2.24 | 2.1 | 0.8 | 2.0 | 0.16 | 0.15 | 0.057 |
| 15 | 46 | 3.68 | 3.45 | 1.314 | 1.7 | 0.136 | 0.128 | 0.049 |
| 16 | 18 | 1.44 | 1.35 | 0.514 | 2.0 | 0.16 | 0.15 | 0.057 |
| 17 | 44 | 3.52 | 3.3 | 1.257 | 1.3 | 0.104 | 0.098 | 0.037 |
| 18 | 44 | 3.52 | 3.3 | 1.257 | 1.9 | 0.152 | 0.143 | 0.054 |
| 19 | 26 | 2.08 | 1.95 | 0.743 | 1.8 | 0.144 | 0.135 | 0.051 |
| 20 | 40 | 3.2 | 3.0 | 1.143 | 1.5 | 0.12 | 0.113 | 0.043 |
| 21 | 36 | 2.88 | 2.7 | 1.029 | 1.3 | 0.104 | 0.098 | 0.037 |
| 22 | 47 | 3.76 | 3.53 | 1.343 | 1.8 | 0.144 | 0.135 | 0.051 |
| 23 | 40 | 3.2 | 3.0 | 1.143 | 1.5 | 0.12 | 0.113 | 0.043 |
| 24 | 43 | 3.44 | 3.23 | 1.229 | 1.7 | 0.136 | 0.128 | 0.049 |
| 25 | 44 | 3.52 | 3.3 | 1.257 | 2.0 | 0.16 | 0.15 | 0.057 |
| 26 | 42 | 3.36 | 3.15 | 1.2 | 1.8 | 0.144 | 0.135 | 0.051 |
| 27 | 5.1 | 0.408 | 0.38 | 0.146 | 1.12 | 0.09 | 0.084 | 0.032 |
| 28 | 2.1 | 0.168 | 0.16 | 0.06 | 1.05 | 0.084 | 0.079 | 0.03 |
| 29 | 3.2 | 0.256 | 0.24 | 0.091 | 0.7 | 0.056 | 0.053 | 0.02 |
| 30 | 4.7 | 0.376 | 0.35 | 0.134 | 0.65 | 0.052 | 0.049 | 0.019 |
| 31 | 4.8 | 0.384 | 0.36 | 0.137 | 0.63 | 0.05 | 0.047 | 0.018 |
| 32 | 1.3 | 0.104 | 0.1 | 0.037 | 0.87 | 0.07 | 0.065 | 0.025 |
| 33 | 0.7 | 0.056 | 0.05 | 0.02 | 0.76 | 0.061 | 0.057 | 0.022 |
| 34 | 2.1 | 0.168 | 0.16 | 0.06 | 0.88 | 0.07 | 0.066 | 0.025 |
| 35 | 1.1 | 0.088 | 0.08 | 0.031 | 0.78 | 0.062 | 0.059 | 0.022 |
| 36 | 2.8 | 0.224 | 0.21 | 0.08 | 0.79 | 0.063 | 0.059 | 0.023 |

Table 4. The hazard quotient (HQ) and hazard index (HI) for fluoride and nitrate in adults, children, and infants.

| S.N. | HQ Nitrates | | | HQ Fluoride | | | HI | | |
|---|---|---|---|---|---|---|---|---|---|
| | Infants | Children | Adults | Infants | Children | Adults | Infants | Children | Adults |
| 1 | 0.9 | 0.844 | 0.321 | 1.227 | 1.15 | 0.438 | 2.127 | 1.994 | 0.76 |
| 2 | 0.8 | 0.75 | 0.286 | 1.147 | 1.075 | 0.41 | 1.947 | 1.825 | 0.695 |
| 3 | 1.3 | 1.219 | 0.464 | 1.147 | 1.075 | 0.41 | 2.447 | 2.294 | 0.874 |
| 4 | 0.8 | 0.75 | 0.286 | 1.467 | 1.375 | 0.524 | 2.267 | 2.125 | 0.81 |
| 5 | 1.6 | 1.5 | 0.571 | 1.76 | 1.65 | 0.629 | 3.36 | 3.15 | 1.2 |
| 6 | 1.2 | 1.125 | 0.429 | 1.16 | 1.088 | 0.414 | 2.36 | 2.213 | 0.843 |
| 7 | 1.0 | 0.938 | 0.357 | 1.213 | 1.138 | 0.433 | 2.213 | 2.075 | 0.79 |
| 8 | 0.9 | 0.844 | 0.321 | 1.267 | 1.188 | 0.452 | 2.167 | 2.031 | 0.774 |
| 9 | 0.95 | 0.891 | 0.339 | 1.4 | 1.313 | 0.5 | 2.35 | 2.203 | 0.839 |
| 10 | 1.4 | 1.313 | 0.5 | 1.48 | 1.388 | 0.529 | 2.88 | 2.7 | 1.029 |
| 11 | 1.1 | 1.031 | 0.393 | 1.493 | 1.4 | 0.533 | 2.593 | 2.431 | 0.926 |
| 12 | 0.65 | 0.609 | 0.232 | 1.173 | 1.1 | 0.419 | 1.823 | 1.709 | 0.651 |
| 13 | 1.65 | 1.547 | 0.589 | 2 | 1.875 | 0.714 | 3.65 | 3.422 | 1.304 |
| 14 | 1.4 | 1.313 | 0.5 | 2.667 | 2.5 | 0.952 | 4.067 | 3.813 | 1.452 |
| 15 | 2.3 | 2.156 | 0.821 | 2.267 | 2.125 | 0.81 | 4.567 | 4.281 | 1.631 |
| 16 | 0.9 | 0.844 | 0.321 | 2.667 | 2.5 | 0.952 | 3.567 | 3.344 | 1.274 |
| 17 | 2.2 | 2.063 | 0.786 | 1.733 | 1.625 | 0.619 | 3.933 | 3.688 | 1.405 |
| 18 | 2.2 | 2.063 | 0.786 | 2.533 | 2.375 | 0.905 | 4.733 | 4.438 | 1.69 |
| 19 | 1.3 | 1.219 | 0.464 | 2.4 | 2.25 | 0.857 | 3.7 | 3.469 | 1.321 |
| 20 | 2.0 | 1.875 | 0.714 | 2.0 | 1.875 | 0.714 | 4.0 | 3.75 | 1.429 |
| 21 | 1.8 | 1.688 | 0.643 | 1.733 | 1.625 | 0.619 | 3.533 | 3.313 | 1.262 |
| 22 | 2.35 | 2.203 | 0.839 | 2.4 | 2.25 | 0.857 | 4.75 | 4.453 | 1.696 |
| 23 | 2.0 | 1.875 | 0.714 | 2.0 | 1.875 | 0.714 | 4.0 | 3.75 | 1.429 |
| 24 | 2.15 | 2.016 | 0.768 | 2.267 | 2.125 | 0.81 | 4.417 | 4.141 | 1.577 |
| 25 | 2.2 | 2.063 | 0.786 | 2.667 | 2.5 | 0.952 | 4.867 | 4.563 | 1.738 |
| 26 | 2.1 | 1.969 | 0.75 | 2.4 | 2.25 | 0.857 | 4.5 | 4.219 | 1.607 |
| 27 | 0.255 | 0.239 | 0.091 | 1.493 | 1.4 | 0.533 | 1.748 | 1.639 | 0.624 |
| 28 | 0.105 | 0.098 | 0.038 | 1.4 | 1.313 | 0.5 | 1.505 | 1.411 | 0.538 |
| 29 | 0.16 | 0.15 | 0.057 | 0.933 | 0.875 | 0.333 | 1.093 | 1.025 | 0.39 |
| 30 | 0.235 | 0.22 | 0.084 | 0.867 | 0.813 | 0.31 | 1.102 | 1.033 | 0.393 |
| 31 | 0.24 | 0.225 | 0.086 | 0.84 | 0.788 | 0.3 | 1.08 | 1.013 | 0.386 |
| 32 | 0.065 | 0.061 | 0.023 | 1.16 | 1.088 | 0.414 | 1.225 | 1.148 | 0.438 |
| 33 | 0.035 | 0.033 | 0.013 | 1.013 | 0.95 | 0.362 | 1.048 | 0.983 | 0.374 |
| 34 | 0.105 | 0.098 | 0.038 | 1.173 | 1.1 | 0.419 | 1.278 | 1.198 | 0.457 |
| 35 | 0.055 | 0.052 | 0.02 | 1.04 | 0.975 | 0.371 | 1.095 | 1.027 | 0.391 |
| 36 | 0.14 | 0.131 | 0.05 | 1.053 | 0.988 | 0.376 | 1.193 | 1.119 | 0.426 |

The hazard index (HI) ranged from 0.37 to 1.74 (average of 0.99) for adults, 0.98 to 4.56 (average of 2.59) for children, and 1.05 to 4.87 (average of 2.77) for infants (Table 4, Figure 4). Groundwater samples surpassed the safety level of 1 by 44.44 (16 out of 36), 97.22 (35 out of 36), and 100% for adults, children, and infants, respectively (Figure 4). According to the study's findings, drinking water in most of the study areas, particularly in water samples from Jubailah (boreholes 13–26) and a few from Wadi Nisah (boreholes 5 and 10), could expose infants, children, and adults to non-cancer health concerns. Additionally, our findings show that infants and children are more vulnerable to non-carcinogenic health risks than adults.

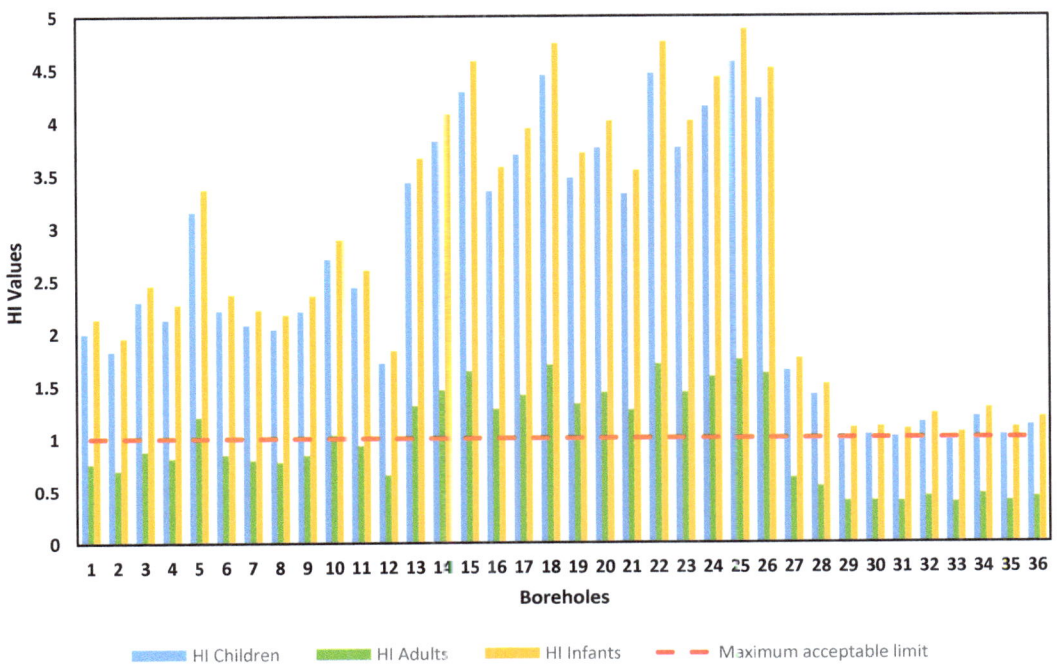

**Figure 4.** Non-carcinogenic risks induced by fluoride and nitrate in drinking water.

## 5. Discussion

The Al Jubailah area has a shallow unconfined aquifer, with a groundwater depth ranging from 25–100 m, that is recharged by flood waters from nearby mountains. The aquifer has high hydraulic conductivity and porosity due to the presence of fractures and karst features in the carbonate rocks. These factors may influence the transport and attenuation of nitrate and fluoride in groundwater [22]. Moreover, the Al Jubailah region is a fertile valley that sustains extensive cultivation of palm trees, vegetables, and fruits. These crops demand significant quantities of water and nutrients, which can result in excessive use of fertilizers and pesticides that contain nitrate and fluoride compounds [47]. Some of these pollutants may also be due to the region's rocky geology. Sedimentary rocks, such as limestone, dolomite, and gypsum, dominate Al Jubailah. These rocks may contain nitrate and fluoride minerals that can dissolve into groundwater under certain hydrogeochemical conditions, such as pH, temperature, redox potential, and salinity.

On the contrary, the region of the Wasia Well Field is located far away from agricultural areas, which are known to be the main sources of nitrate and fluoride contamination caused by fertilizers and pesticides. Unlike other study areas where the groundwater is shallow (25–100 m), the groundwater in the Wasia Well Field is deeper (18–300 m), making it less vulnerable to surface runoff and infiltration of pollutants from human activities. This groundwater is considered fossil water [18], which means it has a low recharge rate and a long residence time, reducing the likelihood of nitrate and fluoride minerals leaching from rocks and sediments. Additionally, the Wasia aquifer contains fine sandstones with low porosity and permeability, which could limit the movement of nitrate and fluoride ions in groundwater.

The Wadi Nisah area has moderate groundwater quality in terms of nitrate and fluoride concentrations. Compared to the Wadi Al Jubailah area, the Wadi Nisah area has lower nitrate and fluoride concentrations in groundwater. This could be due to the low extent and intensity of agricultural activities in Wadi Nisah as well as the low recharge rate [19]. The Wadi Al Jubailah area has more intensive farming of palm trees, vegetables, and fruits, which require more fertilizers and pesticides that increase nitrate and fluoride concentration.

Fluoride and nitrate are two of the most frequent and pervasive contaminants in many groundwater supplies, making water contamination a major environmental problem. Excessive nitrate levels in drinking water have been linked to various human health issues, including gastrointestinal cancers, methemoglobinemia, Alzheimer's disease, vascular dementia, and multiple sclerosis [48]. The amount of fluoride ions released into groundwater is controlled by the degree of saturation of fluorite and calcite and the concentrations of calcite, sodium, and bicarbonate ions in groundwater [43]. According to the study's findings from HQ and HI, drinking water in most of the study areas could expose infants, children, and adults to non-cancer health concerns. Infants and children are more vulnerable to non-carcinogenic health risks than adults, possibly due to their lower body weights. Many researchers worldwide have obtained similar results assessing health risks caused by nitrate and fluoride in groundwater, e.g., in northwest China, western Khorasan Razavi, Iran, the Medchal area, South India, the Noyyal basin, India, the Khyber district, north-western Pakistan, and central India [2,11,40,43,49,50].

However, to reduce the high concentrations of fluoride and nitrate in groundwater, there are three effective techniques: adsorption, electrocoagulation, and reverse osmosis. Adsorption involves using solid materials to capture fluoride and nitrate ions from water. It is a simple and low-cost method, but it requires regular adsorbent replacement [51]. Electrocoagulation involves using electric currents to generate metal hydroxides that precipitate fluoride and nitrate ions from water. It is a fast and eco-friendly method, but it requires high energy and proper sludge disposal [52]. Reverse osmosis involves using high pressure and a membrane to filter fluoride and nitrate ions from water. It is a highly effective and selective method, but it can be costly and requires proper brine disposal [53].

## 6. Conclusions

The present work highlighted the non-carcinogenic risk of drinking nitrate and fluoride-contaminated groundwater collected from central Saudi Arabia. Our findings indicate that none of the 36 studied boreholes had nitrate levels above WHO guidelines (50.00 mg/L). Differently, 11 out of 36 areas had fluoride levels exceeding the acceptable limit for drinking water (1.5 mg/L). The lowest nitrate and fluoride levels were recorded in the Wasia Well Field area, whereas the highest concentrations were recorded in the Jubailah area. The increased nitrate and fluoride concentrations in Jubailah may be attributed to the rock–water interaction, the intensive palm farms, and the region's heavy use of fertilizers and sanitation. The average values of the HI for adults, children, and infants were 0.99, 2.59, and 2.77, respectively. According to the USEPA method, the water samples had HI values exceeding the safety level for adults, children, and infants at 44.44%, 97.22%, and 100%, respectively. Infants and children are more vulnerable to non-carcinogenic health risks due to consumption of groundwater than adults.

In the future, it would be helpful to study different methods of removing fluoride and nitrate from groundwater, choose the best method for a particular location, and enhance drinking water quality for the local population.

**Author Contributions:** Conceptualization, T.A. and A.S.E.-S.; methodology, T.A. and A.S.E.-S.; software, T.A. and A.S.E.-S.; writing—original draft preparation, T.A. and A.S.E.-S.; writing—review and editing, T.A. and A.S.E.-S. All authors have read and agreed to the published version of the manuscript.

**Funding:** Researchers Supporting Project number (IFKSUOR3-157-2), King Saud University, Riyadh, Saudi Arabia.

**Data Availability Statement:** All data generated or analyzed during this study are included in this published article.

**Acknowledgments:** The authors extend their appreciation to the Deputyship for Research & Innovation, Ministry of Education in Saudi Arabia for funding this research work through the project no. (IFKSUOR3-157-2). Moreover, the authors thank the anonymous reviewers for their valuable suggestions and constructive comments.

**Conflicts of Interest:** The authors declare no conflict of interest.

# References

1. Patel, P.M.; Saha, D.; Shah, T. Sustainability of groundwater through community-driven distributed recharge: An analysis of arguments for water scarce regions of semi-arid India. *J. Hydrol. Reg. Stud.* 2020, 29, 100680. [CrossRef]
2. Kom, K.P.; Gurugnanam, B.; Bairavi, S. Non-carcinogenic health risk assessment of nitrate and fluoride contamination in the groundwater of Noyyal basin, India. *Geodesy Geodyn.* 2022, 13, 619–631. [CrossRef]
3. Dippong, T.; Mihali, C.; Hoaghia, M.-A.; Cical, E.; Cosma, A. Chemical modeling of groundwater quality in the aquifer of Seini town—Someș Plain, Northwestern Romania. *Ecotoxicol. Environ. Saf.* 2018, 168, 88–101. [CrossRef] [PubMed]
4. Gao, Y.; Qian, H.; Ren, W.; Wang, H.; Liu, F.; Yang, F. Hydrogeochemical characterization and quality assessment of groundwater based on integrated-weight water quality index in a concentrated urban area. *J. Clean Prod.* 2020, 260, 121006. [CrossRef]
5. Nawale, V.; Malpe, D.; Marghade, D.; Yenkie, R. Non-carcinogenic health risk assessment with source identification of nitrate and fluoride polluted groundwater of Wardha sub-basin, central India. *Ecotoxicol. Environ. Saf.* 2021, 208, 111548. [CrossRef] [PubMed]
6. Toolabi, A.; Bonyadi, Z.; Paydar, M.; Najafpoor, A.A.; Ramavandi, B. Spatial distribution, occurrence, and health risk assessment of nitrate, fluoride, and arsenic in Bam groundwater resource, Iran. *Groundw Sustain. Dev.* 2021, 12, 100543. [CrossRef]
7. Briki, M.; Zhu, Y.; Gao, Y.; Shao, M.; Ding, H.; Ji, H. Distribution and health risk assessment to heavy metals near smelting and mining areas of Hezhang, China. *Environ. Monit. Assess* 2017, 189, 458. [CrossRef] [PubMed]
8. Jinling, W.; Yin, Y.; Wang, J. Hydrogen-based membrane biofilm reactors for nitrate removal from water and wastewater. *Int. J. Hydrog. Energy* 2018, 43, 1–15.
9. Eggers, M.J.; Doyle, J.T.; Lefthand, M.J.; Young, S.L.; Moore-Nall, A.L.; Kindness, L.; Ford, T.E.; Dietrich, E.; Parker, A.E.; Hoover, J.H.; et al. Community engaged cumulative risk assessment of exposure to inorganic well water contaminants, crow reservation, montana. *Int. J. Environ. Res. Public Health* 2018, 15, 76. [CrossRef] [PubMed]
10. Amiri, V.; Berndtsson, R. Fluoride occurrence and human health risk from groundwater use at the west coast of Urmia Lake, Iran. *Arab. J. Geosci.* 2020, 13, 921. [CrossRef]
11. Duvva, L.K.; Panga, K.K.; Dhakate, R.; Himabindu, V. Health risk assessment of nitrate and fluoride toxicity in groundwater contamination in the semi-arid area of Medchal, South India. *Appl. Water Sci.* 2022, 12, 1–21. [CrossRef]
12. Su, H.; Kang, W.; Li, Y.; Li, Z. Fluoride and nitrate contamination of groundwater in the Loess Plateau, China: Sources and related human health risks. *Environ. Pollut.* 2021, 286, 117287. [CrossRef]
13. Khan, A.F.; Srinivasamoorthy, K.; Prakash, R.; Gopinath, S.; Saravanan, K.; Vinnarasi, F.; Babu, C.; Rabina, C. Human health risk assessment for fluoride and nitrate contamination in the groundwater: A case study from the east coast of Tamil Nadu and Puducherry, India. *Environ. Earth Sci.* 2021, 80, 1–17. [CrossRef]
14. Kimambo, V.; Bhattacharya, P.; Mtalo, F.; Ahmad, A. Fluoride occurrence in groundwater systems at global scale and status of defluoridation-State of the art. *Groundw. Sustain. Dev.* 2019, 9, 100223. [CrossRef]
15. Vithanage, M.; Bhattacharya, P. Fluoride in the environment: Sources, distribution and defluoridation. *Environ. Chem. Lett.* 2015, 13, 131–147. [CrossRef]
16. Suresh, M.; Gurugnanam, B.; Vasudevan, S.; Dharanirajan, K.; Raj, N.J. Drinking and irrigational feasibility of groundwater, GIS spatial mapping in upper Thirumanimuthar sub-basin, Cauvery River, Tamil Nadu. *J. Geol. Soc. India* 2010, 75, e518–e525. [CrossRef]
17. Tanwer, N.; Deswal, M.; Khyalia, P.; Laura, J.S.; Khosla, B. Fluoride and nitrate in groundwater: A comprehensive analysis of health risk and potability of groundwater of Jhunjhunu district of Rajasthan, India. *Environ Monit Assess* 2023, 195, 267. [CrossRef] [PubMed]
18. Alharbi, T. Hydrochemical Evaluation of Wasia Well Field in Riyadh Area. Master Thesis, King Saud University, Riyadh, Saudi Arabia, 2005; p. 130p.
19. Alharbi, T.G. Identification of hydrogeochemical processes and their influence on groundwater quality for drinking and agricultural usage in Wadi Nisah, Central Saudi Arabia. *Arab. J. Geosci.* 2018, 11, 359. [CrossRef]
20. Alharbi, T.G.; Zaidi, F.K. Hydrochemical classification and multivariate statistical analysis of groundwater from Wadi Sahba area in central Saudi Arabia. *Arab. J. Geosci.* 2018, 11, 643. [CrossRef]
21. Aly, A.A.; Alomran, A.; Alharby, M.M. The water quality index and hydrochemical characterization of groundwater resources in Hafar Albatin, Saudi Arabia. *Arab. J. Geosci.* 2014, 8, 4177–4190. [CrossRef]
22. Alharbi, T.; El-Sorogy, A.S.; Qaysi, S.; Alshehri, F. Evaluation of groundwater quality in central Saudi Arabia using hydrogeochemical characteristics and pollution indices. *Environ. Sci. Pollut. Res.* 2021, 28, 53819–53832. [CrossRef] [PubMed]
23. Mallick, J.; Singh, C.K.; AlMesfer, M.K.; Singh, V.P.; Alsubih, M. Groundwater Quality Studies in the Kingdom of Saudi Arabia: Prevalent Research and Management Dimensions. *Water* 2021, 13, 1266. [CrossRef]

24. Alharbi, T.; El-Sorogy, A.S. Quality and groundwater contamination of Wadi Hanifa, central Saudi Arabia. *Environ. Monit. Assess* **2023**, *195*, 525. [CrossRef] [PubMed]
25. Powers, R.W.; Ramirez, L.F.; Redmond, C.D.; Elberg, E.L. *Geology of the Arabian Peninsula Sedimentary Geology of Saudi Arabia*; United States Geological Survey: Reston, VA, USA, 1966.
26. Al Husseini, M.I.; Mathews, R. Stratigraphic note: Orbital calibration of the Arabian plate second order sequence stratigraphy. *GeoArabia* **2006**, *11*, 161–170. [CrossRef]
27. Hussein, M.T.; Al Yousif, M.M.; Awad, H.S. Potentiality of secondary aquifers in Saudi Arabia: Evaluation of groundwater quality in Jubaila limestone. *Int. J. Geosci.* **2012**, *3*, 71–80. [CrossRef]
28. Youssef, M.; El Sorogy, A.S. Palaeoecology of benthic foraminifera in coral reefs recorded in the Jurassic Tuwaiq Mountain Formation of the Khashm Al-Qaddiyah area, central Saudi Arabia. *J. Earth Sci.* **2015**, *26*, 224–235. [CrossRef]
29. Gameil, M.; El-Sorogy, A.S. Gastropods from the Campanian–Maastrichtian Aruma Formation, Central Saudi Arabia. *J. Afr. Earth Sci.* **2015**, *103*, 128–139. [CrossRef]
30. El-Asmar, H.M.; Assal, E.M.; El-Sorogy, A.S.; Youssef, M. Facies analysis and depositional environments of the Upper Jurassic Jubaila Formation, Central Saudi Arabia. *J. Afr. Earth Sci.* **2015**, *110*, 34–51. [CrossRef]
31. El-Sorogy, A.S.; Al-Kahtany, K.M. Contribution to the scleractinian corals of Hanifa Formation, Upper Jurassic, Jabal al-Abakkayn central Saudi Arabia. *Hist. Biol.* **2015**, *27*, 90–102. [CrossRef]
32. Tawfik, M.; Al-Dabbagh, M.E.; El-Sorogy, A.S. Sequence stratigraphy of the late middle Jurassic open shelf platform of the Tuwaiq Mountain Limestone Formation, central Saudi Arabia. *Proc. Geol. Assoc.* **2016**, *127*, 395–412. [CrossRef]
33. El-Sorogy, A.S.; Al-Kahtany, K.H.M.; El-Asmar, H. Marine benthic invertebrates of the Upper Jurassic Tuwaiq Mountain Limestone, Khashm Al-Qaddiyah, Central Saudi Arabia. *J. Afr. Earth Sci.* **2014**, *97*, 161–172. [CrossRef]
34. El-Sorogy, A.S.; Almadani, S.A.; Al-Dabbagh, M.E. Microfacies and diagenesis of the reefal limestone, Callovian Tuwaiq Mountain Limestone Formation, central Saudi Arabia. *J. Afr. Earth Sci.* **2016**, *115*, 63–70. [CrossRef]
35. Özer, S.; El-Sorogy, A.S.; Al-Dabbagh, M.; Al-Kahtany, K. Campanian-Maastrichtian unconformities and rudist diagenesis, Aruma Formation, Central Saudi Arabia. *Arab. J. Geosci.* **2019**, *12*, 34–45. [CrossRef]
36. El-Sorogy, A.; Al-Kahtany, K.; Almadani, S.; Tawfik, M. Depositional architecture and sequence stratigraphy of the Upper Jurassic Hanifa Formation, central Saudi Arabia. *J. Afr. Earth Sci.* **2018**, *139*, 367–378. [CrossRef]
37. El-Sorogy, A.S.; Gameil, M.; Youssef, M.; Al-Kahtany, K.M. Stratigraphy and macrofauna of the Lower Jurassic (Toarcian) Marrat Formation, central Saudi Arabia. *J. Afr. Earth Sci.* **2017**, *134*, 476–492. [CrossRef]
38. Al-Dabbagh, M.E.; El-Sorogy, A.S. Diagenetic alterations of the Upper Jurassic scleractinian corals, Hanifa Formation, Jabal Al-Abakkayn, Central Saudi Arabia. *J. Geol. Soc. India* **2016**, *87*, 337–344. [CrossRef]
39. Khalifa, M.; Al-Kahtany, K.; Farouk, S.; El-Sorogy, A.S.; Al Qahtani, A. Microfacies architecture and depositional history of the Upper Jurassic (kimmeridgian) Jubaila Formation in central Saudi Arabia. *J. Afr. Earth Sci.* **2021**, *174*, 104076. [CrossRef]
40. Qasemi, M.; Afsharnia, M.; Zarei, A.; Farhang, M.; Allahdadi, M. Non-carcinogenic risk assessment to human health due to intake of fluoride in the groundwater in rural areas of Gonabad and Bajestan, Iran: A case study. *Hum. Ecol. Risk Assess. Int. J.* **2018**, *25*, 1222–1233. [CrossRef]
41. Vaiphei, S.P.; Kurakalva, R.M. Hydrochemical characteristics and nitrate health risk assessment of groundwater through seasonal variations from an intensive agricultural region of upper Krishna River basin, Telangana, India. *Ecotoxicol. Environ. Saf.* **2021**, *213*, 112073. [CrossRef]
42. USEPA. *Human Health Evaluation Manual, Supplemental Guidance: Update of Standard Default Exposure Factors, OSWER Directive 9200.1–120*; United States Environmental Protection Agency: Washington, DC, USA, 2014.
43. Chen, J.; Wu, H.; Qian, H.; Gao, Y. Assessing Nitrate and Fluoride Contaminants in Drinking Water and Their Health Risk of Rural Residents Living in a Semiarid Region of Northwest China. *Expo. Health* **2016**, *9*, 183–195. [CrossRef]
44. Arora, R.P.; Sachdev Sud, Y.K.; Luthra, V.K.; Subbiah, B.V. Fate of fertilizer nitrogen in a multiple cropping system. In *Soil Nitrogen as Fertilizer or Pollution*; International Atomic Energy Agency: Vienna, Austria, 1980.
45. Zhao, B.; Li, X.; Liu, H.; Wang, B.; Zhu, P.; Huang, S.M.; Bao, D.; Li, Y.; So, H. Results from long-term fertilizer experiments in China: The risk of groundwater pollution by nitrate. *NJAS-Wageningen J. Life Sci.* **2011**, *58*, 177–183. [CrossRef]
46. Kalpana, L.; Brindha, K.; Elango, L. FIMAR a new fluoride index to mitigate geogenic contamination by managed aquifer recharge. *Chemosphere* **2019**, *220*, 381–390. [CrossRef] [PubMed]
47. Alharbi, T.; Abdelrahman, K.; El-Sorogy, A.S.; Ibrahim, E. Contamination and health risk assessment of groundwater along the Red Sea coast, Northwest Saudi Arabia. *Mar. Pollut. Bull.* **2023**, *192*, 115080. [CrossRef]
48. Narsimha, A.; Rajitha, S. Spatial distribution and seasonal variation in fluoride enrichment in groundwater and its associated human health risk assessment in Telangana State, South India. *Hum. Ecol. Risk Assess.* **2018**, *24*, 2119–2132. [CrossRef]
49. Ather, D.; Muhammad, M.; Ali, W. Fluoride and nitrate contaminations of groundwater and potential health risks assessment in the Khyber district, North-Western Pakistan. *Int. J. Environ. Anal. Chem.* **2022**. [CrossRef]
50. Sarkar, N.; Kandekar, A.; Gaikwad, S.; Kandekar, S. Health risk assessment of high concentration of fluoride and nitrate in the groundwater–A study of central India. *Transactions* **2022**, *44*, 13.
51. Zhou, H.; Tan, Y.; Gao, W.; Zhang, Y.; Yang, Y. Selective nitrate removal from aqueous solutions by a hydrotalcite-like absorbent FeMgMn-LDH. *Sci. Rep.* **2020**, *10*, 16126. [CrossRef]

52. Sandoval, M.A.; Fuentes, R.; Thiam, A.; Salazar, R.; van Hullebusch, E.D. Arsenic and fluoride removal by electrocoagulation process: A general review. *Sci. Total Environ.* **2021**, *753*, 142108. [CrossRef] [PubMed]
53. Epsztein, R.; Nir, O.; Lahav, O.; Green, M. Selective nitrate removal from groundwater using a hybrid nanofiltration–reverse osmosis filtration scheme. *Chem. Eng. J.* **2015**, *279*, 372–378. [CrossRef]

**Disclaimer/Publisher's Note:** The statements, opinions and data contained in all publications are solely those of the individual author(s) and contributor(s) and not of MDPI and/or the editor(s). MDPI and/or the editor(s) disclaim responsibility for any injury to people or property resulting from any ideas, methods, instructions or products referred to in the content.

Article

# The Sources of Sedimentary Organic Matter Traced by Carbon and Nitrogen Isotopes and Environmental Effects during the Past 60 Years in a Shallow Steppe Lake in Northern China

Hongbin Gao [1,2], Yanru Fan [1,2,*], Gang Wang [1,2], Lin Li [3], Rui Zhang [1,2], Songya Li [1,2], Linpei Wang [1,2], Zhongfeng Jiang [1,2], Zhekang Zhang [1,2], Junfeng Wu [1,2] and Xinfeng Zhu [1,2]

1. College of Municipal and Environmental Engineering, Henan University of Urban Construction, Pingdingshan 467000, China; gaohongbin0922@126.com (H.G.); 15103752123@163.com (G.W.); 17530809062@163.com (R.Z.); 20201013@huuc.edu.cn (S.L.); 20202017@huuc.edu.cn (L.W.); jiangzf@huuc.edu.cn (Z.J.); wjf8047@163.com (J.W.); zhuxf780@163.com (X.Z.)
2. Henan Province Key Laboratory of Water Pollution Control and Rehabilitation Technology, Pingdingshan 467000, China
3. Pingdingshan Ecological Environment Monitoring Center of Henan Province, Pingdingshan 467000, China; 20171004@huuc.edu.cn
* Correspondence: 20171019@huuc.edu.cn; Tel.: +86-185-3755-7690

**Abstract:** The organic matter of lake sediment plays an important role in paleolimnological reconstruction. Here, we report a detailed study of organic matter components ($C_{org}$%, N%, $\delta^{13}C$, $\delta^{15}N$) in a dated sediment core of Hulun Lake in northern China. Multiple mixing models based on the stoichiometric ratios and stable isotopic compositions were applied to quantify the contributions of organic matter sources in lake sediment. The results show that the organic matter in the sediments from Hulun Lake mainly comes from terrestrial organic matter: the proportion of terrestrial organic matter is more than 80%. The results of the SIAR mixing model further reveal that the proportions of terrestrial $C_3$ plants-derived organic matter, soil organic matter, and lake plankton-derived organic matter were 76.0%, 13.9%, and 10.1%, respectively. The organic matter content of lake sediment from terrestrial sources began to increase significantly from 1980 onward, which is consistent with the growth in overgrazing in the Hulun Lake basin. The content of organic matter from endogenous lake-derived sources began to increase significantly after 2000 due to the nutrients gradually becoming concentrated in lake water, indicating that the reduction in rivers' discharge and the downgrade of the lake water level were the immediate causes of the lake's environmental deterioration during this period.

**Keywords:** sources of sedimentary organic matter; carbon and nitrogen isotopes; end-member model; SIAR mixing model; overgrazing; Hulun Lake

**Citation:** Gao, H.; Fan, Y.; Wang, G.; Li, L.; Zhang, R.; Li, S.; Wang, L.; Jiang, Z.; Zhang, Z.; Wu, J.; et al. The Sources of Sedimentary Organic Matter Traced by Carbon and Nitrogen Isotopes and Environmental Effects during the Past 60 Years in a Shallow Steppe Lake in Northern China. *Water* 2023, *15*, 2224. https://doi.org/10.3390/w15122224

Academic Editor: Daniel D. Snow

Received: 10 May 2023
Revised: 7 June 2023
Accepted: 8 June 2023
Published: 13 June 2023

**Copyright:** © 2023 by the authors. Licensee MDPI, Basel, Switzerland. This article is an open access article distributed under the terms and conditions of the Creative Commons Attribution (CC BY) license (https://creativecommons.org/licenses/by/4.0/).

## 1. Introduction

The organic matter of lake sediment plays an important role in paleolimnological reconstruction. The vertical profiles of organic matter components, including abundance, elemental content, isotopic composition, and molecular ratio, were used to identify and differentiate between the effects of changes in natural conditions and anthropogenic activities on the environment of lake and its surrounding area [1–3]. In general, the organic matter in lake sediment is commonly derived from terrestrial (allochthonous) and aquatic organic materials (autochthonous). Over the past century, because of the growth in human activities, the terrestrial sources may be not only be controlled by natural conditions, such as precipitation, surface runoff, and vegetation cover, but also impacted by anthropogenic factors, such as use of agricultural fertilizers, land clearing, and cropping. Studies about sources of organic matter are of high importance for understanding how natural environmental changes and human activities affect the aquatic environment, biological productivity, and the global cycle of carbon [4–6].

The stoichiometric ratios and stable isotopic compositions of C (carbon) and N (nitrogen) in organic matter were developed as effective methods to trace the organic matter sources and identify predominant processes [7,8]. Moreover, the sediment profiles of stoichiometric ratios and stable isotopes of organic matter can reflect the historical changes in aquatic ecosystem productivity and terrigenous organic matter transportation processes, providing important information for the interpretation of paleoenvironmental conditions [4,5,9,10].

With the development of the techniques in tracing organic matter sources, the quantitative mathematical models were developed to analyze the proportions of different sources of the organic matter in lake sediment. In particular, the quantitative models based on the stoichiometric ratios and stable isotopic compositions of C and N were widely used to quantify the contributions of organic matter sources, such as the end-member mixing models and Bayesian mixing models [11–14]. The end-member mixing models are linear mixing models based on the mass balance equation that calculate the contributions of different organic matter sources in mixtures [14,15]. The average values of potential sources and mixture samples were used for the calculation of the end-member mixing models, and these models were only suitable for calculating the contribution ratios of no more than three major pollution sources [14]. Bayesian mixing models use Bayesian statistical theory to quantify source contributions. The contributions of potential sources in the model were estimated using the probability distribution of the proportional contribution of each source, which is determined via the logistic distribution and posterior distribution [16]. The models developed include mixing sample-importance resampling (MixSIR, R indicates the R Programming Language), stable isotope analysis in R (SIAR), mixing stable isotope analysis in R (MixSIAR), and compound-specific stable isotopes analysis in R (CSSIR) [13,17,18]. Compared to the end-member models, Bayesian mixing models incorporate all sources (i.e., more than three potential sources) and mixture sample values to account for the uncertainties in the sample data. The output of Bayesian mixing models are reported as probability distributions of the source contributions, rather than as a single value in end-member mixing models, which define the uncertainty in the experimental process [16].

The Hulun Lake is located in a sparsely populated, mildly farmed, and slightly industrialized steppe area in the northeastern part of Inner Mongolia, China (Figure 1), which has relatively limited anthropogenic factors affecting the lake water ecosystem. Notably, the Hulun Lake is in an active, NE–SW-trending, translithospheric fault zone [19,20]. However, it has recently experienced severe environmental deterioration due to the high concentrations of nutrients in lake water [21]. It is important to know the sources and processes of nutrients loading to understand the past environmental evolution of Hulun Lake. Unfortunately, available data about nutrient level of water, pollutant loading, and human activities in Hulun Lake and its basin are rare. Furthermore, since few instrumental and documentary records are detailed, it is difficult to identify whether the changes over a long time-scale are caused by natural condition changes and/or human activities. It is necessary to carry out paleolimnological reconstruction using the paleoenvironmental proxies archived in lake sediment, which were successfully used to trace the changes in sources of nutrients in and environmental evolution of lakes [22,23].

In this study, we report detailed studies of organic matter components, including the contents of organic carbon ($C_{org}\%$), the contents of nitrogen ($N\%$), and the stable isotopic compositions of carbon ($\delta^{13}C$) and nitrogen ($\delta^{15}N$), which were conducted in a dated sediment core. Two end-member mixing models and a Bayesian mixing model based on the stoichiometric ratios and stable isotopic compositions of carbon and nitrogen were used to quantify the contributions of organic matter sources to lake sediment. The objective was to trace the sources of sedimentary organic matter and quantify the contributions of different sources of organic matter within Hulun Lake's sediment. Furthermore, the proxies of $C_{org}$%, N%, $\delta^{13}C$, and $\delta^{15}N$ as paleolimnological indicators were used to further trace the historical environmental evolution of Hulun Lake in the past 60 years, as well as infer the environmental effects of climate change and human activities on the recent environmental deterioration in Hulun Lake. Due to its importance for the region, this information can be used to provide baseline data for the environmental management of Hulun Lake basin.

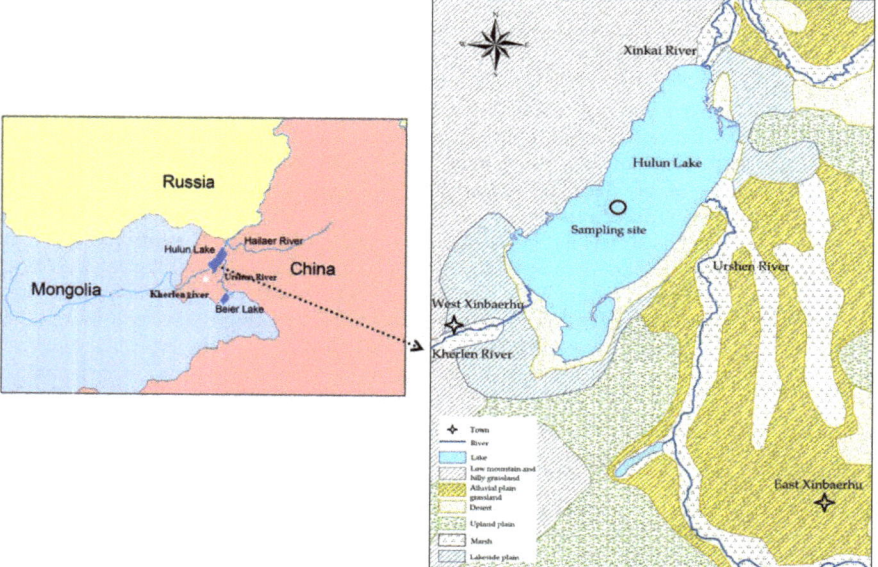

**Figure 1.** Location and geography of study area and sedimentary core sampling site from Hulun Lake (black circle).

## 2. Materials and Methods

### 2.1. Study Area

Hulun Lake (48°31′–49°20′ N, 116°58′–117°48′ E), which is located in a steppe area in the northeastern part of Inner Mongolia, China (Figure 1), is the fifth largest lake in China [24]. Although the lake is located in China, about 63.7% of the total basin areas of 256,000 km² are located in Mongolia. In addition, two rivers (Kherlen River and Urshen River) controlling the main input sources of Hulun Lake both originate from Mongolia [25]. Most areas of the lake basin are covered by the steppe grassland and are used for grazing [26]. There were relatively large changes in land use type in the Hulun Lake basin in recent years, which indicated that grassland degradation became increasingly serious [27]. Recent studies indicated that Hulun Lake suffered from eutrophication, and, sometimes, a cyanobacterial bloom occurs in certain areas of the lake [21]. Monitoring data gathered from Lake Hulun over the past 20 years (1994–2015) show that the TN (Total Nitrogen) and TP (Total Phosphorus) concentrations ranged from 0.80 to 3.30 mg/L and 0.04 to 0.25 mg/L (Figure 2), respectively; these values greatly exceed the National Grade IV Standards for Surface Water [28,29].

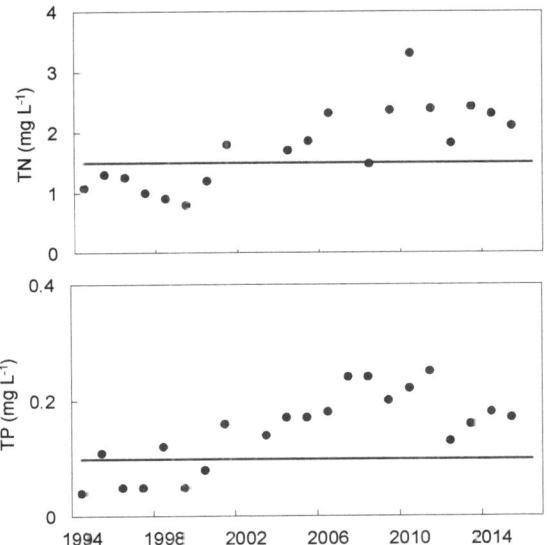

**Figure 2.** Changes in aqueous TN (Total Nitrogen) and TP (Total Phosphorus) concentrations in Hulun Lake during 1994–2015 (black lines denote limiting values of TN (Total Nitrogen) and TP (Total Phosphorus) of National Grade IV Standards for Surface Water [30]).

## 2.2. Sampling

A 41-centimeter-long sedimentary core was obtained at the deepest site in the center of Hulun Lake (Figure 1), China, in July 2015. Core samples were sliced immediately in 1 cm intervals on board the vessel. Sub-samples were stored in the sealing bags in an ice cooler and then transferred to the refrigerator (<4 °C) after transportation to the laboratory.

## 2.3. Experiments and Methods

All sediment sub-samples were measured in 1-centimeter intervals. Sediments used in carbon, nitrogen and isotopic analyses were ground in a mortar and homogenized. The $C_{org}$ and N contents (% of dry weight) and their corresponding stable isotope compositions ($\delta^{13}C_{org}$ and $\delta^{15}N$) were determined using a CN Automatic Elemental Analyzer and an isotope ratio mass spectrometer (DELTA plus Advantage), respectively. Analytical accuracy and precision were compared with known isotopic standards (Vienna Pee Dee Belemnite (VPDB) for carbon and atmospheric $N_2$ for nitrogen). The analytical precision for standards was within ±0.2‰ for $\delta^{13}C_{org}$ and ±0.3‰ for $\delta^{15}N$. The results are expressed innovatively as deviations in per milliliter (‰) differences relative to standard values of international standards (VPDB), as shown below:

$$\delta[‰ \text{vs. VPDB}] = [R_{sample}/R_{standard} - 1] \times 1000 \qquad (1)$$

To determine the age of the sediment core, the radioactive elements of $^{137}Cs$ and $^{210}Pb$ for 41 sub-samples in one-centimeter intervals were conducted via gamma spectrometry at Nanjing Institute of Geography and Limnology Chinese Academy of Sciences. The profile of $^{210}Pb$ dating for 41 samples was calculated through the constant initial concentration model (CIC). Combined with the $^{137}Cs$ activity data, a chronology framework for the whole core was established, which corresponded to a 57-year series from 1958 to 2015. The detailed results were described in a previous study [31].

*2.4. Calculations*

Organic matter in lake sediment is often described as a binary mixture of aquatic and terrestrial end members [32]. The binary model proposed by Qian, et al. can be employed to quantify the amount and percentage of allochthonous and autochthonous organic matter [11]. This model is designed as follows for carbon and nitrogen:

$$C(i) = C_{al}(i) + C_{au}(i) \tag{2}$$

$$N(i) = N_{al}(i) + N_{au}(i) \tag{3}$$

$$R_{al}(i) = C_{al}(i)/N_{al}(i) \tag{4}$$

$$R_{au}(i) = C_{au}(i)/N_{au}(i) \tag{5}$$

where $C(i)$ and $N(i)$ are the measured values of $C_{org}$ and N in sample ($i$), respectively; $C_{al}(i)$ and $N_{al}(i)$ are the content of $C_{org}$ and N derived from allochthonous organic matter, respectively; $C_{au}(i)$ and $N_{au}(i)$ are the content of $C_{org}$ and N derived from autochthonous organic matter respectively; and $R_{al}$ and $R_{au}$ are $C_{org}$/N ratios derived from allochthonous and autochthonous sources, respectively. Thus, the results are as follows:

$$N_{al}(i) = [C(i) - R_{au}(i) \cdot N(i)] / [R_{al}(i) - R_{au}(i)] \tag{6}$$

$$N_{au}(i) = [C(i) - R_{al}(i) \cdot N(i)] / [R_{au}(i) - R_{al}(i)] \tag{7}$$

$$C_{al}(i) = R_{al}(i)[C(i) - R_{au}(i) \cdot N(i)] / [R_{al}(i) - R_{au}(i)] \tag{8}$$

$$C_{au}(i) = R_{au}(i)[C(i) - R_{al}(i) \cdot N(i)] / [R_{au}(i) - R_{al}(i)] \tag{9}$$

Therefore, if the values of $R_{al}$ and $R_{au}$ are provided, the amounts and relative proportions of allochthonous and autochthonous sources can be calculated via this model. In this study, the $C_{org}$/N weight ratios for allochthonous ($R_{al}$) and autochthonous ($R_{au}$) sources of organic matter are given as 20 and 6, respectively [33].

The terrestrial (allochthonous) and lake (autochthonous) organic carbon fractions in sediment can also be estimated using a reliable two-end-member isotope-mixing model based on $\delta^{13}C$ [14]:

$$\delta^{13}C(i) = f_{al}\delta^{13}C_{al}(i) + f_{au}\delta^{13}C_{au}(i) \tag{10}$$

$$f_{al} + f_{au} = 1 \tag{11}$$

$$C_{al}(i) = f_{al}C(i) \tag{12}$$

$$C_{au}(i) = f_{au}C(i) \tag{13}$$

where $\delta^{13}C(i)$ is the measured value of $\delta^{13}C$ in sample ($i$); $f_{al}$ and $f_{au}$ are the proportions of allochthonous organic matter and autochthonous organic matter, respectively; $\delta^{13}C_{al}(i)$ and $\delta^{13}C_{au}(i)$ are end-members of allochthonous organic matter and autochthonous organic matter, respectively; and $C_{al}(i)$ and $C_{au}(i)$ are the content of organic matter derived from autochthonous organic matter and autochthonous organic matter, respectively.

To quantify the relative contributions of organic carbon from multiple sources, the potential sources were considered. Since the low temperatures recorded throughout the year in the study area are not conducive to the growth of aquatic plants, the macrophytes are absent in Lake Hulun, and most of the autochthonous organic matter in Lake Hulun was derived from algae [21]. Combined with previous research reports, the main plant type in the Hulun Lake basin is $C_3$ plants [34]. Thus, the allochthonous organic matter of Hulun Lake may mainly come from terrestrial $C_3$ plants. In this study, the end-members $\delta^{13}C$ of allochthonous organic matter ($-28.11 \pm 0.12$‰) were obtained from the $\delta^{13}C$ values measured in $C_3$ plants around the lake [35,36]. Since the $\delta^{13}C$ values of lake plankton in Hulun Lake were not measured in this study, the end-members $\delta^{13}C$ of autochthonous organic matter ($-21.37 \pm 2.84$‰) in Hulun Lake were obtained from Liang's surveys, which provide an average $\delta^{13}C$ value for plankton sourced from 10 lakes in Eastern Yunnan, China [37]. To assess the uncertainties in differentiating between the contributions of allochthonous organic matter and autochthonous organic matter associated with the range in $\delta^{13}C$ values for the different sources, we implemented three sets of calculations for each sample. Our "best" estimates were based on the average $\delta^{13}C$ values ($-28.11$‰ for allochthonous organic matter and $-21.37$‰ for autochthonous organic matter). The upper limit $\delta^{13}C$ for allochthonous organic matter contributions was calculated using $\delta^{13}C = -27.99$‰, while the upper limit for autochthonous organic matter concentrations was calculated using $\delta^{13}C = -24.21$‰.

Finally, a Bayesian mixing model (Stable Isotope Analysis in R, SIAR) based on $\delta^{13}C$ and $\delta^{15}N$ was run to determine the potential sources of sediment organic matter in more detail [18]. The SIAR model can be expressed as follows:

$$X_{ij} = \sum_{k=1}^{k} P_k \left( S_{jk} + C_{jk} \right) + \varepsilon_{ij} \tag{14}$$

$$S_{jk} \sim N(\mu_{jk}, \omega_{jk}^2) \tag{15}$$

$$C_{jk} \sim N(\lambda_{jk}, \tau_{jk}^2) \tag{16}$$

$$\varepsilon_{ij} \sim N(0, \sigma_j^2) \tag{17}$$

where $X_{ij}$ is the observed isotope value $j$ of the mixture $i$, in which $i = 1, 2, 3$, etc., $I$, and $j = 1, 2, 3$, etc., $J$; $P_k$ is the proportion of source $k$, which needs to be estimated via SIAR model; $S_{jk}$ is the source value $k$ on isotope $j$ ($k = 1, 2, 3$, etc., $K$) under normal distribution with mean $\mu_{jk}$ and variance $\omega_{jk}^2$; $C_{jk}$ is the isotopic fractionation factor for isotope $j$ ($k = 1, 2, 3$, etc., $K$) under normal distribution with mean $\lambda_{jk}$ variance $\tau_{jk}^2$; and $\varepsilon_{ij}$ is the residual error representing the additional unquantified variation between individual mixtures under normal distribution, with mean 0 and standard deviation $\sigma_j$ being estimated through the model.

Compared to the two-end-member model, the SIAR model can incorporate more potential sources to account for the contributions for each source. Combined with previous analysis of the possible sources of sediment organic matter in Hulun Lake, the potential four sources, including pasture ($C_3$), soil organic matter, lake phytoplankton, and lake zooplankton, were considered in the model's calculations. Due to the limitation of sampling conditions, the samples of plankton were not collected for elemental and isotopic determination. The values of $\delta^{13}C$ and $\delta^{15}N$ as the end-members for the SIAR model were instead cited from the results as being within the typical ranges of previous studies (Table 1). The fractionation factors for all sources were set to zero [$C_{jk} = 0$ in Equation (15)] because corresponding experiments for determining enrichment factors were not conducted, and no significant isotope fraction signals were observed in this study.

**Table 1.** Data for two-end-member model and SIAR model (‰).

| Source | $\delta^{13}C$ | SD | $\delta^{15}N$ | SD | Reference |
|---|---|---|---|---|---|
| Terrestrial $C_3$ plants | −28.11 | ±0.12 | 6.20 | ±0.50 | [34,35] |
| Lake plankton | −21.37 | ±2.84 | 9.14 | ±3.51 | [36] |
| Soil organic matter | −26.04 | ±0.29 | 4.74 | ±1.64 | [21,35] |
| Lake phytoplankton | −21.88 | ±2.97 | 7.26 | ±3.83 | [36] |
| Lake zooplankton | −20.85 | ±2.70 | 11.02 | ±3.18 | [36] |

Note: lake plankton was divided into lake phytoplankton and lake zooplankton.

## 3. Results and Discussion

### 3.1. The Characteristic of $C_{org}$%, N%, $C_{org}/N$, $\delta^{13}C$ and $\delta^{15}N$ Distribution in Sediment Profile

The results of $C_{org}$%, N%, $C_{org}/N$, $\delta^{13}C$, and $\delta^{15}N$ are plotted as profiles with core depth (left y-axis) and sediment ages (right y-axis) in Figure 3. The mean value, maximum value, minimum value, and standard deviation (SD) for values of $C_{org}$%, N%, $C_{org}/N$, $\delta^{13}C$, and $\delta^{15}N$ are shown in Table 2. The content of organic carbon ($C_{org}$%) shows a trend of gradual increase, ranging from 3.06 to 5.07%, with an average value of 3.57%, and the maximum value appears at the surface of the sediment core. The vertical variation trend for nitrogen content (N%) is consistent with that of organic carbon, which increases gradually from the bottom to the surface of the sediment profile, ranging from 0.41 to 0.20%, with an average value of 0.25%. The maximum value of N% also appears at the surface of the sediment core. The lake's organic matter mainly comes from the input by aquatic organisms in the lake itself and land sources in the basin. The changes in $C_{org}$% and N% in lake sediments reflect the primary productivity of the lake and its surrounding area, and the higher $C_{org}$% and N% indicate the improvement in the primary productivity [33]. The distribution of $C_{org}$% and N% in the sediment profile changed, displaying a rapid growth trend approximately after the year 2000 (corresponding above the depth of 12 cm), indicating that the organic matter in Hulun Lake increased in this period. This outcome may have been caused by either the increased input of terrestrial substances or the increase in the number of endogenous organisms in the lake.

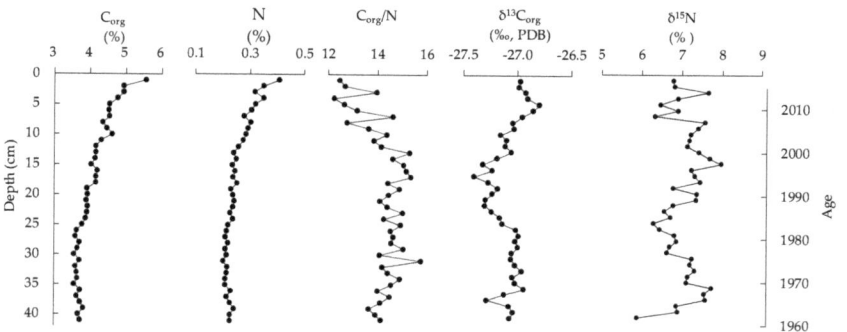

**Figure 3.** Profiles of $C_{org}$%, N%, $C_{org}/N$, $\delta^{13}C$, and $\delta^{15}N$ in lake sediment core.

Due to the different characteristics of the $C_{org}/N$ ratios of aquatic plants and land plants, the method of using $C_{org}/N$ ratio is widely used to determine the source of organic matter in lake sediments. Generally, the $C_{org}/N$ ratio of aquatic plankton is 4–10, that of aquatic plants ranges from 2.80 to 3.40, and that of terrestrial vascular plants is 20 or greater [33]. If the $C_{org}/N$ ratio in the sediment exceeds eight, it is usually considered that the composition of organic matter includes both endogenous and exogenous sources. The increase in $C_{org}/N$ ratio in the vertical depth of sediment is often considered to represent an increase in the proportion of terrestrial materials received by lakes during this period, while the proportion of aquatic plankton decreased. In contrast, the decreasing $C_{org}/N$ ratio in the vertical depth of the sediment is often considered to represent an increase in the proportion of aquatic plankton in the lake, while the proportion of terrestrial materials received decreased. The $C_{org}/N$ ratio increased vertically in Hulun Lake's sediment profile, ranging from 12.25 to 15.79, with a mean value of 14.25, and there is an obvious turning point at the depth of 12 cm approximately corresponding to the year 2000. The decrease in $C_{org}/N$ may reflect the increase in the proportion of endogenous organic matter relative to the total organic matter in the lake; thus, the primary productivity of plankton in the lake was relatively high during this period.

**Table 2.** Observed values of $C_{org}\%$, $N\%$, $C_{org}/N$, $\delta^{13}C$, and $\delta^{15}N$ in lake sediment core.

| Samples | $C_{org}\%$ | $N\%$ | $C_{org}/N$ | $\delta^{13}C$ | $\delta^{15}N$ |
|---|---|---|---|---|---|
| 1 | 5.07 | 0.41 | 12.46 | −26.97 | 6.78 |
| 2 | 4.44 | 0.35 | 12.70 | −26.98 | 6.82 |
| 3 | 4.45 | 0.32 | 13.98 | −26.92 | 7.65 |
| 4 | 4.28 | 0.35 | 12.25 | −26.90 | 6.89 |
| 5 | 4.05 | 0.32 | 12.66 | −26.79 | 6.47 |
| 6 | 4.03 | 0.31 | 13.17 | −26.85 | 6.89 |
| 7 | 4.06 | 0.28 | 14.65 | −26.95 | 6.32 |
| 8 | 3.87 | 0.30 | 12.78 | −27.04 | 7.56 |
| 9 | 3.98 | 0.29 | 13.64 | −27.03 | 7.39 |
| 10 | 4.12 | 0.29 | 14.40 | −27.15 | 7.22 |
| 11 | 3.82 | 0.28 | 13.88 | −27.09 | 7.18 |
| 12 | 3.67 | 0.26 | 14.17 | −27.11 | 7.13 |
| 13 | 3.68 | 0.24 | 15.32 | −27.05 | 7.41 |
| 14 | 3.66 | 0.25 | 14.64 | −27.18 | 7.68 |
| 15 | 3.54 | 0.24 | 15.06 | −27.31 | 7.97 |
| 16 | 3.72 | 0.25 | 15.17 | −27.22 | 7.22 |
| 17 | 3.67 | 0.24 | 15.37 | −27.39 | 7.32 |
| 18 | 3.67 | 0.25 | 14.44 | −27.26 | 7.45 |
| 19 | 3.43 | 0.23 | 14.91 | −27.18 | 6.77 |
| 20 | 3.46 | 0.24 | 14.46 | −27.23 | 7.36 |
| 21 | 3.42 | 0.24 | 14.12 | −27.29 | 7.35 |
| 22 | 3.45 | 0.24 | 14.42 | −27.29 | 6.77 |
| 23 | 3.43 | 0.23 | 15.04 | −27.23 | 6.56 |
| 24 | 3.39 | 0.24 | 14.29 | −27.16 | 6.72 |
| 25 | 3.29 | 0.22 | 14.97 | −27.13 | 6.29 |
| 26 | 3.14 | 0.22 | 14.56 | −27.01 | 6.44 |
| 27 | 3.11 | 0.21 | 14.67 | −26.98 | 6.82 |
| 28 | 3.22 | 0.22 | 14.57 | −27.01 | 6.86 |
| 29 | 3.16 | 0.21 | 15.06 | −26.99 | 6.69 |
| 30 | 3.06 | 0.22 | 14.12 | −27.04 | 6.63 |
| 31 | 3.22 | 0.20 | 15.79 | −27.05 | 7.25 |
| 32 | 3.10 | 0.22 | 14.21 | −27.01 | 7.20 |
| 33 | 3.14 | 0.22 | 14.45 | −26.95 | 7.31 |
| 34 | 3.17 | 0.21 | 14.95 | −27.04 | 7.15 |
| 35 | 3.07 | 0.21 | 14.57 | −27.01 | 7.11 |
| 36 | 3.24 | 0.23 | 14.03 | −26.93 | 7.73 |
| 37 | 3.14 | 0.22 | 14.52 | −27.11 | 7.55 |
| 38 | 3.24 | 0.23 | 14.14 | −27.27 | 7.58 |
| 39 | 3.34 | 0.24 | 13.67 | −27.07 | 6.86 |
| 40 | 3.18 | 0.23 | 13.95 | −27.03 | 6.90 |
| 41 | 3.25 | 0.23 | 14.18 | −27.06 | 5.89 |
| Mean | 3.57 | 0.25 | 14.25 | −27.08 | 7.05 |
| Maximum | 5.07 | 0.41 | 15.79 | −26.79 | 7.97 |
| Minimum | 3.06 | 0.20 | 12.25 | −27.39 | 5.89 |
| SD | 0.46 | 0.05 | 0.82 | 0.13 | 0.45 |

Note: Mean, maximum, minimum, and SD indicate mean value, maximum value, minimum value, and Standard Deviation (SD) for values of $C_{org}\%$, $N\%$, $C_{org}/N$, $\delta^{13}C$, and $\delta^{15}N$ in lake sediment core.

The carbon isotopic composition of organic matter in lake sediments is important in identifying organic matter sources and reconstructing the changes in past productivity [33]. The carbon isotopic composition of organic matter in Hulun Lake sediments ($\delta^{13}$C) had a relatively large fluctuation range, varying from −27.39 to −26.79‰, with an average value of −27.09‰. The $\delta^{13}$C values had a small change range from the bottom to 26 cm of the sediment profile, beyond which there is an obvious decreasing trend from a minimum value at 17 cm, and, finally, a rapid increase from 18 cm to the surface layer of the sediment profile. The shift in $\delta^{13}$C values indicates that the productivity of the lake or the surrounding area of the lake changed during this period.

The nitrogen isotopic compositions ($\delta^{15}$N) can similarly help to identify sources of organic matter in lakes and reconstruct past productivity rates [38]. However, additional factors besides source discrepancy complicate interpretations of the nitrogen isotopic composition of organic matter in lake sediments, such as denitrification DIN in anoxic bottom water, seasonal changes in phytoplankton, and nitrogen fixation [16]. Thus, $\delta^{15}$N-assisted $\delta^{13}$C in organic matter tracing will provide more reliable results [39]. The nitrogen isotopic compositions ($\delta^{15}$N) in the Hulun Lake sediment profile show a large fluctuation range, varying from 5.89 to 7.97‰, with an average value of 7.05‰. The $\delta^{15}$N values change irregularly throughout the depth of the sediment profile, indicating that the processes of nitrogen isotopic fractionation could be affected by complicated factors that compare the carbon isotopes during the transportation and deposition of organic matter.

The correlation between $C_{org}$% and N% in the sediment core is shown in Figure 4. The changes in $C_{org}$% and N% at different depths were extremely consistent and in significant correlation with a correlation coefficient 0.92, indicating that organic carbon and nitrogen in lake sediment cores originated from the same source, while most nitrogen may exist as organic form in sediment.

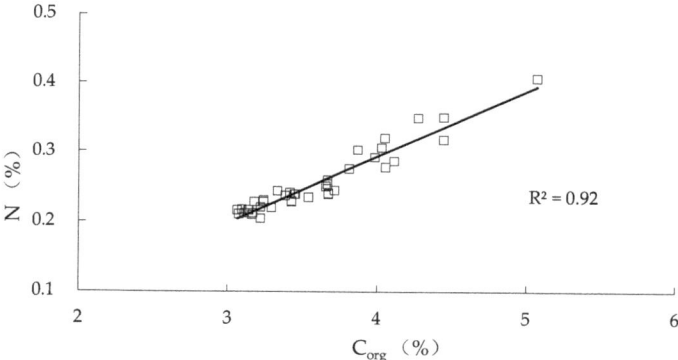

**Figure 4.** Relationships between $C_{org}$% values and N% values in lake sediment.

*3.2. The Sources of Sedimentary Organic Matter in Hulun Lake*

The diagram of the relationship between $C_{org}$/N ratios and $\delta^{13}$C values of sediment organic matter was successfully used to distinguish between the different organic matter sources in the sediment, as proposed by Meyers [33]. In this study, the diagram of the relationship between $C_{org}$/N ratios and $\delta^{13}$C values was plotted in Figure 5. Generally, plankton in fresh aquatic have low $C_{org}$/N ratios between 4 and 10, whereas vascular land plants with cellulose-rich and protein-poor traits usually have $C_{org}$/N ratios of 20 and greater. In contrast, soil organic matter have intermediate $C_{org}$/N ratios, ranging from 8 to 15 [40]. In the lake ecosystem, $\delta^{13}$C is another effective tracer used to identify the autochthonous and allochthonous organic matter sources of sediment. Autochthonous organic matter sources are produced by the biota within aquatic ecosystems, such as aquatic plants ($\delta^{13}$C values range from −20 to −12‰) and plankton ($\delta^{13}$C values range from −32 to −24‰) [41], while allochthonous organic matter sources are derived from sources found in areas surrounding the lake, such as C$_3$ terrestrial plants ($\delta^{13}$C values range from $-33$ and $-24$‰), C$_4$ terrestrial plants ($\delta^{13}$C values range from $-16$ to $-10$‰), and soil organic matter ($\delta^{13}$C values range from $-32$ to $-20$‰) [33,40].

The $\delta^{13}$C values between $-27.39$ and $-26.79$‰ in lake sediment fall within a typical range for C$_3$ terrestrial plants, soil organic matter, and plankton, while the $\delta^{13}$C values fall outside of the values for organic matter produced by C$_4$ terrestrial plants. As the previous section described, since the low temperatures are not conducive to the growth of aquatic plants, the macrophytes are absent in Lake Hulun; thus, the C$_{org}$/N ratios range from 12.25 to 15.79 and $\delta^{13}$C values range from $-27.39$‰ to $-26.79$‰, suggesting that the contribution of organic matter to Hulun Lake sediment may mainly derived from a mixture of C$_3$ terrestrial plants, soil organic matter, and lake plankton. Moreover, the C$_{org}$/N ratios and $\delta^{13}$C values were closer to the range for C$_3$ terrestrial plants and soil organic matter, indicating that a greater proportion of organic matter in Hulun Lake sediment was derived from allochthonous sources.

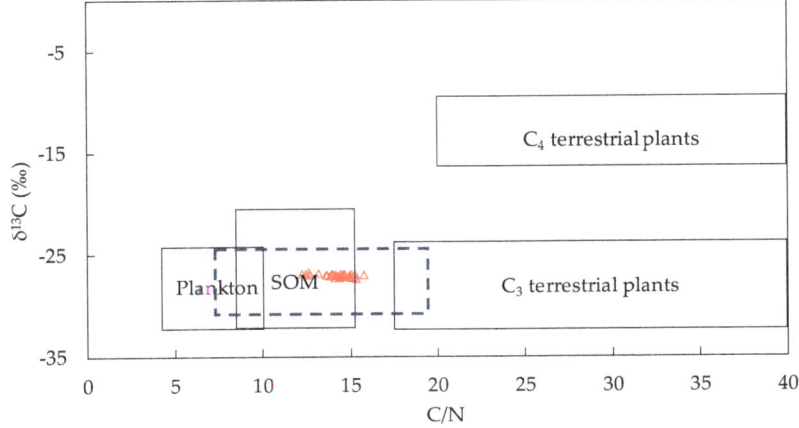

**Figure 5.** Distributions of $\delta^{13}$C values and C$_{org}$/N values in sediment cores from Hulun Lake (red triangle) and diagram of potential identification of sources of organic matter using $\delta^{13}$C values and C$_{org}$/N values for sediment samples. (SOM denotes soil organic matter).

Based on the binary models of C$_{org}$/N and $\delta^{13}$C, the relative proportions of allochthonous and autochthonous organic carbon are calculated, as shown in Figures 6 and 7. The results show that the proportions of allochthonous organic carbon in the sediment core calculated via the C$_{org}$/N model varied from 72.9 to 88.6%, with an average value of 82.5%, and the proportions of autochthonous organic carbon calculated via the C$_{org}$/N model varied from 11.4 to 27.1%, with an average value of 17.5%. The results of $\delta^{13}$C model show that the proportions of allochthonous organic carbon in the sediment core varied from 80.5 to 89.4%, with an average value of 84.7%, and the proportions of autochthonous organic carbon varied from 10.6 to 19.5%, with an average value of 15.3%. Comparing the results calculated via the two models, it can be seen that the proportions of allochthonous and autochthonous organic carbon are relatively consistent. Furthermore, the organic matter in the sediments of Hulun Lake mainly comes from terrestrial organic matter, of which the proportion is more than 80%, while the proportion of endogenous plankton organic matter is less than 20%. The binary models' results support the findings depicted in the diagram of the relationship between C$_{org}$/N ratios and $\delta^{13}$C values shown in Figure 5.

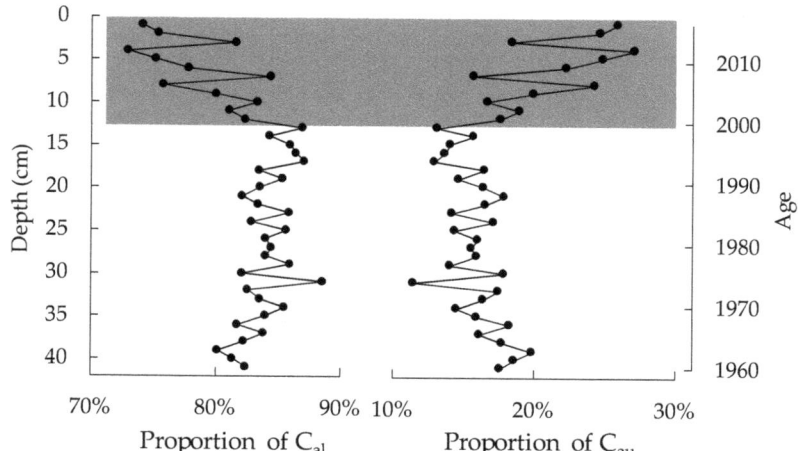

**Figure 6.** Proportions of allochthonous ($C_{al}$) and autochthonous ($C_{au}$) organic carbon in lake sediment calculated via binary models of $C_{org}/N$.

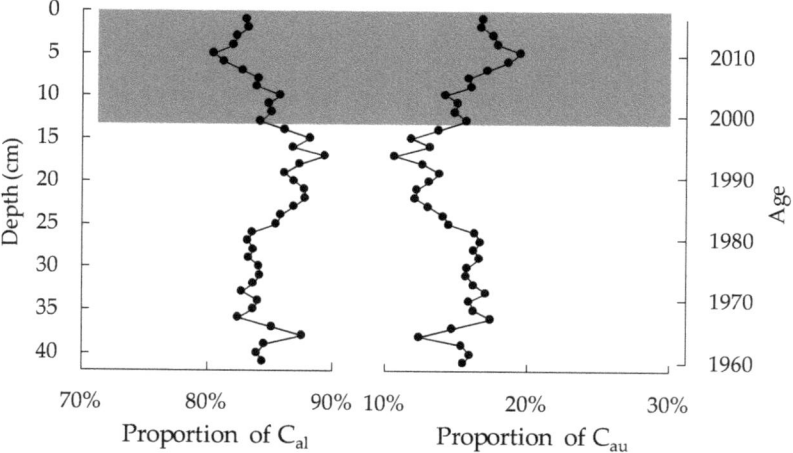

**Figure 7.** Proportions of allochthonous ($C_{al}$) and autochthonous ($C_{au}$) organic carbon in lake sediment calculated using $\delta^{13}C$ values.

As shown in the vertical profiles of proportions of allochthonous and autochthonous organic carbon based on the binary models of $C_{org}/N$ and $\delta^{13}C$ (Figures 6 and 7), obvious alterations occur in the upper 12 cm (corresponding to the approximate period after the year 2000) of the sediment core. The relative constant proportion of $C_{au}$ began to increase and reached the maximum, while a continuous increase in proportion of $C_{al}$ since 2000 was recorded, indicating that the productivity of phytoplankton in the lake increased during this period.

The SIAR mixing model outputs regarding the proportional contributions are presented in Table 3. The results show that terrestrial $C_3$ plants-derived organic matter (average proportion of 76.0%) was the predominant source, while soil organic matter from the lake basin was the second source, with an average proportion of 13.9%. The lake phytoplankton- and lake zooplankton-derived organic matter were the smaller contributors, with average values of 6.9% and 3.2%, respectively. Therefore, the total proportion of allochthonous and autochthonous organic matter in Hulun Lake sediment can be calculated as being 89.9%

and 10.1%, respectively. The values are close to the calculated results based on the binary models of $C_{org}/N$ and $\delta^{13}C$, indicating the reliability of the model results based on $\delta^{13}C$ and $\delta^{15}N$. In addition, these results also reveal that allochthonous organic matter input was the predominant source of sediment in Hulun Lake.

**Table 3.** Relative contributions of putative sources of sedimentary organic matter in Hulun Lake calculated via SIAR mixing model.

| Source | Mean | SD | 25% | 50% | 75% | 95% |
|---|---|---|---|---|---|---|
| Terrestrial C$_3$ plants | 0.760 | 0.081 | 0.711 | 0.775 | 0.823 | 0.868 |
| Soil organic matter | 0.139 | 0.100 | 0.061 | 0.119 | 0.197 | 0.334 |
| Lake phytoplankton | 0.069 | 0.029 | 0.050 | 0.071 | 0.091 | 0.115 |
| Lake zooplankton | 0.032 | 0.024 | 0.013 | 0.027 | 0.046 | 0.079 |

Note: SD denotes standard deviation, and contributions are designated as estimated region mode with probability distribution ranging from 25% to 95%.

### 3.3. The Environmental Effects during the Past 60 Years in Hulun Lake

#### 3.3.1. Allochthonous Organic Matter

The concentrations of allochthonous organic matter in Hulun Lake sediment cores calculated via $C_{org}/N$ ratio and $\delta^{13}C$ values are shown in Figure 8. The results calculated via the two different methods are very consistent, which show that the organic matter from terrestrial sources began to increase significantly at the sediment depth of 25 cm (the corresponding year is about 1980). The contents of organic matter from terrestrial sources increased from 25 cm to the top of the sediment core, which range from 2.82 to 3.76% calculated via $C_{org}/N$ ratio and 2.82 to 4.21% calculated via $\delta^{13}C$ values.

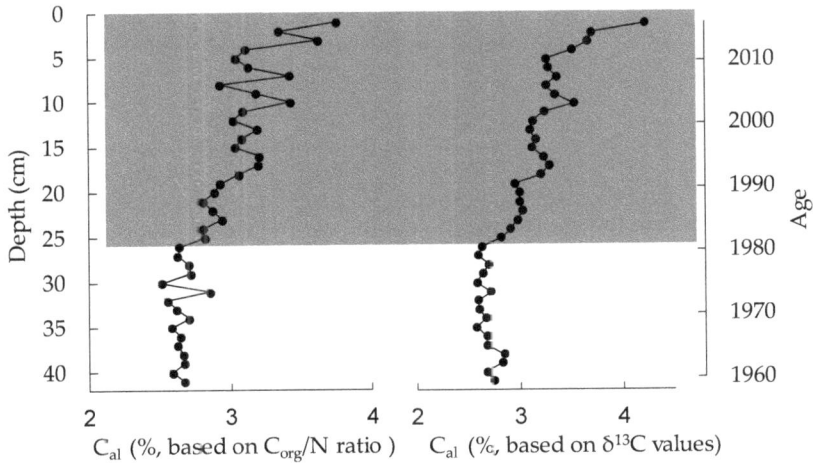

**Figure 8.** Concentrations of allochthonous ($C_{al}$) organic carbon in Hulun Lake sediment calculated via binary models of $C_{org}/N$ ratio and $\delta^{13}C$ values, respectively.

The changes in organic matter content in the lake are mainly controlled through the external input and the internal change in the lake. Generally, the increase in exogenous organic matter may be due to the changes in land use in the basin, as well as the domestic and industrial pollution generated via direct human activities. Hulun Lake is located in Hulunbeier Grassland in the north of China, and is, thus, surrounded by grassland. Human activities mainly include grazing without industrial pollution and large-scale urban domestic sewage discharge, and there is no strong non-point source pollution caused by livestock manure [21]. In addition, the upstream catchment areas of the lake's main discharge rivers are located in the sparsely populated mountainous areas, which have no direct discharge of pollutants. Thus, the sources of organic matter in Hulun Lake are mainly plant debris, hay, soil, etc. from the surrounding grassland, which are carried into the lake by rivers and winds.

The vegetation coverage condition is the main factor that controls the loss of surface materials caused by soil erosion or wind erosion. Hulun Lake basin is mainly covered by grassland, and most of the area is used by local herdsmen for grazing. Due to the lack of awareness of grassland environmental protection rules, herdsmen usually adopt the most primitive grazing system. This problematic grazing system may cause serious damage to the grassland and aggravate the water and soil loss. Researchers evaluated the grassland soil loss caused by different grazing intensities, indicating that grassland degradation will occur if the grazing intensity reaches 0.5–0.6 sheep per hectare [42]. In recent years, some researchers also carried out research on the impact of different grazing systems and grazing amounts on the grassland in the Hulun Lake basin. Onda Y. et al. conducted a survey on the grassland in the Klulan River basin in Mongolia, which is a sub-basin of the Hulun Lake basin, showing that the number of grazing livestock in the region increased from the 1980s, and the number of sheep converted from grazing intensity was 0.8 sheep per hectare in the 2000s [26]. Furthermore, the grazing intensity in parts of the Hulun Lake basin even reached to 1.7 sheep per hectare in recent years, according to the survey of the grassland in Hulunbeier City [21]. It can be seen that the grazing intensity in the Hulun Lake basin exceeded the reasonable grazing range (0.5–0.6 sheep per hectare), which could have an impact on its grassland ecosystem. Thus, overgrazing system may be a critical factor affecting the degradation of grassland in Hulun Lake basin, resulting in an increase in soil and water loss as the materials are carried into the lake.

As discussed previously, the organic matter of lake sediment from terrestrial sources began to increase significantly from the year 1980, which is consistent with the time when grazing intensity started to increase in the basin, indicating that the increase in grazing intensity in the lake basin may be the main reason for the increase in organic matter entering the lake.

### 3.3.2. Autochthonous Organic Matter

The concentrations of autochthonous organic matter in Hulun Lake sediment cores calculated via the $C_{org}/N$ ratio and $\delta^{13}C$ values are shown in Figure 9. The results calculated via the two different methods are also very consistent, which shows that the content of organic matter from endogenous lake-derived sources remained stable below the sediment depth of 12 cm, before beginning to increase significantly at the sediment depth of 12 cm (the corresponding period is about 2000), with a range from 0.65 to 1.31% calculated via the $C_{org}/N$ ratio and 0.55 to 0.86% calculated via $\delta^{13}C$ values. These results show that the nutrients in the lake were sufficient in the period after 2000, which was conducive to the growth of plankton and improved the productivity of the lake.

For closed lakes in arid and semi-arid regions, their hydrochemistry is very sensitive to climate change and hydrological processes. Due to the reduction in precipitation and possible upstream artificial closure, the water supply from two discharge rivers of Hulun Lake decreased rapidly since 2000, from $17.5 \times 10^8$ m$^3$ in 1999 to to $2.5 \times 10^8$ m$^3$ in 2011. Due to cold and dry climate conditions, the water level of Hulun Lake dropped sharply since 2000. Compared to the highest water level, the water level dropped by a maximum of 4 m, making it unable to flow out through the outlet; thus, the lake became a closed lake. Furthermore, the substances in the lake could not be exchanged with outside sources, and the amount of water replenished by rivers was far from balanced with the strong evaporation loss experiencd in the lake, which made the substances in the lake become gradually more concentrated. Monitoring data gathered from Lake Hulun over the past 20 years (1994–2015) also show that the nutrient (TN and TP) concentrations increased and the lake experienced eutrophication from the year 2000 (Figure 2). This period of time is very consistent with the results of sediment records. In this period, the high concentration of nutrients in the lake could have benefitted the growth of lake plankton, resulting in increased autochthonous organic matter content being deposited in Hulun Lake sediment. In addition, this finding also reveals that the reduction in rivers' discharge and the downgrading of lake water level were the immediate causes of the lake's environmental deterioration during this period.

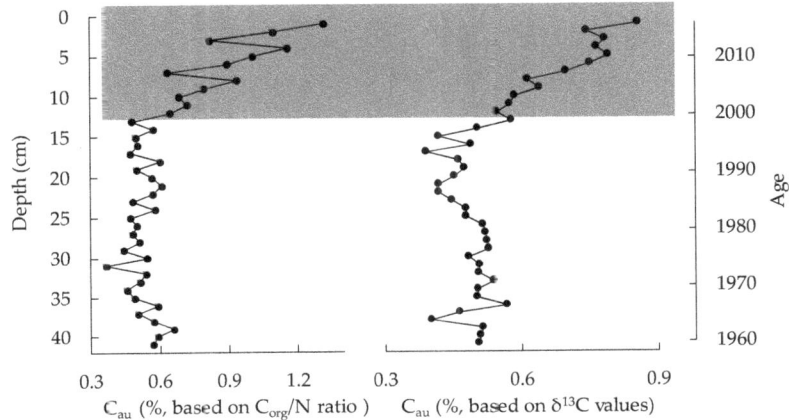

**Figure 9.** Concentrations of autochthonous ($C_{au}$) organic carbon in Hulun Lake sediment calculated via binary models of $C_{org}/N$ ratio and $\delta^{13}C$ values, respectively.

## 4. Conclusions

The variation patterns of organic matter components and isotope signatures of C and N were exhibited in a dated sediment core of Hulun Lake in this study. Multiple models based on the stoichiometric ratios and stable isotopic compositions revealed that terrestrial $C_3$ plants-derived organic matter was the predominant source of sediment in Hulun Lake. The variation patterns of organic matter in the sediment were associated with the impact of human activities and climatic changes, especially those related to grazing, inflow discharge, and the lake water level. The organic matter of lake sediment from terrestrial sources began to increase significantly from the year 1980, which is consistent with the time when grazing intensity started to increase in the basin, indicating that overgrazing in the lake basin may be the main reason for the increase in organic matter entering the lake. The content of organic matter from endogenous lake-derived sources began to increase significantly after 2000, indicating that high concentrations of nutrients in the lake could be beneficial to the growth of lake plankton, resulting in increased autochthonous organic matter content being deposited in Hulun Lake sediment. In addition, it also revealed that the reduction in rivers' discharge and the downgrading of the lake water level was the immediate cause of the lake's environmental deterioration during this period. These results highlight the need to pay attention to the inputs of terrestrial organic matter in Hulun Lake and take measures to control the decline in the lake's water level.

**Author Contributions:** H.G. and Y.F. designed and performed research. H.G., R.Z. and G.W. wrote the paper. Z.Z., L.L., L.W. and S.L. assisted experiment, Z.J., X.Z. and J.W. provided comments. All authors have read and agreed to the published version of the manuscript.

**Funding:** This research was supported by the Natural Science Foundation of Henan (No. 222300420106), the National College Students' Innovation and Entrepreneurship Training Program (No. 202211765003), the Henan Provincial Department of Education Key Project (No. 19A210008, 19B570001, 21B610002), the Science and Technology Project of Henan Province (No. 212102310277, No. 232102320132, No. 232102320115), and the Project of Young Core Instructors of Henan University of Urban Construction (No. YCJQNGGJS202103).

**Data Availability Statement:** The data that support the findings of this study are available from the authors upon reasonable request.

**Acknowledgments:** Comments from the anonymous reviewers are appreciated.

**Conflicts of Interest:** The authors declare no conflict of interest.

## References

1. Liu, C.; Dong, Y.; Li, Z.; Chang, X.; Nie, X.; Liu, L.; Xiao, H.; Bashir, H. Tracing the source of sedimentary organic carbon in the Loess Plateau of China: An integrated elemental ratio, stable carbon signatures, and radioactive isotopes approach. *J. Environ. Radioactiv.* **2016**, *167*, 201–210. [CrossRef]
2. Rumolo, P.; Barra, M.; Gherardi, S.; Marsella, E.; Sprovieri, M. Stable isotopes and C/N ratios in marine sediments as a tool for discriminating anthropogenic impact. *J. Environ. Monit.* **2011**, *13*, 3399–3408. [CrossRef] [PubMed]
3. Feng, W.; Wang, T.; Zhu, Y.; Sun, F.; Giesy, J.P.; Wu, F. Chemical composition, sources, and ecological effect of organic phosphorus in water ecosystems: A review. *Carbon Res.* **2023**, *2*, 12. [CrossRef]
4. Fornace, K.L.; Whitney, B.S.; Galy, V.; Hughen, K.A.; Mayle, F.E. Late quaternary environmental change in the interior South American tropics: New insight from leaf wax stable isotopes. *Earth Planet. Sci. Lett.* **2016**, *438*, 75–85. [CrossRef]
5. Zhuo, Y.; Zeng, W.; Cui, D.; Ma, B.; Xie, Y.; Wang, J. Spatial-temporal variation, sources and driving factors of organic carbon burial in rift lakes on Yunnan-Guizhou plateau since 1850. *Environ. Res.* **2021**, *201*, 111458. [CrossRef]
6. Chen, F.X.; Fang, N.F.; Wang, Y.X.; Tong, L.S.; Shi, Z.H. Biomarkers in sedimentary sequences: Indicators to track sediment sources over decadal timescales. *Geomorphology* **2017**, *278*, 1–11. [CrossRef]
7. Märki, L.; Lupker, M.; Gajurel, A.P.; Gies, H.; Haghipour, N.; Gallen, S. Molecular tracing of riverine soil organic matter from the central Himalaya. *Geophys. Res. Lett.* **2020**, *47*, e2020GL087403. [CrossRef]
8. Chen, J.; Yang, H.; Zeng, Y.; Guo, J.; Song, Y.; Ding, W. Combined use of radiocarbon and stable carbon isotope to constrain the sources and cycling of particulate organic carbon in a large freshwater lake, China. *Sci. Total Environ.* **2018**, *625*, 27–38. [CrossRef]
9. Lu, L.; Cheng, H.G.; Pu, X.; Wang, J.; Cheng, Q.; Liu, X. Identifying organic matter sources using isotopicratios in a watershed impacted by intensive agricultural activities in Northeast China. *Agric. Ecosyst. Environ.* **2016**, *222*, 48–59. [CrossRef]
10. Tue, N.T.; Hamaoka, H.; Sogabe, A.; Quy, T.D.; Nhuan, M.T.; Omori, K. The application of $\delta^{13}C$ and C/N ratios as indicators of organic carbon sources and paleoenvironmental change of the mangrove ecosystem from Ba Lat Estuary, Red River, Vietnam. *Environ. Earth Sci.* **2011**, *64*, 1475–1486. [CrossRef]
11. Qian, J.L.; Wang, S.M.; Xue, B.; Chen, R.S.; Ke, S.Z. A method of quantitatively calculating amount of allochthonous organic carbon in lake sediments. *Chin. Sci. Bull.* **1997**, *42*, 1821–1823. [CrossRef]
12. Nosrati, K.; Govers, G.; Semmens, B.X.; Ward, E.J. A mixing model to incorporate uncertainty in sediment fingerprinting. *Geoderma* **2014**, *217–218*, 173–180. [CrossRef]
13. Stock, B.C.; Jackson, A.L.; Ward, E.J.; Parnell, A.C.; Phillips, D.L.; Semmens, B.X. Analyzing mixing systems using a new generation of Bayesian tracer mixing models. *PeerJ* **2018**, *6*, e5096. [CrossRef] [PubMed]
14. Phillips, D.L. Mixing models in analyses of diet using multiple stable isotopes: A critique. *Oecologia* **2001**, *127*, 166–170. [CrossRef]
15. Li, S.; Xia, X.; Zhang, S.; Zhang, L. Source identification of suspended and deposited organic matter in an alpine river with elemental, stable isotopic, and molecular proxies. *J. Hydrol.* **2020**, *590*, 125492. [CrossRef]
16. Li, Z.; Wang, S.; Nie, X.; Sun, Y.; Ran, F. The application and potential non-conservatism of stable isotopes in organic matter source tracing. *Sci. Total Environ.* **2022**, *838*, 155946. [CrossRef]
17. Moore, J.W.; Semmens, B.X. Incorporating uncertainty and prior information into stable isotope mixing models. *Ecol. Lett.* **2008**, *11*, 470–480. [CrossRef]
18. Parnell, A.; Inger, R.; Bearhop, S.; Jackson, A.L. SIAR: Stable Isotope Analysis in R. 2008. Available online: http://cran.rproject.org/web/packages/siar/index.html (accessed on 2 May 2009).
19. Zhang, K.J. Genesis of the Late Mesozoic Great Xing'an Range Large Igneous Province: A Mongol–Okhotsk slab window model. *Int. Geol. Rev.* **2014**, *56*, 1557–1583. [CrossRef]
20. Zhang, K.J.; Yan, L.L.; Ji, C. Switch of NE Asia from extension to contraction at the mid-Cretaceous: A tale of the Okhotsk oceanic plateau from initiation by the Perm Anomaly to extrusion in the Mongol–Okhotsk ocean? *Earth-Sci. Rev.* **2019**, *198*, 102941. [CrossRef]
21. Chen, X.; Chuai, X.; Yang, L.; Zhao, H. Climatic warming and overgrazing induced the high concentration of organic matter in Lake Hulun, a large shallow eutrophic steppe lake in northern China. *Sci. Total Environ.* **2012**, *431*, 332–338. [CrossRef]
22. Li, Z.; Li, X.; Wang, X.; Ma, J.; Xu, J.; Xu, X. Isotopic evidence revealing spatial heterogeneity for source and composition of sedimentary organic matters in Taihu Lake, China. *Ecol. Indic.* **2020**, *109*, 105854.1–105854.10. [CrossRef]
23. Fang, J.; Wu, F.; Xiong, Y.; Li, F.; Du, X.; An, D.; Wang, L. Source characterization of sedimentary organic matter using molecular and stable isotopic composition of n-alkanes and fatty acids in sediment core from Lake Dianchi, China. *Sci. Total Environ.* **2014**, *473–474*, 410–421. [CrossRef] [PubMed]
24. Xu, Z.J.; Jiang, F.Y.; Zhao, H.W.; Zhang, Z.B.; Sun, L. *Annals of Hulun Lake*; Jilin Literature and History Publishing House: Jilin, China, 1989.

25. Sun, E.; Li, C.Y.; Zhang, S.; Zhao, S.N. Analysis of Hulun digital basin based on spatial information technology. *Environ. Pollut. Control.* **2010**, *32*, 5–19.
26. Onda, Y.; Kato, H.; Tanaka, Y.; Tsujimura, M.; Davaa, G.; Oyunbaatar, D. Analysis of runoff generation and soil erosion processes by using environmental radionuclides in semiarid areas of Mongolia. *J. Hydrol.* **2007**, *333*, 124–132. [CrossRef]
27. Wang, G. Land Use and Land Cover Change of Hulun Lake Nature Reserve in Inner Mongolia, China: A Modeling Analysis. *Exp. Parasitol.* **2012**, *131*, 210–214. [CrossRef]
28. Liang, L.E.; Li, C.Y.; Shi, X.H.; Zhao, S.N.; Tian, Y.; Zhang, L.J. Analysis on the eutrophication trends and affecting factors in Lake Hulun, 2006–2015. *J. Lake Sci.* **2016**, *28*, 1265–1273.
29. Chuai, X.; Chen, X.; Yang, L.; Zeng, J.; Miao, A.; Zhao, H. Effects of climatic changes and anthropogenic activities on lake eutrophication in different ecoregions. *Int. J. Environ. Sci. Technol.* **2012**, *9*, 503–514. [CrossRef]
30. GB3838-2002; Environmental Quality Standard for Surface Water. State Environmental Protection Administration: Beijing, China, 2002.
31. Gao, H.; Zhang, R.; Wang, G.; Fan, Y.; Zhu, X.; Wu, J.; Wu, L. A dam construction event recorded by high-resolution sedimentary grain size in an outflow-controlled lake (Hulun Lake, China). *Water* **2022**, *14*, 3878. [CrossRef]
32. Lu, C.; He, J.; Sun, H.; Xue, H. Application of allochthonous organic carbon and phosphorus forms in the interpretation of past environmental conditions. *Environ. Geol.* **2008**, *55*, 1279–1289. [CrossRef]
33. Meyers, P.A. Applications of organic geochemistry to paleolimnological reconstructions: A summary of examples from the Laurentian Great Lakes. *Org. Geochem.* **2003**, *34*, 261–289. [CrossRef]
34. Kiyokazu, K.; Kurosu, M.; Cheng, Y.; Tsendeekhuu, T.; Wu, Y.; Nakamura, T. Floristic composition, grazing effects and aboveground plant biomass in the Hulunbeier Grasslands of Inner Mongolia, China. *J. Ecol. Field Biol.* **2008**, *31*, 297–307.
35. Wang, Y.T.; Wang, X.; Zhu, G.D.; Zhao, L.Q.; Jia, L.X.; Zhao, M.L. Responses of water and nitrogen use efficiency to stocking rates in $C_3$, $C_4$ plants of stipa-breviflora steppe in Inner Mongolia. *Chin. J. Grassl.* **2017**, *39*, 59–64.
36. Liu, X.Z.; Zhang, Y.; Su, Q.; Li, Z.; Feng, T.; Song, Y. Gradient variation of $\delta^{15}N$ values in herbs and its indication to environmental information in the agro-pastoral ecotone in the north of China. *J. Univ. Chin. Acad. Sci.* **2018**, *35*, 749–760.
37. Liang, H. Study on Geographical Distribution Characteristics of Carbon and Nitrogen Stable Isotopes and Elemental of Lake Primary Producers and Zooplankton in Eastern Yunnan. Master's Thesis, Yunnan Normal University, Kunming, China, 2019
38. Last, W.M.; Smol, J.P. *Tracking Environmental Change Using Lake Sediments: Physical and Geochemical Methods*; Springer: Dordrecht, The Netherlands, 2001; pp. 401–439.
39. Liu, C.; Li, Z.; Berhe, A.A.; Hu, B.X. The isotopes and biomarker approaches for identifying eroded organic matter sources in sediments: A review. *Adv. Agron.* **2020**, *162*, 257–303.
40. Kendall, C.; Silva, S.R.; Kelly, V.J. Carbon and nitrogen isotopic compositions of particulate organic matter in four large river systems across the United States. *Hydrol. Process.* **2001**, *15*, 1301–1346. [CrossRef]
41. Lee, Y.; Hong, S.; Kim, M.S.; Kim, D.; Choi, B.H.; Hur, J. Identification of sources and seasonal variability of organic matter in Lake Shwa and surrounding inland creeks, South Korea. *Chemosphere* **2017**, *177*, 109–119. [CrossRef]
42. Evans, R. Overgrazing and soil erosion on hill pastures with particular reference to the Peak District. *Grass Forage Sci.* **1977**, *32*, 65–76. [CrossRef]

**Disclaimer/Publisher's Note:** The statements, opinions and data contained in all publications are solely those of the individual author(s) and contributor(s) and not of MDPI and/or the editor(s). MDPI and/or the editor(s) disclaim responsibility for any injury to people or property resulting from any ideas, methods, instructions or products referred to in the content.

Article

# Analysis of the Driving Mechanism of Water Environment Evolution and Algal Bloom Warning Signals in Tai Lake

Cuicui Li [1,2,3,4,5] and Wenliang Wu [1,5,*]

1. College of Resources and Environment, China Agricultural University, Beijing 100193, China
2. Anrong Credit Rating Co., Ltd., Beijing 100032, China
3. Guangzhou Institute of Geochemistry, Chinese Academy of Sciences, Guangzhou 510640, China
4. State Key Laboratory of Environmental Criteria and Risk Assessment, Chinese Research Academy of Environmental Sciences, Beijing 100012, China
5. Beijing Key Laboratory of Biodiversity and Organic Agriculture, Beijing 102208, China
* Correspondence: wuwenl@cau.edu.cn

**Abstract:** Understanding the evolution characteristics and driving mechanisms of eutrophic lake ecosystems, especially over long time scales, remains a challenge. Little research on lake ecosystem mutation has been conducted using long-term time series data. In this study, long-term water quality indicators, as well as ecological indexes, natural meteorological factors, and socio-economic indexes, were collected for Tai Lake to enable us to study the environmental evolution of the lake ecosystem. The key time nodes and early warning signals of the steady-state transformation of Tai Lake were also identified, which could provide a theoretical basis for early indication of the transformation of lake ecosystems. Furthermore, the characteristics and driving mechanism of the lake's ecosystem evolution were analyzed based on the physical and chemical indexes of its sediments and its long-term water quality indexes. The results show that the early warning signals (variance, autocorrelation, and skewness) of ecosystem mutation included abnormal changes 10 years before the steady-state change, and the evolution of Tai Lake was driven by the complex nonlinear effects of biological, physical, chemical, and socio-economic factors in the lake basin. These results have important theoretical and practical value for pollution control and the management of eutrophic lakes.

**Keywords:** water ecology; Tai Lake; tipping point; driving mechanism

Citation: Li, C.; Wu, W. Analysis of the Driving Mechanism of Water Environment Evolution and Algal Bloom Warning Signals in Tai Lake. Water 2023, 15, 1245. https://doi.org/10.3390/w15061245

Academic Editors: Fang Yang and Jing Liu

Received: 6 January 2023
Revised: 3 March 2023
Accepted: 7 March 2023
Published: 22 March 2023

Copyright: © 2023 by the authors. Licensee MDPI, Basel, Switzerland. This article is an open access article distributed under the terms and conditions of the Creative Commons Attribution (CC BY) license (https://creativecommons.org/licenses/by/4.0/).

## 1. Introduction

Lake ecosystems have a variety of ecological functions and have provided ecosystem services for humans for thousands of years. However, most lakes are experiencing serious ecosystem degradation, which directly leads to a loss of biodiversity and an imbalance in the ecosystem structure and function [1,2]. With the changing climate and increasingly intensive nature of human activity, multiple driving forces interacting with each other have exerted serious damage upon the global natural ecosystem. Therefore, it is important to predict sudden and nonlinear changes in the system [3–6]. Ecologists have long recognized that ecosystems can exist in one steady state and operate within a predictable range for a long period of time and then suddenly switch to another state [7–10]. For example, the desertification of grassland, the expansion of shrubs in the Arctic region, the eutrophication of lakes, the acidification of seawater, and the degradation of coral reefs are all real or potential ecosystem mutations that are indicated by tipping points or system thresholds and driven by one or more external driving forces, as well as the internal control variables of the system. This leads to changes in the structure, function, and dynamics of the system [11]. Many scientific studies have shown that the Earth's system is currently developing along an unsustainable trajectory [12], with shallow lake ecosystems being relatively fragile. The lakes' low pollution loading capacity and the vulnerability of their water–soil interfaces to external disturbances result in strong material exchange and instability. In addition, in shallow lakes located in areas with intense human activity, it is difficult to fundamentally control external discharge. When the cycle characteristics of the nutrients in lake sediments, the food web structure, and the aquatic environment in the lower layer of a lake are destroyed, changes in the hydrodynamic conditions and the physical and chemical properties of the surface sediments will form a harmful positive feedback loop, thus hindering the process of ecological restoration in the lake [13].

Many studies have been conducted on the basic principles, driving mechanisms, statistical methods, and early indicators of steady-state transition in shallow lakes. Crawford S Holling (1973) studied the ability of ecosystems to respond to stress and proposed the term "resilience" to clarify the nonlinear characteristics of ecosystems to external stress [10]. Robert M May (1977) proposed the multistable state and threshold theory of ecosystems, pointing out that persistent external stress will weaken the resilience of an ecosystem and trigger a homeostatic transition [14]. Carpenter (2011) confirmed that methods such as experimental observation, statistical analysis, and model simulation can be used to identify the driving factors of the steady-state transition of lake ecosystems [15]. Wang Rong (2012) took Erhai Lake in Yunnan Province as an example to discuss the early warning signals of lake ecosystem mutation and revealed that under the strong interference of environmental change, the ecosystem state fluctuated frequently before the occurrence of steady-state transition; he termed this the "flickering" phenomenon [16]. However, due to the relatively scattered nature of the quantitative and long-term observation and research data, and the use of different research scales, key areas, research directions or starting points, and methods, his research conclusions were not completely consistent. Therefore, there is an urgent need for a long-term comprehensive analysis based on a large number of monitoring data. This paper takes the large, shallow Tai Lake as the key research object. We collected long-term ecological data and data on natural meteorological factors and the social economy, as well as other physical, chemical, biological, and comprehensive data, to explore the driving mechanism of water environment evolution.

The continuous monitoring of the water environment in Tai Lake began in the 1980s when it first entered a eutrophication state. Therefore, in order to study the characteristics of ecosystem mutation in Tai Lake, monitoring data from before the 1980s are needed for analysis. There is insufficient research on the evolution characteristics of the lake water environment before the 1980s due to the lack of water quality monitoring data in the 1970s. However, based on the "amplification effect" of environmental information on lake sediments and the orderliness of time records, physical, chemical, and biological information from different historical periods can be obtained by studying lake sediments; moreover, the historical process of ecological environment changes in lakes can be indirectly obtained by analyzing this information. Among a series of lacustrine sediment indicators, paleoecological indicators have been widely used in the study of lake eco-environmental evolution, among which diatoms in sediments are especially suitable for the study of high-resolution environmental change events because of their short life cycle, rapid reproduction, extreme sensitivity to environmental changes in water, and easy preservation in strata [17].

Therefore, in this paper, the evolution of diatom community structure in Tai Lake sediments was used to characterize the evolution process of Tai Lake and identify the key time nodes of steady-state transition. Furthermore, the characteristics of structural transformation and the driving mechanism during the steady-state transition were analyzed in order to provide a scientific basis for determining the steady-state transformation of large shallow lake ecosystems and management of the lake water environment [18–20].

## 2. Materials and Methods

### 2.1. Study Area

Tai Lake is the third largest freshwater lake in China [21] and is located near the Yangtze River Delta in a subtropical monsoon zone (Figure 1). The catchment area is 36,500 km$^2$, and the lake shoreline is 393.2 km long [22]. The average annual temperature and precipitation are between 16.0 and 18.0 °C and 1100 and 1150 mm, respectively. The lake supplies humans with vital ecosystem services such as agricultural grain production, flood control, fish, tourist tours, shipping, etc. In addition, Tai Lake also acts as a repository for a large quantity of industrial and domestic sewage discharge from nearby cities, villages, and industries due to the rapidly growing economy [23].

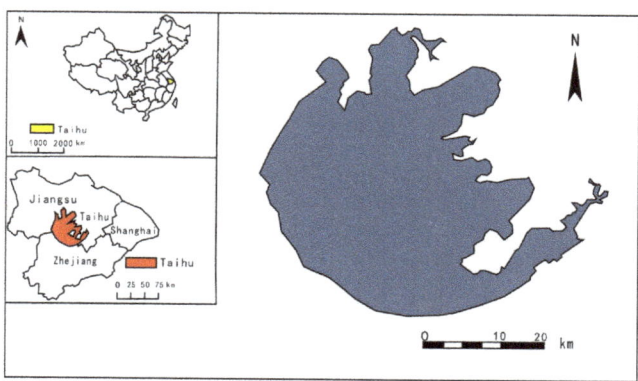

**Figure 1.** Location of Tai Lake in China.

*2.2. Data Source*

2.2.1. Physical and Chemical Indicators of the Water Environment

In this study, water quality indexes such as total phosphorus [TP], total nitrogen [TN], ammonia nitrogen [$NH_3$-N], chlorophyll [Chl-a], 5-day BOD [$BOD_5$], chemical oxygen demand [COD], dissolved oxygen [DO], pH, water temperature [WT], and transparency [SD] were collected from Tai Lake from 1980 to 2012 and sorted. Among them, water quality index data from 1980 to 2000 were obtained from water environmental monitoring data (GEMS/Water) collected by the Global and Regional Environmental Monitoring Coordination Center under the United Nations Environment Program. Data from 2000 to 2012 were collected from the National Ecosystem Observation and Research Network (CNERN). Water quality monitoring data from 2012 to 2017 were obtained from literature published in related fields [24,25], China's Environmental Status Bulletin of the Ministry of Environmental Protection, and the Tai Lake Health Report of the Taihu Basin Administration of the Ministry of Water Resources.

2.2.2. Aquatic Ecological Indexes

The aquatic ecological indicators were collected from the National Ecosystem Observation and Research Network (CNERN: http://www.cnern.org.cn/data/iitDRsearch?classcode=SYC_A01 (access on 20 October 2022); it is important to note that the time range of the data may change over time) and the Tai Lake Health Report of the Tai Lake Network. The aquatic ecological indicators of Tai Lake in this study mainly include phytoplankton, zooplankton, benthic animals, macro-aquatic plants, and bacteria. The first principle component analysis (PCA1) results on the presence of algae in Tai Lake sediments were obtained from the literature [26]; the TOC (%), TN (%), and C/N data on sediments were obtained from the literature [27]; and data on the abundance of Bosimina spp were obtained from the literature [28].

2.2.3. Meteorological Factors

Meteorological element data were collected from the National Ecosystem Observation and Research Network (CNERN) and provided by the Meteorological Data Room of the National Meteorological Information Center, China Meteorological Administration. These data included rainfall, wind speed, and temperature indicators.

2.2.4. Hydrological Data

The data on hydrological elements (which mainly refer to water level) were collected from the National Ecosystem Observation and Research Network (CNERN), and daily water level data were recorded at five water stations in the Tai Lake area: Wangting, Dapukou, Jiapu, Xiaomeikou, and Dongting, Xishan.

2.2.5. Socio-Economic Indicators

Data on China's annual grain production and fertilizer use from 1960 to 2013 were downloaded from: http://www.earthpolicy.org/data_center/C24.html (access on 20 October 2022).

2.3. Data Analysis and Processing

In this paper, STARS (Sequential T-test Algorithm for Analyzing), proposed by Rodionov [29], was used to analyze the statistical mutation of the mean level and amplitude of fluctuation of the evolution of the diatom community structure PCA1, and the abundance of the clean-tolerant species *Bosmina*, using a 10-year cutoff length ($p < 0.01$). The abundance of PCA1 and *Bosmina* spp in the sediments was calculated using Z-scores, which further supported the detection results regarding the mutation point of the lake ecosystem. Gaussian kernel density estimation of the PCA1 time series was conducted using R software to obtain a bimodal curve, which confirmed the existence of bistability in the ecosystem. Furthermore, an Autoregressive Integrated Moving Average (ARIMA (p, d, q)) model of the time series before PCA1 mutation was fitted using R software using the Stats program package, which was downloaded from http://www.r-project.org/ (access on 20 October 2022). Based on the minimum value of the Akaike Information Criterion (AIC), the optimal model was selected to predict the evolution process of the PCA1 time series from 1850 to 1960, and the predicted values were compared with the actual observed values. In addition, single exponential smoothing was conducted using Minitab software to detrend the PCA1 time series and lake sediments before the mutation. Then, the first-order lag autocorrelation coefficient, standard deviation, and skewness were calculated based on the detrended residuals with a 4-year moving window to assess changes in the elasticity of the lake ecosystem. Furthermore, SPSS statistical software was used to create a simple regression model and a multiple linear regression model and to verify the hypothesis that non-stationary driving factors lead to changes in the ecosystem.

## 3. Results and Discussion

*3.1. Historical Records of Response Variables and Environmental Driving Factors in the Sediment of the Tai Lake Ecosystem*

Continuous monitoring of the water environment in Tai Lake began in the 1980s, and during this time, the understanding of water environment evolution was insufficient. In view of the "amplification" effect of environmental information on sediments, spatial statistical representativeness, and chronological orderliness, this paper analyzes the evolution processes of the sedimentary environment and anthropogenic pollution in Tai Lake by combining geochemical records of sedimentary cores with dating data. Based on this analysis, we were able to reconstruct and analyze the historical evolution of the water environment in Tai Lake. The time series of the abundance of the diatom community PCA1 showed a downward trend from around 1960, with a sharp decline in 1970 (Figure 2a). The abundance of *Bosmina* spp increased 1.3-fold from the 1950s to the 1970s and began to decrease rapidly in the 1970s; following this, the abundance of *Bosmina* spp in the sediments decreased sharply and nearly disappeared in the 2000s, which indicates that the water environment of Tai Lake deteriorated quickly from the 1970s (Figure 2b). Furthermore, the amount of TOC in the sediments of Tai Lake increased rapidly from 1970 to 2000 and nearly doubled. Meanwhile, the content of TN increased slowly from 1950 to 1970 and then decreased by 5% from 1970 to 1979, followed by a rapid increase of 10% from 1979 to 2000 (Figure 2c); this is consistent with the strengthening of agricultural economic activity in the Tai Lake basin and indicates that the extensive application of farmland fertilizers is an important contributor to the total nitrogen in lake sediments [22]. Since the 1970s, the nutrient contents in the sediments of Tai Lake have increased rapidly, and the quality of the water environment has deteriorated remarkably. In addition, since 1970, the C/N ratio in the sediment has also increased significantly (Figure 2d); combined with the variation in TOC and TN, this indicates that before the 1970s, the carbon and nitrogen cycles were in a balanced state, and occurred mainly in the internal cycle of the system. During this time, the trophic status of the lake was still within the threshold before the occurrence of its mutation. After the 1970s, with the significant increase in the proportion of exogenous organic matter in the lake sediments, the balance of carbon and nitrogen was disturbed, and Tai Lake gradually became eutrophic.

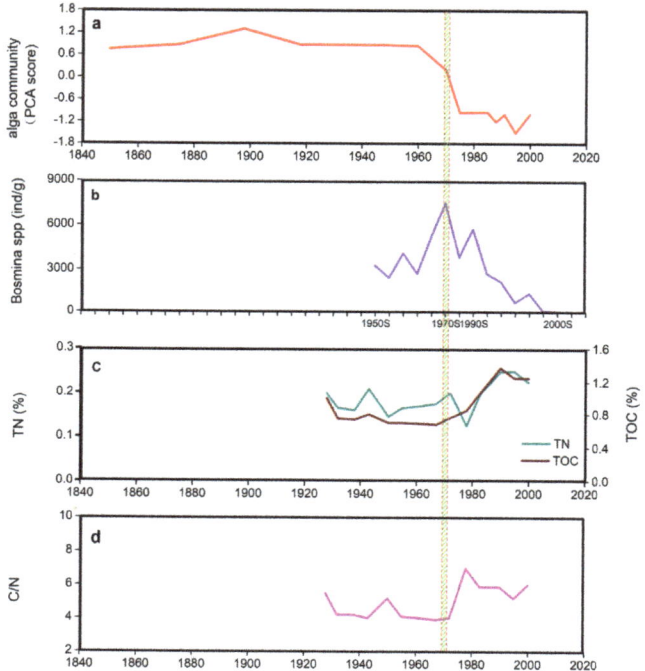

**Figure 2.** Historical records of lake sediment-based aquatic system response variables (The dotted box represents tipping points of the lake sediment-based aquatic system response variables). (**a**) Silicon alga composition in sediment; (**b**) density of *Bosmina* spp. species in sediment; (**c**) TN and TOC in sediment(%); (**d**) C/N in sediment.

From 1951 to 1970, historical records of natural driving factors in the Tai Lake basin show that the temperature and rainfall decreased from 15.1 °C and 1650 mm to 14.9 °C and 970 mm, respectively (Figure 3a), while from 1970 to 2017, the temperature and rainfall increased significantly ($p < 0.01$). Since 1970, the 2-min average wind speed, which can affect the chemical and biological processes in Tai Lake, has decreased year by year (from 3.1 m/s to 1.72 m/s in 2018) (Figure 3b). With wind speed gradually decreasing, vertical disturbance in the lake has gradually weakened, which has slowed the increase in nutrients in the water body, thus alleviating the eutrophication process. In addition, the lake water level decreased from 3.58 m in 1954 to 3.1 m in 1970 and then gradually increased (Figure 3b) to 3.58 m in 2016. Furthermore, the annual average grain output and fertilizer use in China has increased rapidly since 1970. By 2011, fertilizer use had increased 49.5-fold (Figure 3c), and by 2013, grain output had increased 4.42-fold. The intensive development of agriculture has led to the discharge of a large number of nutrients into the lake, thus causing continuous deterioration of its environment.

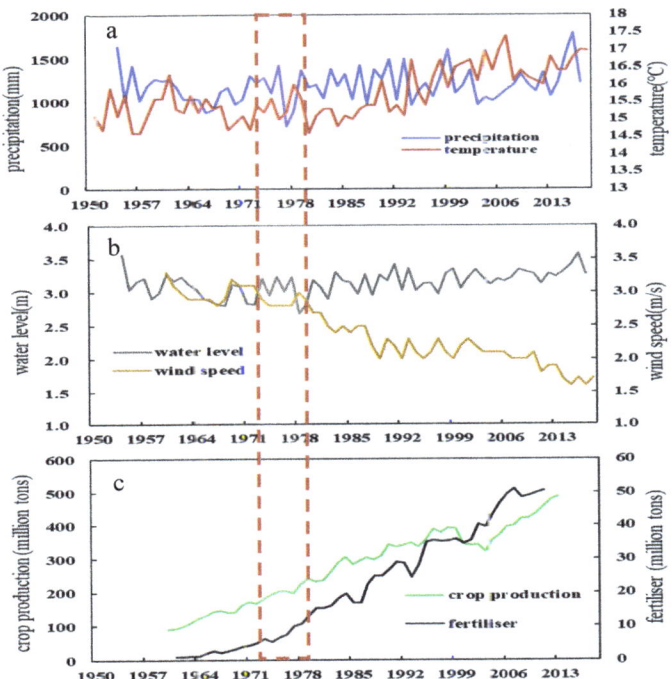

**Figure 3.** Historical records of environmental drivers during the period 1840–2017 (The dotted box represents tipping points of the lake sediment-based aquatic system response variables). (**a**) Temperature; (**b**) water level and wind speed; (**c**) grain production and fertilizer.

### 3.2. Detection and Evidence of Mutation in the Tai Lake Ecosystem

Based on the comprehensive analysis of sediment data, water level, meteorology, and other natural factors in the Tai Lake basin (Figures 2 and 3), we can deduce that the algal community structure changed significantly after the 1970s. The exogenously driven historical record (the 1950s) shows that the structural evolution of the algal community, which began in the 1960s, has changed in line with the nutrient loading of the lake due to agricultural intensification, which is a slow-driving variable that takes effect over decades (Figure 3c). In addition, the rapid driving variables, comprising the short-term regulation of lake water levels and short-term water volume changes caused by low rainfall from 1950 to 1970, have jointly triggered the transformation of the Tai Lake ecosystem. With an increased nutrient concentration in the lake (Figure 2c), its productivity is increased, and the dissolved oxygen in the water body is decreased; this results in the release of bioavailable phosphorus into the water body of Tai Lake from its upper sediments, which further aggravates eutrophication [30]. Although the water level of Tai Lake increased in 1980 (Figure 3b), the diatom community did not show signs of recovery until 1998 due to the occurrence of positive feedback in the eutrophication process.

The results of STARS mutation detection show that there was a downward mutation in the PCA1 time series in 1974 (Figure 4a). In order to further prove the existence of the mutation point, a set of autoregressive integrated moving average (ARIMA (p, d, q)) models were fitted using the Stats package of R software, where $p$ is the autoregressive (AR) order, q is the moving average (MA) order, and d is the differential part of the 1850–1960 PCA1 time series. Based on the minimum Akaike information criterion (AIC), the evolution process of the PCA1 time series from 1970 to 2000 was predicted, and ARIMA (1,1,2) was selected as the optimal model for the PCA1 time series (Table 1). The predicted values for 1970–2000 (red lines represent 95% confidence intervals), observed values (solid black lines), and predicted values (circles) are significantly different within the 95% probability level (Figure 4b); this indicates that the abrupt change observed in the 1970s cannot be predicted by a linear model, further proving the existence of an ecosystem break point. Furthermore, the value of PCA1 decreases dramatically from about 0.9 to −0.9 (Figure 4a). Compared with Figure 4b, it can be seen that in the absence of strong external interference, the value should have been within the predicted range instead of the mutation range. In addition, the probability density function (Gaussian kernel density estimate) from 1850 to 2000 indicates the existence of bistability in the ecosystem, with short vertical bars indicating the density of individual points (Figure 4b). This indicates that the ecosystem of Tai Lake has undergone dramatic mutation in the past 150 years, with a significant increase in eutrophic planktonic diatoms in its sediments [26] and a rapid decrease in the clean-tolerant species *Bosmina* spp. (Figure 4c). Overall, these results indicate that the ecosystem of Tai Lake underwent large-scale mutation and rapid reorganization in the 1970s.

**Table 1.** The selection details of the ARIMA (1,1,2) model of PCA1, including model type (AR and MA), coefficient, standard deviation of coefficient, T-test statistic, and probability level (P).

| | PCA1_ARIMA (1,1,2) | | | $R^2 = 0.98$ |
|---|---|---|---|---|
| Type | Cofficient | SE Coef | T | P |
| AR(1) | 0.533 | 0.375 | 1.422 | 0.193 |
| AR(2) | 0.466 | 0.368 | 1.265 | 0.242 |
| MA(1) | 0.888 | 0.171 | 2.185 | 0.0008 |

In addition, the PCA1 and *Bosmina* spp. data were normalized based on their means and standard deviation using the function Z-score (X) in MATLAB software. The obtained Z-score values indicate the variability of the PCA1 and *Bosmina* spp. time series data. Before the 1970s, the Z-score value fluctuates within a relatively low range of one standard deviation, and a sudden jump occurs in the 1970s, exceeding the range of one standard deviation, indicating that the state of the Tai Lake ecosystem suddenly changed in the 1970s (Figure 5).

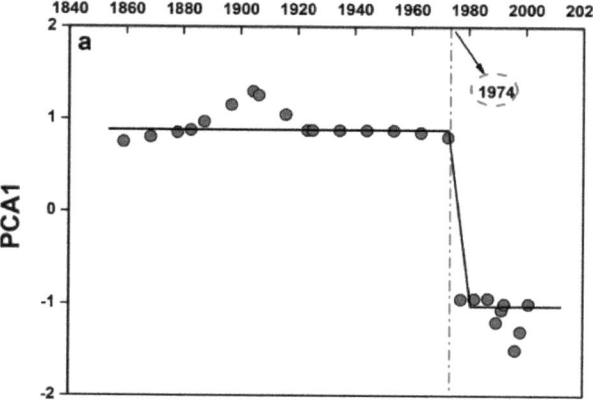

**Figure 4.** *Cont.*

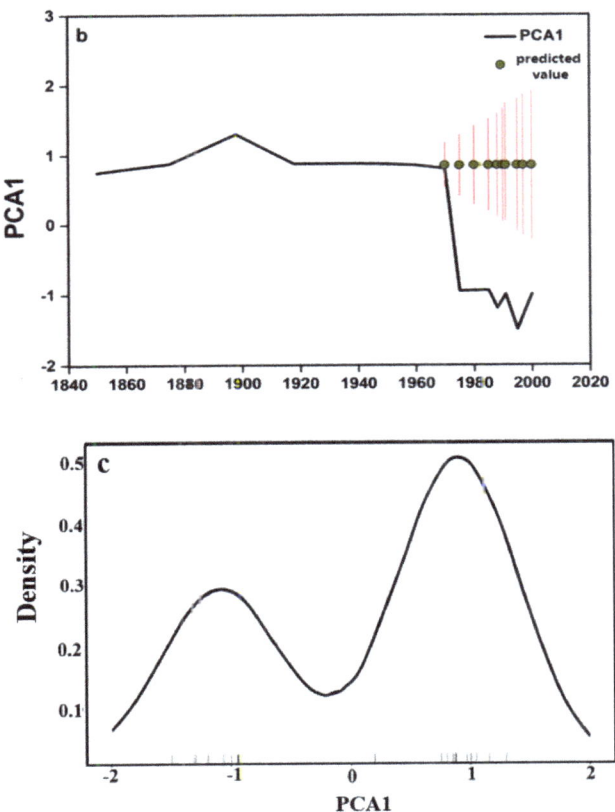

**Figure 4.** Evidence of bistability and regime shift in time series of PCA1.

In 1960, the transparency of Tai Lake was 1.57 m, which decreased to 0.45 m in 1980. At the same time, the area covered by submerged macrophytes decreased from 252 km² to 146 km², while the chlorophyll concentration increased from 0.003 mg/L to 0.004 mg/L. Meanwhile, COD content increased from 1.9 mg/L to 2.83 mg/L. This shows that in 1960, the water quality of Tai Lake was good, the ecological environment was not affected by large-scale external stress, and the water environment was still undergoing natural evolution. By 1980, water quality monitoring results showed that Tai Lake had entered a stable state of eutrophication (Figure 6). This indicates that the water environment of Tai Lake underwent a steady-state transition during the 1970s, changing from a natural evolution state to a disordered state. By 1980, the ecosystem of Tai Lake had completed its steady-state transition (Figure 6).

In order to further exclude the possibility of PCA1 time series mutation caused by non-stationary external stress, a regression model was used to test our hypothesis. The results of the multiple linear regression analysis for 1950–2000 (including the potential mutation point of the 1970s) (Table 2) show that rainfall, temperature, wind speed, fertilizer, water level, and grain yield do not provide a clear explanation for the changes in linear relationships with the PCA1 time series (none were significant at a probability level of $p \leq 0.05$). In our simple linear regression model (Table 2), only wind speed, fertilizer, and grain yield showed significant statistical significance, but the relationship between wind speed and PCA1 was a counterintuitive negative correlation, which was difficult to explain using a simple causal relationship. Therefore, the mutation of the Tai Lake ecosystem could be a critical transition phenomenon caused by a series of complex nonlinear interactions, internal and external feedback of the ecosystem, and internal threshold crossing.

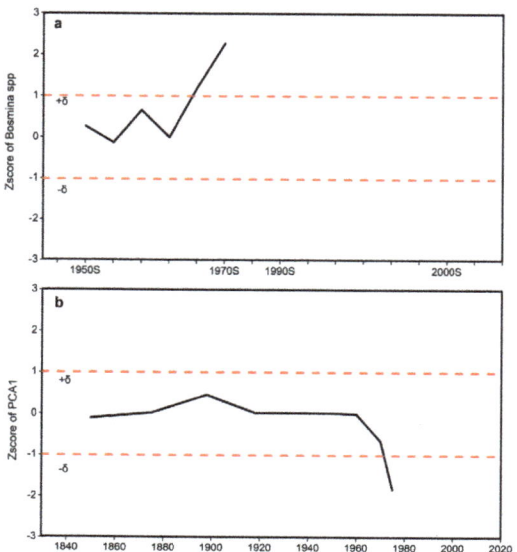

**Figure 5.** Z-scores of *Bosmina* spp. species and PCA1. (**a**) Z-score of *Bosmina* spp.; (**b**) Z-score of PCA1.

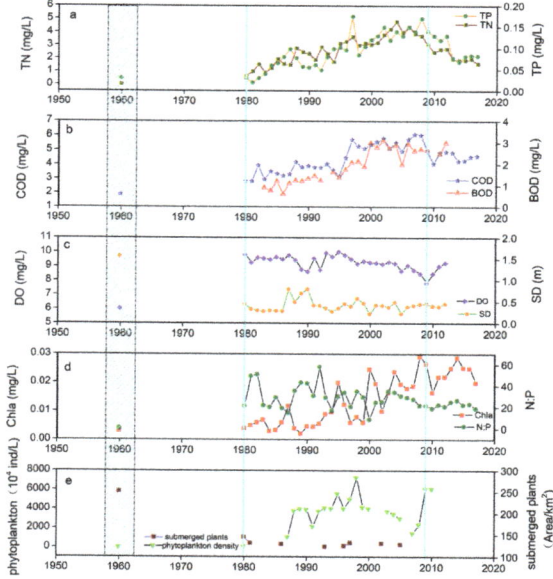

**Figure 6.** Time series of water quality indicators between 1960 and 2011~2017 in Tai Lake (The dotted box was used to highlight 1960, and the blue and green lines were used to mark time nodes of water environment change). (**a**) Total nitrogen and total phosphorous; (**b**) chemical oxygen demand and biological oxygen demand; (**c**) dissolved oxygen and transparency; (**d**) chla, ratio of nitrogen to phosphorous; (**e**) phytoplankton and submerged plants.

**Table 2.** A simple, multivariate regression analysis of potential external drivers (rainfall, air temperature, wind speed, fertilizer, water level, and crop production) and algal response (PCA1) in the Tai Lake ecosystem from 1950 to 2000.

| | Multiple regression model standardized coefficients | |
|---|---|---|
| | | PCA1 |
| precipitation | | −0.046 |
| air temperature | | −0.039 |
| wind speed | | 0.942 |
| fertilizer | | 0.586 |
| crop production | | −0.438 |
| Water level | | 0.046 |
| Notes: $p \leq 0.05$ insignificant. | | |
| | Simple regression model R value | |
| | | PCA1 |
| precipitation | | −0.476 |
| air temperature | | −0.476 |
| wind speed * | | 0.834 |
| Fertilizer * | | −0.739 |
| water level | | 0.077 |
| crop production * | | −0.773 |
| Notes: $p < 0.05$ * (2-tailed test) significant, the rest are insignificant. | | |

### 3.3. Extraction of Potential Early Warning Signals of Ecosystem Mutation

R software and Minitab software were adopted to perform Gaussian kernel density estimation and trend decomposition to obtain the residuals of the PCA1 time series (Figure 7b), and the red line represents the smooth curve fitted after the time series before the mutation point was detrended (Figure 7a). The first-order autocorrelation coefficient (lag1-autocorrelation), standard deviation (SD), and skewness (skew) of the residuals were calculated using a semi-time series sliding window, and the early warning signal of ecosystem mutation was extracted. The results show that the first-order autocorrelation coefficient and standard deviation of the PCA1 time series residuals have increased significantly since 1960, and the skewness has decreased since 1960. Compared with the mutation that occurred in the 1970s, the change in the trend of warning signals occurred about 10 years earlier, in 1960. Previous studies have found that the responses of shallow lake ecosystems to stress drivers will change suddenly with changes or increases in disturbance intensity, which will lead to a series of changes in the structure or function of the ecosystem, that is, the homeostatic transition of the ecosystem [7,31]. The results of the PCA1 long-time series analysis of the diatom community structure in the sediments of Tai Lake show that the standard deviation, autocorrelation coefficient, and skewness of the Tai Lake ecosystem increased significantly about 10 years before the threshold point of steady-state transition in the 1970s. Previous studies have shown that when driven or stressed by the external environment, the rate of change of a dynamic system will slow down when the tipping point is approached, and in an ecosystem, the autocorrelation of the system will increase in the short term [31,32]. Scheffer et al. studied changes in the characteristics of the variable autocorrelation coefficient in the transition of a lake from an oligotrophic state to a eutrophic state and found that the autocorrelation of variables was significantly enhanced before the steady-state transition of the lake ecosystem [31] Owing to the increasing intensity of external stress factors, the number of species showed an obvious downward trend. In this process, the autocorrelation of statistics that indicate the number of species was significantly enhanced, suggesting that the enhancement of the autocorrelation of a large number of statistics can be used for early indication of the steady-state transition of ecosystems [31,33]. This is consistent with this study's conclusion that the system autocorrelation of Tai Lake was enhanced 10 years before the tipping point in the 1970s. This indicates that the change rate of the ecosystem in Tai Lake gradually slowed down 10 years before the threshold mutation occurred under the influence of external interference. Therefore, the long-term monitoring data of Tai Lake can be used for analysis; moreover, the autocorrelation coefficient of the system state residual can be extracted so as to provide an early indication of the threshold mutation of the ecosystem and determine the time node of steady-state transition.

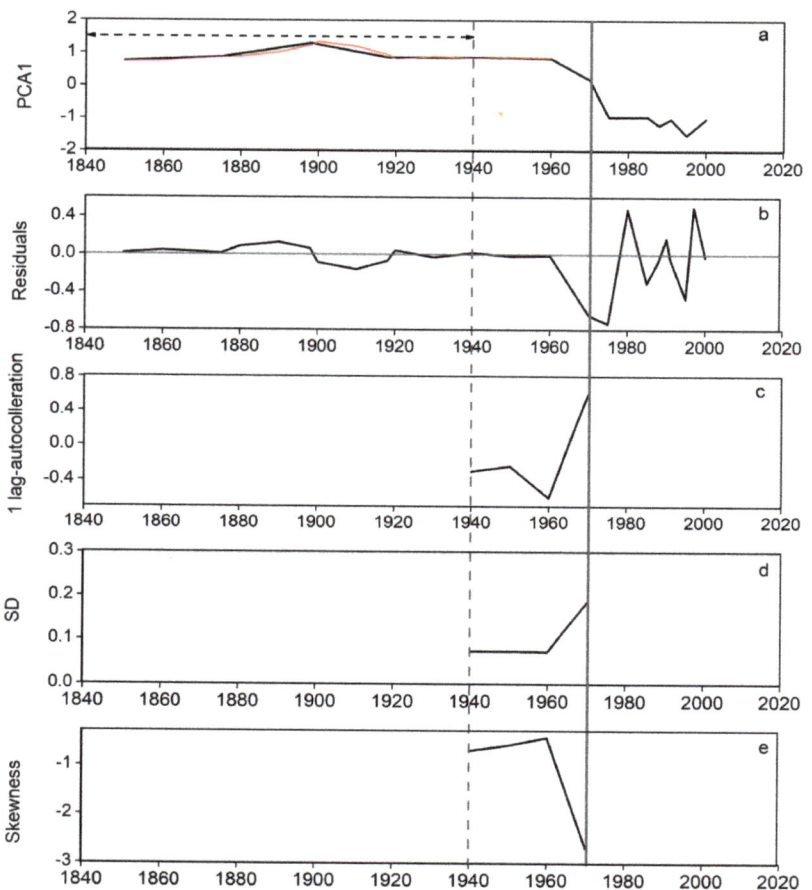

**Figure 7.** Potential early warning signals of a regime shift in the lake's trophic state for PCA time series. (**a**) PCA1 scores of sediment diatom composition (derived from Figure 2) (The red line is the Minitab curve after filtering out the slow trend, and the arrow indicates the width of the moving window used to calculate the warning signal); (**b**) residuals of PCA1 (the gray line indicates zero); (**c**) lag-autocorrelation of (**b**); (**d**) SD of (**b**); (**e**) skewness of (**b**) (the solid gray vertical line indicates the 1970s).

Secondly, the early warning signals extracted in this study also include an increase in the standard deviation of the residuals. Carpenter (2011) successfully obtained advanced warning of a sudden change in the food web of an ecosystem by introducing competitive predators to destroy the food web of the aquatic ecosystem and analyzing significant changes in statistical data, such as a sudden increase in standard deviation or a sudden decrease in recovery rate. It was proven that a regular change in long-term time series data can be used to judge the occurrence of steady-state transition in an ecosystem [13]. When Scheffer (2007) studied the community structure of aquatic plants in shallow lakes, he simulated a "large aquatic plant quantity model" over a long time scale and analyzed the change of the systems upon approaching the mutation point of a multi-stable curve and found that the variance increased significantly [34]. The above results are consistent with those of the Tai Lake ecosystem in this paper, which show a significant increase in variance 10 years before approaching the threshold point of the multistable system. As a result of the strong evidence of exogenous driving forces (Figures 2 and 3), we can reject the hypothesis that the increase in variance is explained by internal noise generated only by internal changes in the ecosystem. Thus, the rising variance is most likely indicative of a phenomenon caused by multiple exogenously driven interactions and transitions across thresholds within the ecosystem that amplify the system response.

In addition, in the field of ecosystem steady-state transitions, sudden changes in the skewness of statistical data can be used to accurately indicate whether the symmetry of the target long-term time series data has changed [35]. It has been confirmed that when the system is close to the threshold point where the mutation is about to occur, the nonlinear influence of a large number of external stress factors is gradually strengthened, which leads to an increasing trend of asymmetry in the statistical data density distribution. Therefore, the gradual strengthening of the law of asymmetry of statistical data can be used to identify the occurrence of steady-state transition in the system. Thus, through the research and analysis of long-term time series data of an ecosystem, if it is confirmed that the skewness of the statistical data curve has a sudden change, it can be predicted that the system will cross the mutation point of the steady-state transition, and then, transform to other steady states. Guttal (2008) regarded the oligotrophic and eutrophic states of a lake as two stable states of an ecosystem [36]. Our research results show that in the process of the system approaching the threshold point of the multistable curve, with the increasing intensity of external interference, the symmetry of the curve describing the system state variables decreased significantly, and the skewness value of statistical data changed abruptly. This trend appeared 10 years before the steady-state transition of the lake ecosystem, which confirmed that the sudden change in skewness could be used as an important early indicator of lake eutrophication. The results of this study show that the skewness of residuals has become increasingly left-aligned since 1960 (the absolute value of the negative number has become increasingly larger), and this appeared 10 years before the steady-state transition of the Tai Lake ecosystem; this is consistent with Guttal's results [36]. Therefore, the feasibility of developing an early warning system for lake ecosystem eutrophication based on the increased skewness of statistical data is further verified.

At present, the driving mechanisms for the steady-state transition of lake ecosystems are mainly divided into six types [37], and the characteristics of external force and the nature of random disturbance in the ecosystem determine the type of steady-state transition and whether it can be identified [38]. Whether or not the warning factor for steady-state transition is effective can determine the type of mechanism that drives steady-state transition. In our study, external environmental driving forces, such as climate and nutrient load, slowly pushed the ecosystem to the threshold of steady-state transition, which represents a typical slow environmental driving mechanism (Figure 8). Therefore, the potential early warning signals (variance, first-order, autocorrelation, and skewness) extracted in this paper were generated by the slow environmental driving mechanism and the reorganization and feedback of the lake's internal system [39]. These factors can be identified and used for the early identification of water bloom in Tai Lake, with an early warning time scale of as long as 10 years.

### 3.4. Dynamic Change Process of the Socio-Lake System

Ecosystems can be driven by sudden environmental events or by slowly changing environmental variables that cause internal reorganization and feedback [30,40,41]. The research in this paper shows that the degradation of the Tai Lake ecosystem has experienced three different stages corresponding to different climate–socioeconomic regimes in the last two centuries (Figure 9). Before the 1960s, the ecosystem of Tai Lake was in an oligotrophic state (Figure 6), and the lake ecosystem maintained a relatively stable state, which can be seen from the relatively stable natural variability of the PCA1 time series (Figure 5). Although human beings began long-term reclamation and development of the Tai Lake basin thousands of years ago, due to the limitations of farming technology and the sparse population density [42], this reclamation was limited to a local area of the Tai Lake basin; thus, the impact of human activity was much smaller than that of modern society. Therefore, during this period, the ecosystem maintained a high recovery capacity after external disturbance, that is, it had high system elasticity (upper part of Figure 9), which enabled the maintenance of an ecological balance between the lake ecosystem and human disturbance.

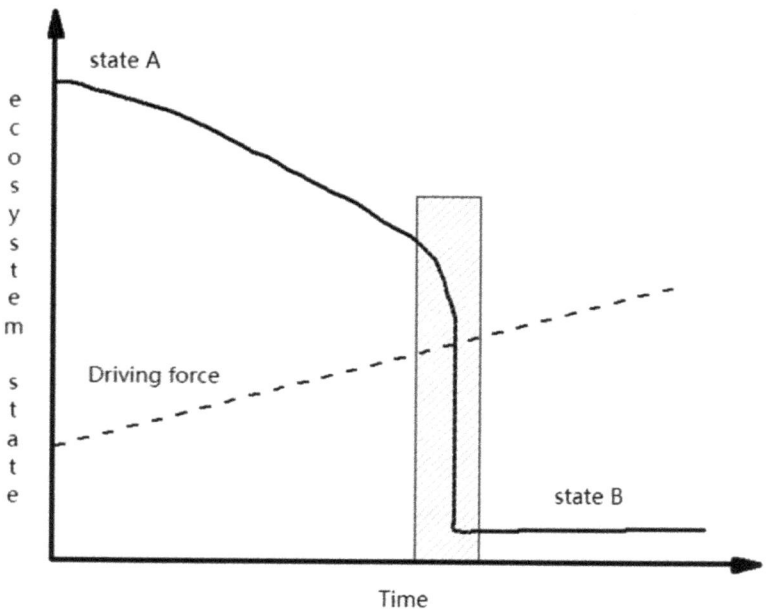

**Figure 8.** The mechanism of driving forces for regime shift in Tai Lake (the dotted line represents the time range of ecosystem mutation).

However, after 1960, the balance between the ecosystem of Tai Lake and the social system surrounding the basin has gradually been broken. The pressure of population growth and the increasing demand for food has led to an increase in the intensity of agricultural activity in the Tai Lake basin, such as the Great Leap Forward in 1958–1961 and the People's Commune Movement in 1958–1982 [40], which were marked by large-scale wetland reclamation and a campaign to build farmland around the lake. As a result, the lake area has shrunk, and nutrients have been discharged into the lake. At the same time, the water level of Tai Lake declined from 1965 to 1970, which further aggravated the degradation of the ecological environment. In the 1970s, fluctuation in the environmental, chemical, and physical processes of Tai Lake led to significant changes in its species composition (Figure 7a) and food web structure. The system elasticity continuously decreased, and steady-state mutation of the ecosystem from an oligotrophic state to a eutrophic state occurred (upper part of Figure 9).

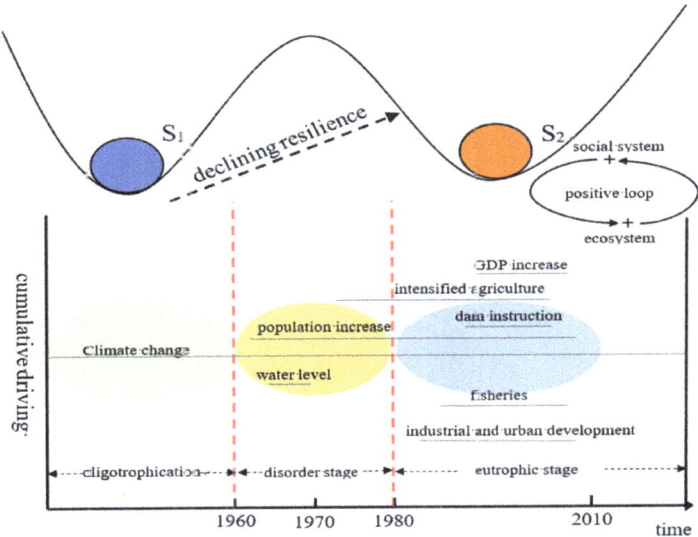

**Figure 9.** A time axis of multiple cumulative driving factors in the Tai Lake basin (the bottom figure) over the past 60 years, which have led to degradation of the ecosystem and reduced its resilience (the top figure). The three ellipses of different colors represent the different development stages of the Lake Tai ecosystem, representing the oligotrophic stage (before 1960S), the disordered mutation stage (1960–1980), and the eutrophication stage (1980–2010).

Since the 1980s, with the deepening of the reform and opening-up policy, the impact of human activity on the ecosystem of Tai Lake has increased exponentially. In 2010, the urbanization rate of the Tai Lake basin in Jiangsu Province reached 67.9% [42]. Large-scale and high-intensity human socio-economic activity has greatly accelerated the eutrophication process of Tai Lake [43,44]. With the development of society, people obtain increasing amounts of natural resources from lakes, which leads to problems such as overfishing, the extensive use of chemical fertilizers and pesticides, and the excessive excavation of sediment. Meanwhile, the temperature of the Tai Lake basin keeps rising (Figure 3), which further aggravates the ecological and environmental effects of pollution and overfishing, such as the outbreak of cyanobacteria in Tai Lake in 2007 and the "black water mass" event in 2008, both of which were extreme ecological disasters caused by the synergistic effects of external driving factors. Under these conditions, the response of the Tai Lake ecosystem has been much greater than if it were under the influence of a single external driving force. In addition, there is a positive feedback loop of interaction and mutual reinforcement between the development of the Tai Lake ecosystem and its social system, which further accelerates the lack of elastic resilience of the Tai Lake ecosystem. For example, a decline in the fishing rate in Tai Lake would lead to the strengthening of fishing activity, and a long-term decline in fishing would stimulate the economic value of rare goods, which would lead to a further increase in fishing activity. Under the strong influence of social activity [45] and the positive feedback loop of the ecosystem, we believe that after the 1980s, the external stress exerted on the Tai Lake ecosystem exceeded the elastic adjustment ability of the system itself, which has led to great changes in the composition, structure, and function of its species (Figure 10).

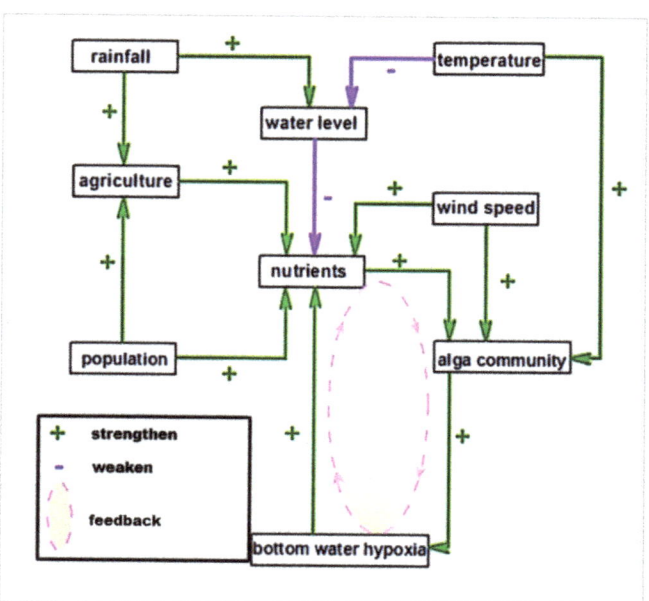

**Figure 10.** The figure describes the occurrence of eutrophication (Figure 6) after the 1980s. At this time, a positive feedback loop (dotted ellipse) of phosphorus in the sediments circulating in the lake water was firmly established. The figure shows the diversity and interconnectivity of the main external drivers (bold) of the algal community in Lake Tai, with typical relationships (positive and negative) shown for major interactions. This figure is modified from Wang et al., 2012 [16].

## 4. Conclusions

Our study offers new insight into research on lake ecosystems. In this research, the sediment environmental index and water quality monitoring data were used to analyze the abrupt change in the Tai Lake ecosystem, which successfully fills the research gap caused by the lack of environmental monitoring records for Tai Lake in the 1970s.

In order to further understand the evolutionary process and driving mechanism of the Tai Lake ecosystem, we considered the evolutionary process of diatom community structure in the sediments to represent the evolutionary process of Tai Lake. Time series with long time scales were used for inversion, and early indications (variance, autocorrelation, and skewness) of ecosystem mutation showed the occurrence of abnormal changes ten years before the steady-state transition; moreover, we found that the evolution of the Tai Lake ecosystem is driven by complex nonlinear interactions between biophysical, ecological and socio-economic factors.

In addition, this study also shows that the evolution of the ecological structure of the diatom community in the Tai Lake ecosystem is driven by complex interactions between biophysics, ecology, and the social economy. Therefore, in the processes of protecting and restoring a lake's ecological environments, the tipping point and threshold of ecosystem mutation should be included in their management; moreover, on this basis, the driving forces of the further collapse of lake ecosystems should be studied in depth.

**Author Contributions:** Conceptualization, W.W. and C.L.; methodology, C.L.; software, C.L.; validation, W.W.; formal analysis, C.L.; investigation, C.L.; data curation, C.L.; writing—original draft preparation, C.L.; writing—review and editing, C.L.; visualization, C.L.; supervision, W.W.; project administration, W.W.; funding acquisition, W.W. All authors have read and agreed to the published version of the manuscript.

**Funding:** This research was funded by [Soil carbon sequestration potential of typical land use under climate change] grant number [31170489].

**Data Availability Statement:** Not applicable.

**Acknowledgments:** This research was supported by National Natural Science Foundation of China, International Cooperation Project of Land Use Change and its Control in the Context of Climate Change.

**Conflicts of Interest:** The authors declare no conflict of interest.

# References

1. Xu, X.; Zhang, Y.B.; Chen, Q.; Li, N.; Shi, K.; Zhang, Y.L. Regime shifts in shallow lakes observed by remote sensing and the implications for management. *Ecol. Indic.* 2020, *113*, 2–9. [CrossRef]
2. Petrescu, A.M.R.; Lohila, A.; Tuovinen, J.-P.; Baldocchi, D.D.; Desai, A.R.; Roulet, N.T.; Vesala, T.; Dolman, A.J.; Oechel, W.C.; Marcolla, B.; et al. The uncertain climate footprint of wetlands under human pressure. *Proc. Natl. Acad. Sci. USA* 2015, *112*, 4594–4599. [CrossRef] [PubMed]
3. Kuiper, J.J.; van Altena, C.; de Ruiter, P.C.; van Gerven, J.H.; Janse, J.H.; Mooij, W.M. Food-web stability signals critical transitions in temperate shallow lakes. *Nat. Commun.* 2015, *6*, 7727. [CrossRef] [PubMed]
4. Carpenter, S.R.; Mooney, H.A.; Agard, J.; Capistrano, D.; Defries, R.S.; Díaz, S.; Dietz, T.; Duraiappah, A.K.; Oteng-Yeboah, A.; Pereira, H.M.; et al. Science for managing ecosystem services: Beyond the Millennium Ecosystem Assessment. *Proc. Natl. Acad. Sci.* 2009, *106*, 1305–1312. [CrossRef]
5. Rockström, J.; Steffen, W.; Noone, K.; Persson, A.; Chapin, F.S., III; Lambin, E.F.; Lenton, T.M.; Scheffer, M.; Folke, C.; Schellnhuber, H.J.; et al. A safe operating space for humanity. *Nature* 2009, *461*, 472–475. [CrossRef]
6. Lenton, T.M.; Held, H.; Kriegler, E.; Hall, J.W.; Lucht, W.; Rahmstorf, S. Tipping elements in the Earth's climate system. *Proc. Natl. Acad. Sci. USA* 2008, *105*, 1786–1793. [CrossRef]
7. Hastings, A.; Wysham, D.B. Regime shifts in ecological systems can occur with no Warning. *Ecol. Lett.* 2010, *13*, 464–472. [CrossRef]
8. Andersen, T.; Carstensen, J.; Hernández-García, E.; Duarte, C.M. Ecological thresholds and regime shifts: Approaches to identification. *Trends Ecol. Evol.* 2009, *24*, 49–57. [CrossRef]
9. Scheffer, M.; Carpenter, S.R. Catastrophic regime shifts in ecosystems: Linking theory to observations. *Trends Ecol. Evol.* 2003, *18*, 648–656. [CrossRef]
10. Holling, C.S. Resilience and stability of ecological systems. *Annu. Rev. Ecol. Syst.* 1973, *4*, 1–23. [CrossRef]
11. Hoegh-Guldberg, O.; Mumby, P.J.; Hooten, A.J.; Steneck, R.S.; Greenfield, P. Coral reefs under rapid climate change and ocean acidification. *Science* 2007, *318*, 1737–1742. [CrossRef]
12. Wu, J. Landscape sustainability science: Ecosystem services and human well-being in changing landscapes. *Landsc. Ecol.* 2013, *28*, 999–1023. [CrossRef]
13. Batt, R.D.; Carpenter, S.R.; Cole, J.J.; Pace, M.L.; Johnson, R.A. Changes in ecosystem resilience detected in automated measures of ecosystem metabolism during a whole-lake manipulation. *Proc. Natl. Acad. Sci. USA* 2013, *110*, 17398–17403. [CrossRef]
14. Robert, M.M. Thresholds and breakpoints in ecosystems with a multiplicity of stable states. *Nature* 1977, *269*, 471–477. [CrossRef]
15. Carpenter, S.R.; Cole, J.J.; Pace, M.L.; Batt, R.; Brock, W.A.; Cline, T.; Coloso, J.; Hodgson, J.R.; Kitchell, J.F.; Seekell, D.A.; et al. Early Warnings of Regime Shifts: A Whole-Ecosystem Experiment. *Science* 2011, *332*, 1079–1082. [CrossRef]
16. Wang, R.; Dearing, J.A.; Landon, P.G.; Zhang, E.L.; Yang, X.D.; Dakos, V.; Scheffer, M. Flickering gives early warning signals of a critical transition to a eutrophic lake state. *Nature* 2012, *492*, 419–422. [CrossRef]
17. Fan, X.; Cheng, F.J.; Yu, Z.M.; Song, X.X. The environmental implication of diatom fossils in the surface sediment of the Changjiang River estuary (CRE) and its adjacent area. *J. Oceanol. Limnol.* 2019, *37*, 552–567. [CrossRef]
18. Feng, W.Y.; Wang, T.K.; Zhu, Y.R.; Sun, F.H.; Giesy, J.P.; Wu, F.C. Chemical composition, sources, and ecological effect of organic phosphorus in water ecosystems: A review. *Carbon Res.* 2023. [CrossRef]
19. Feng, W.Y.; Yang, F.; Zhang, C.; Liu, J.; Song, F.H.; Chen, H.Y.; Zhu, Y.R.; Liu, S.S.; Giesy, J.P. Composition characterization and biotransformation of dissolved, particulate and algae organic phosphorus in eutrophic lakes. *Environ. Pollut.* 2020, *265*, 114838. [CrossRef]
20. Wang, X.R.; Liang, D.Y.; Wang, Y.; Peijnenburg, W.J.; Monikh, F.A.; Zhao, X.L.; Dong, Z.M.; Fan, W.H. A critical review on the biological impact of natural organic matter on nanomaterials in the aquatic environment. *Carbon Res.* 2022, *1*, 13. [CrossRef]
21. Li, C.C.; Feng, W.Y.; Chen, H.Y.; Li, X.F.; Song, F.H.; Guo, W.J.; Giesy, J.P.; Sun, F.H. Temporal variation in zooplankton and phytoplankton community species composition and the affecting factors in Lake Taihu—A large freshwater lake in China. *Environ. Pollut.* 2019, *245*, 1050–1057. [CrossRef] [PubMed]
22. Qin, B.Q.; Hu, W.M.; Chen, W.M. *Water Environment Evolution Process and Mechanism of Tai Lake*; Science Press: Beijing, China, 2004; ISBN 9787030124081. (In Chinese)
23. Li, C.C.; Feng, W.Y.; Song, F.H.; He, Z.Q.; Wu, F.C.; Zhu, Y.R.; Giesy, J.P.; Bai, Y.C. Three decades of changes in water environment of a large freshwater Lake and its relationship with socio-economic indicators. *J. Environ. Sci.* 2019, *77*, 156–166. [CrossRef] [PubMed]

24. Zhu, G.W.; Qin, B.Q.; Zhang, Y.L.; Xu, H.; Zhu, M.Y.; Yang, H.W.; LI, K.Y.; Min, S.; Shen, R.J.; Zhong, C.N. Changes of chlorophyll a and nutrient salts in northern Taihu Lake and their influencing factors from 2005 to 2017. *J. Lake Sci.* **2018**, *30*, 279–295. (In Chinese) [CrossRef]
25. Dai, X.L.; Qian, P.Q.; Ye, L.; Song, T. Evolution trend of nitrogen and phosphorus concentrations in Taihu from 1985 to 2015. *J. Lake Sci.* **2016**, *28*, 935–943. (In Chinese) [CrossRef]
26. Zhang, K.; Dong, X.H.; Yang, X.D.; Kattel, G.; Zhao, Y.J.; Wang, R. Ecological shift and resilience in China's lake systems during the last two centuries. *Glob. Planet Chang.* **2018**, *165*, 147–159. [CrossRef]
27. Liu, E.F.; Shen, J.; Zhu, Y.X. Geochemical records and comparative study of sediment pollution in western Tai Lake. *Sci. Geogr. Sin.* **2005**, *25*, 102–107. (In Chinese)
28. Liu, G.M.; Chen, F.J.; Liu, Z.W. A preliminary study on cladoceran microfossils from Tai Lake. *J. Lake Sci.* **2008**, *20*, 470–476. (In Chinese) [CrossRef]
29. Rodionov, S.N. Use of prewhitening in climate regime shift detection. *Geophys. Res. Lett.* **2006**, *33*, L12707. [CrossRef]
30. Dong, X.H.; Yang, X.D.; Chen, X.; Liu, Q.; Yao, M.; Wang, R.; Xu, M. Using sedimentary diatoms to identify reference conditions and historical variability in shallow lake ecosystems in the Yangtze floodplain. *Mar. Freshw. Res.* **2016**, *67*, 803–815. [CrossRef]
31. Ives, A.R. Measuring resilience in stochastic-systems. *Ecol. Monogr.* **1995**, *65*, 217–233. [CrossRef]
32. Seekell, D.A.; Carpenter, S.R.; Pace, M.L. Conditional heteroscedasticity as a leading indicator of ecological regime shifts. *Am. Nat.* **2011**, *178*, 1–11. [CrossRef]
33. Liu, Y.; Evans, M.A.; Scavia, D. Gulf of mexico hypoxia: Exploring increasing sensitivity to nitrogen loads. *Environ. Sci. Technol.* **2010**, *44*, 5836–5841. [CrossRef]
34. Scheffer, M.; van Nes, E.H. Shallow lakes theory revisited:various alternative regimes driven by climate, nutrients, depth and lake size. *Hydrobiologia* **2007**, *584*, 455–466. [CrossRef]
35. Yu, R.H.; Zhang, X.X.; Liu, T.X.; Hao, Y.L. Forewarned is forewarmed:Limitations and prospects of early warning indicators of regime shifts in shallow lakes. *Acta Ecol. Sin.* **2017**, *37*, 3619–3627. [CrossRef]
36. Guttal, V.; Jayaprakash, C. Changing skewness-an early warning signal of regime shifts in ecosystems. *Ecol. Lett.* **2008**, *11*, 450–460. [CrossRef]
37. Drake, J.M.; Griffen, B.D. Early warning signals of extinction in deteriorating environments. *Nature* **2010**, *467*, 456–459. [CrossRef]
38. Seddon, A.W.R.; Froyd, C.A.; Witkowski, A.; Willis, K.J. A quantitative framework for analysis of regime shifts in a Galápagos coastal lagoon. *Ecology* **2014**, *95*, 3046–3055. [CrossRef]
39. Zhao, Y.J.; Wang, R.; Yang, X.D.; Dong, X.H.; Xu, M. Regime shifts revealed by paleoecological records in Lake Taibai's ecosystem in the middle and lower Yangtze River Basin during the last century. *J. Lake Sci.* **2016**, *28*, 1381–1390. (In Chinese) [CrossRef]
40. Scheffer, M.; Bascompte, J.; Brock, W.A.; Brovkin, V.; Carpenter, S.R.; Dakos, V.; Held, H.; van Nes, E.H.; Rieterk, M.; Sugihara, G. Early-warning signals for critical transitions. *Nature* **2009**, *461*, 53–59. [CrossRef]
41. Wang, H.D. *Lakes of China*; Commercial Press: Beijing, China, 1995; ISBN 9787100014847. (In Chinese)
42. Su, W.Z.; Chen, W.X.; Guo, W.; Ru, J.J. A preliminary study on the spatial occupation mechanism of river network by urban-rural land expansion in Tai Lake Basin. *J. Nat. Resour.* **2016**, *31*, 128–1301. (In Chinese) [CrossRef]
43. Yang, X.J. China's rapid urbanization. *Science* **2013**, *342*, 310. [CrossRef] [PubMed]
44. Liu, J.; Mooney, H.; Hull, V.; Davis, S.J.; Gaskell, J.; Hertel, T.; Lubchenco, J.; Seto, K.C.; Gleick, P.; Kremen, C.; et al. Systems integration for global sustainability. *Science* **2015**, *347*, 125883. [CrossRef] [PubMed]
45. Wu, F.; Li, F.; Zhao, X.; Bolan, N.S.; Fu, P.; Lam, S.S.; Mašek, O.; Ong, H.; Pan, B.; Qiu, X.; et al. Meet the challenges in the "Carbon Age". *Carbon Res.* **2022**, *1*, 1–2. [CrossRef]

**Disclaimer/Publisher's Note:** The statements, opinions and data contained in all publications are solely those of the individual author(s) and contributor(s) and not of MDPI and/or the editor(s). MDPI and/or the editor(s) disclaim responsibility for any injury to people or property resulting from any ideas, methods, instructions or products referred to in the content.

Article

# New Green and Sustainable Tool for Assessing Nitrite and Nitrate Amounts in a Variety of Environmental Waters

H. R. Robles-Jimarez, N. Jornet-Martínez and P. Campíns-Falcó *

MINTOTA Research Group, Departament de Química Analítica, Facultat de Química, Universitat de València, 46100 Burjassot, Spain
* Correspondence: pilar.campins@uv.es

**Abstract:** This paper aims to provide improved selectivity and sensitivity with a short analysis time of about 10 min and low residues for quantitation of nitrite and nitrate in waters by liquid chromatography. Ion-pair formation and ion exchange retention mechanisms were considered. The optimized option was in-tube solid phase microextraction (IT-SPME) by means of a silica capillary of 14 cm length and 0.32 mm id, coupled online with a capillary anion exchange analytical column (Inertsil AX 150 × 0.5 mm id, 5 μm) and the use of their native absorbance. Precision of the retention times expressed as % relative standard deviation (RSD) were <1% for both, nitrite ($t_R$ = 5.8 min) and nitrate ($t_R$ = 10.5 min). Well, river, channel, lake, sea, tap and bottled waters and several matrices of a drinking water treatment plant were analysed, and no matrix effect was observed for all of them. Inorganic anions and several organic acids were tested as possible interferences and suitable selectivity was obtained. Precision expressed as % relative standard deviation (RSD) was between 0.9 and 3%. Low detection limits of 0.9 and 9 μg/L for nitrite and nitrate were obtained, respectively, and low residue generation near 100 μL per run was also achieved.

**Keywords:** nitrite; nitrate; waters; ion exchange; capillary LC; IT-SPME

**Citation:** Robles-Jimarez, H.R.; Jornet-Martínez, N.; Campíns-Falcó, P. New Green and Sustainable Tool for Assessing Nitrite and Nitrate Amounts in a Variety of Environmental Waters. *Water* 2023, 15, 945. https://doi.org/10.3390/w15050945

Academic Editors: Weiying Feng, Fang Yang and Jing Liu

Received: 30 January 2023
Revised: 25 February 2023
Accepted: 27 February 2023
Published: 1 March 2023

**Copyright:** © 2023 by the authors. Licensee MDPI, Basel, Switzerland. This article is an open access article distributed under the terms and conditions of the Creative Commons Attribution (CC BY) license (https://creativecommons.org/licenses/by/4.0/).

## 1. Introduction

The excessive use of products rich in nitrogen in several fields has generated an alteration of its cycle [1]. From some of them, and due to their solubility in water, nitrate and nitrite concentrations affect markedly the quality of drinking water [2] and contribute to eutrophization. When these compounds are ingested, they can generate N-nitrosamines that according to the International Agency for Research on Cancer (IARC) are probably carcinogenic in humans and were included in group 2A [3]. Current regulations on nitrate and nitrite content in drinking water are based on guidelines established by the World Health Organization (WHO) considering short-term effects such as methemoglobinemia and thyroid effects, but not long-term health effects [4].

European legislation established a maximum concentration of 50 mg/L for nitrate and 0.5 mg/L for nitrite [5] in drinking water, and in groundwater the limit value is 50 mg/L for nitrate [6], but their content in environmental waters is variable for both, nitrate and nitrite [7]. Various spectroscopic and electroanalytical techniques have been proposed for nitrate and/or nitrite determination each having its own merits and demerits [8,9]. However, conventional ion chromatography with electronic suppression of eluent conductivity and conductimetric detection is a standard method (method 4110 C) to measure nitrate and nitrite in water and wastewater chemical analysis [10], with limits of detection (LOD) around 0.1 mg/L. Ion-pair formation from adding to the mobile phase of an ion-pairing agent and using conventional reversed phase analytical columns were also proposed [11–13], but higher LODs than that achieved by method 4110 C are achieved.

A more recent work has proposed the determination of inorganic anions in seawater samples by ion chromatography with UV detection using a monolithic octadecylsilyl column of 150 × 4.6 mm i.d. coated with dodecylammonium cation [14] using a flow rate of 400 µL/min and a sample injection volume of 100 µL and achieving LODs of 0.9 µg/L and 1.9 µg/L for nitrite and nitrate, respectively, although both limits were estimated from signal/noise ratios (S/N = 3). Sedyohutomo et al. [15] described the utilization of triacontyl-bonded silica coated with imidazolium ions for capillary ion chromatographic determination of nitrite and nitrate in both river water and seawater. The dimensions of the column used was 100 × 0.32 mm i.d., 5 µm working at a flow rate of 4 µL/min and the injected sample volume was 2 µL. In these conditions, the LODs were 0.14 and 0.16 mg/L for nitrite and nitrate, respectively.

A portable analyser using two-dimensional ion chromatography with UV light-emitting diode-based absorbance detection for nitrate monitoring within both saline and freshwaters was proposed by Fitzhenry et al. [16]. Two columns (50 × 4 mm, 9 µm) in tandem were used, the flow rate was 700 µL/min and the injected sample volume was 195 µL. The LODs obtained in [16] were 20 and 10 µg/L for nitrate in freshwater and marine water, respectively.

The miniaturization of analytical systems in order to achieve real information from samples, in as short as possible time while taking account of environmental and economic issues, is one of the main goals in modern analytical chemistry. In this sense, miniaturized liquid chromatography (LC) contributes in this way. The main advantages perceived in using miniaturized LC are related to the reduction of column id, which comes along with a reduction in mobile phase flow rate; this promotes reductions in both the solvent consumption and waste from the analysis. Although the sample dilution ratio for capillary (Cap)-LC is lower than that achieved in conventional HPLC, as shown before from dimensions of the analytical columns and flow rates of [15] vs. [14,16], respectively, as examples, sensitivity for several applications is unsatisfactory mainly due to the low injected sample volume [15]. A method that facilitates working with higher sample volumes without losses in resolution is the online in-tube solid-phase microextraction (IT-SPME) [17,18]. IT-SPME is based typically on the use of a fused-silica capillary tube packed or coated on its inner surface with an extractive phase. When the sample is passed through the capillary, the analytes are extracted and concentrated by adsorption/absorption onto or into the internal coating of the capillary. Then, the extracted analytes are desorbed by filling the capillary with a solvent, which are transferred to the capillary LC. Clean-up, preconcentration, separation, and detection of analytes can be carried out online by coupling the IT-SPME to LC.

On the other hand, in the context of current chemistry, the requirements of sustainable and green chemistry also have to be considered. The concept "sustainable" first appeared in the 70's, and is connected to the idea of linking economic development with the preservation of natural ecosystems [19]. Later on, green chemistry emerged from the concern about environmental contamination caused by pollution from chemical industry in 1990s. These interests were expressed by Anastas and Warner in 1998, who established a list of 12 principles as a guide for a good practice in green chemistry [20]. In this context, there is an intersection zone between the mentioned subjects, greenness and sustainability, both included into the wider concept of suitable chemistry [21]. A goal of this paper was also to contribute in this direction [22,23].

This paper demonstrates the feasibility of the online coupling of IT-SPME and ion-exchange capillary liquid chromatography (IE-CapLC) with diode array detection for the first time. The selected analytes were nitrite and nitrate considering their relevant influence in health and environmental scenarios. The paper demonstrates the achievements of this couple for improving parameters in the literature indicated in previous paragraphs such detection limits (LODs), time analysis, and applicability to a variety of water matrices with also different levels of both analytes and trueness, besides decreasing waste and without sample treatment.

## 2. Materials and Methods

### 2.1. Materials and Reagents

Ultra-pure water was obtained for a Nanopure II system (Barnstead, NH, USA). Acetonitrile (99.8%) and methanol (96%) grade LC-MS were supplied by VWR Chemicals (Randnor, PA, USA). Potassium nitrate (99%), potassium nitrite (97%), sodium sulfate (99%), potassium biphosphate (98%), citric acid (99%), and orthophosphoric acid (85%) were purchased from Panreac (Barcelona, Spain). Hexadecyltrimethylammonium hydroxide (TBA-OH, 99%), tetrabutylammonium chloride (TBA-Cl, 99%), and hexadecyltrimethylammonium bromide (HTAB, 99%), gallic acid (97.5%), benzoic acid (99.5%), chlorogenic acid (95%), salicylic acid (99%), and phthalic acid (99.5%) were obtained from Sigma-Aldrich (Darmstadt, Germany).

### 2.2. Equipment and Columns

A capillary LC Agilent 1260 infinity series (Agilent, Waldbronn, Germany) with an injection valve rheodyne 7725i and coupled online in in-valve mode to IT-SPME was used. For this purpose, the loop of the six-port injection valve was substituted by the extractive IT-SPME capillary [17,18]. Several IT-SPME capillaries were tested: fused silica capillary of 0.32 mm id and 15 cm length (Supelco, Bellefonte, Pennsylvania USA) and 50% Diphenyl-50% dimethyl polysiloxane, bonded and crosslinked phase capillary: TRB 50 (Teknokroma, Barcelona, Spain) with the same dimensions and 3 μm of film thickness. A photodiode array detector (DAD. Hewlett-Packard 1040M series II) with a cell volume of 80 nL was employed. The analytical columns tested were: ZORBAX SB-C18 150 × 0.5 mm id, 5 μm (Agilent Technologies Spain), ZORBAX SB-C18 35 × 0.5 mm id, 5 μm (Agilent technologies Spain), and Inertsil AX 150 × 0.5 mm id, 5 μm (GL Sciences Tokyo, Japan). The data were acquired and processed by the Agilent HPLC ChemStation Software. Signals were recorded in the range of 210–400 nm and monitored at 220 nm.

The processed volume of standards or samples was 25 μL and all of them were loaded manually into the IT-SPME system using a 100 μL precision syringe at a flow rate of 50 μL/min. Then, the valve was changed to the injection position, so that the analytes retained in the capillary were desorbed with the mobile-phase and transferred to the analytical column for separation and detection. The chromatographic run was carried out with the valve in the injection position.

### 2.3. Optimization of Chromatographic Conditions

Ion pair capillary liquid chromatography (IP-CapLC) technique was studied employing the capillary reversed analytical columns and several ion pairing agents: TBH-OH, TBH-Cl or HTAB at levels 1, 5, 10 mM in the mobile phase (see Table 1) and 1 mg/L standard solution of nitrate and mixtures of nitrate and sulfate and nitrate and phosphate.

Table 1. Mobile phases for IP-CapLC method, flow rate 8 μL/min. Ionic pairing agent solutions at pH = 3.1 reached with HCl containing 20% of methanol (MeOH) and MeOH compositions.

| Time (min) | Mobile Phase % | | | | | |
|---|---|---|---|---|---|---|
| | TBH-CH | MeOH | TBH-Cl | MeOH | HTAB | MeOH |
| 0 | 100 | 0 | 100 | 0 | 50 | 50 |
| 2 | 50 | 50 | 80 | 20 | | |
| 4 | 40 | 60 | 40 | 60 | | |
| 8 | 30 | 70 | 30 | 70 | | |
| 10 | 20 | 80 | 20 | 80 | | |
| 12 | 100 | 0 | 100 | 0 | | |

For ion-exchange capillary liquid chromatography (IE-CapLC) a column Interstil AX was employed and different compositions of acetonitrile and water at pH 3.1 reached with phosphoric acid were evaluated in the mobile phase and also ionic strength was varied by adding NaCl or $KH_2PO_4$. Three flow rates were tested (8, 10 and 12 μL/min). The figures of merit: linearity (n = 5), LODs, limits of quantitation (LOQs), selectivity and matrix effect were evaluated. Nitrite and nitrate concentrations were assayed up to 1 and 2.5 mg/L, respectively, for obtaining the linear calibration graph. Precision was estimated at LOQ (n = 3) and 0.1 (n = 4) and 1 (n = 4) mg/L for both nitrite and nitrate and the matrix effect at the latter levels of concentrations.

LODs and LOQs were obtained by injecting decreasing values of analyte concentration until obtaining signal/noise ratios of 3 and 10 times, respectively.

*2.4. Analysis of Water Samples*

Natural water samples that were collected in different places in the vicinity of Valencia (well, river, channel, sea and lake) and drinking waters such as tap water obtained in the municipality of Burjassot (Valencian community) and several bottled commercial waters. Samples from a potable water treatment plant provided by the municipality of Gandia (Valencian community) collected in three stages of the treatment process of diminishing nitrate concentration (input, mid-process and output) were also analysed. The samples were passed through nylon filters with a pore size of 0.22 μm and/or diluted if necessary.

A study of confirmation was carried out by using the native absorbance of nitrate measured with a CARY UV Visible spectrophotometer model G6860AAR (Agilent Technologies, Santa Clara, CA, USA) with a quartz cuvette of 1 cm path length. Nitrate analysis was performed by measuring the sample directly at a wavelength of 220 nm, the values obtained were interpolated on a calibration line up to 50 mg/L. For nitrite, the Griess technique from PDMS sensors embedding the reagents [24] was employed. A Griess sensor was placed in a vial containing 0.5 mL of the sample and 0.5 mL of citric acid (330 mM), left to stand for 8 min and the absorbance was measured at a wavelength of 540 nm, the values were interpolated on a calibration graph up to 1.5 mg/L.

## 3. Results and Discussion

*3.1. Assessment of Chromatographic Performance*

Two mechanisms were evaluated for the direct analysis of nitrite and nitrate; the first one was ionic pair chromatography with a capillary C18 column (IP-CapLC), and the second one was anion exchange capillary liquid chromatography with an anionic column (IE-CapLC). In both cases, the IT-SPME silica capillary was used. The IT-SPME capillary column was used as the loop of the injector valve (in-valve, in-tube SPME), and the analytes are extracted in the IT-SPME capillary during sample loading, and then transferred to the capillary analytical column with the mobile phase by changing the valve to the injection position. The sensitivity can be improved by flushing a sample volume higher than that of the capillary internal volume through the capillary [17,18].

To study the chromatographic conditions in the IP-CapLC method, three ion pairing agents were evaluated. TBH-OH and TBH-Cl were analysed in gradient mode (see Table 1) and HTAB in isocratic mode. The more suitable responses were obtained working with concentrations of 5 mM for TBH-OH and TBH-Cl and 1 mM for HTAB. The achieved retention times ($t_R$) for nitrate were: 18.6, 11.2 and 4.7 min for TBH-OH, TBH-Cl and HTAB, respectively, as Figure 1 shows.

Mixtures of nitrate and sulfate and/or phosphate were assayed and overlapped peaks were obtained in all cases (Figure 1). Bearing in mind the obtained results for IP-CapLC, ion-exchange capillary liquid chromatography (IE-CapLC) was assayed. The selected column contained diethylamino group, which is a strong base (pka ≈ 10.6). Z. Kadlecová et al. [25] found that the anion-exchange retention gradually diminishes above pH 5, although it is fully charged below pH < 9. The authors indicated a possible mechanism that is deprotonation of residual silanols of uncapped sorbent, which can result in the formation of zwitter-ions and decrease in diethylamino surface charge. The mobile phases were adjusted at pH 3.1 with orthophosphoric acid. Methanol was discarded as a modifier as used in IP-CapLC because assaying a mixture 60:40 methanol:water as the mobile phase, nitrate remained in the analytical column for more than 200 min. Acetonitrile was employed instead of methanol and water/acetonitrile ratios of 70:30, 55:45, and 50:50 were evaluated in the mobile phase and also the effect of adding NaCl or $KH_2PO_4$.

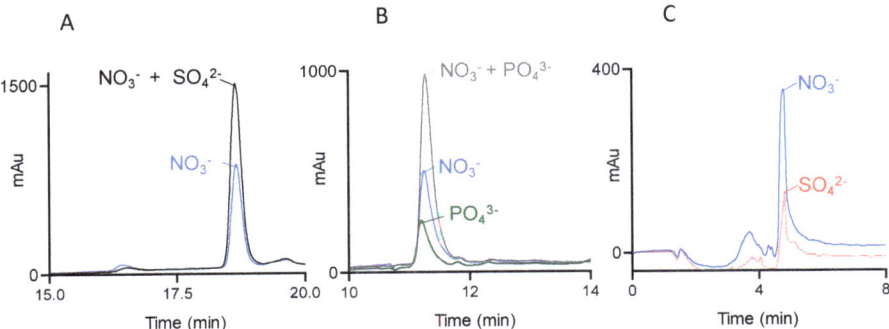

**Figure 1.** Chromatograms obtained by the IP–CapLC method for the standard solutions, using the three ion pairing agents tested: (**A**) TBH–OH (1 mg/L nitrate and 0.5 mg/L Nitrate + 0.5 mg/L Sulfate), (**B**) TBH–Cl (1 mg/L Nitrate, 0.5 mg/L Nitrate + 0.5 mg/L Phosphate and 0.25 mg/L Phosphate) and (**C**) HTA–Br (1 mg/L Nitrate and 0.5 mg/L Sulfate).

As it can be seen in Table 2 the percentage of water had a great influence on the retention time of nitrate for ion-exchange. The influence of the presence of chloride in the mobile phase on $t_R$ of nitrate (36.6 min vs. 46 min without chloride) was not significant compared to the results provided by dihydrogen phosphate ($t_R$ 14 min vs. 46 min without dihydrogen phosphate). The best chromatographic conditions were achieved with mobile phases containing this last anion, which produces the elution of nitrate in short times. The selected conditions are given in Table 2, and it can be seen the retention time for nitrate was 10.5 min. Lower concentrations of dihydrogen phosphate (20 mM and 10 mM) increased $t_R$ of nitrate and they were discarded.

**Table 2.** Optimization of mobile phase in IE-CapLC. For all mobile phases pH = 3.1; [1] 130 mM concentration. * Optimal conditions. For more explanations see text.

| Percentage % Water-Acetonitrile | Salt [1] | Flow Rate µL/min | $t_R$ Nitrate (min) |
|---|---|---|---|
| 70–30 | - | 8 | 75 |
| 55–45 | - | 8 | 57 |
| 50–50 | - | 8 | 46 |
| 50–50 | NaCl | 8 | 36.6 |
| 50–50 | $KH_2PO_4$ | 8 | 14 |
| 50–50 | $KH_2PO_4$ | 10 | 12 |
| * 50–50 | $KH_2PO_4$ | 12 | 10.5 |

Two capillaries with phases with different polarity and with the same internal volume of 12 µL were tested for IT-SPME: fused silica and TRB 50. A sample clean-up step was not necessary after in-valve processing of the samples, which were directly transferred to the analytical column by switching the valve to the injection position as mentioned above. The chromatographic run was achieved in the inject mode of the IT-SPME valve. The chromatographic profile was the same for both IT-SPME capillaries, as it can be seen in Figure 2A; however, similar areas were obtained, although the processed concentrations were 0.1 and 0.5 mg/L of nitrate for silica and TRB 50 capillaries, respectively. The most polar capillary of silica provided a higher level of preconcentration than TRB 50 for a processed volume of 25 µL in accordance with the polar nature of the analytes: nitrite and nitrate. The silica capillary achieved a factor of a preconcentration of five in reference to TRB 50, considering that their internal volumes were the same.

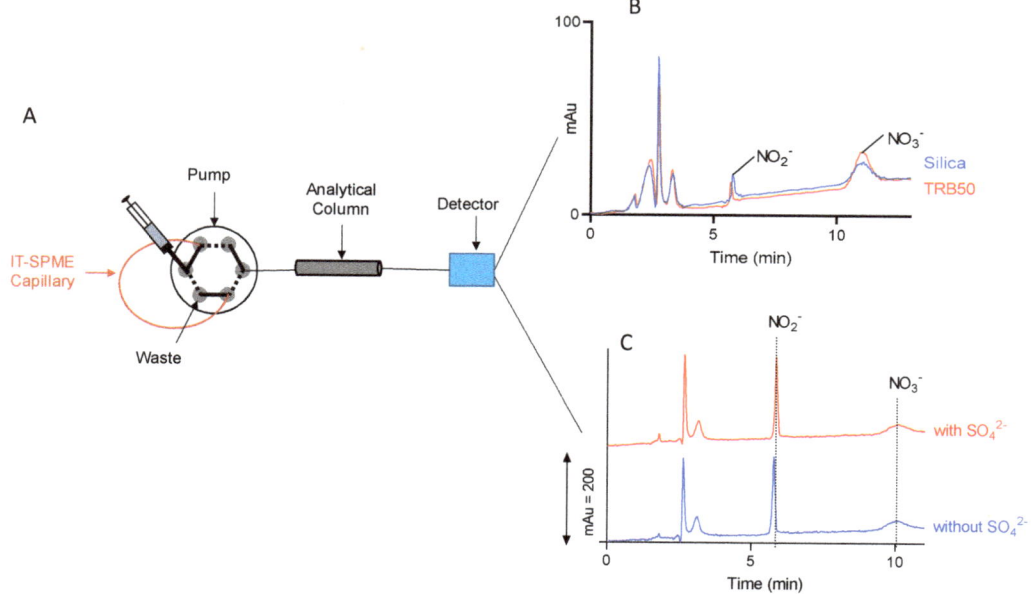

**Figure 2.** (**A**) Scheme of the IT-SPME-CapLC-DAD. (**B**) Influence of the nature of IT-SPME capillary: silica vs. TRB 50 for nitrate concentrations of 0.1 and 0.5 mg/L, respectively. (**C**) Chromatograms obtained by IE-CapLC for mixtures of standard solutions containing 0.5 mg/L of nitrite and 0.1 mg/L of nitrate and 0.5 mg/L of nitrite, 0.1 mg/L of nitrate and 0.5 mg/L of sulfate. The retention times for nitrite and nitrate are 5.8 and 10.5 min, respectively.

Sulpizi et al. [26] gave a detailed understanding of the molecular behaviour of the silica−water interface, indicating that the silanols determine the surface acidity and modulate the water properties. These authors showed how the silanols' orientation and their hydrogen bond properties are responsible for an amphoteric behavior of the surface, which can explain the results obtained here. TRB 50 is a slightly polar stationary phase containing phenyl into methylpolysiloxane, i.e., $-CH_3$ groups with phenyl groups, $-C_6H_5$, providing π-π, dipole–dipole, and dipole-induced dipole interactions and moderate amounts of hydrogen bonding. Figure 2B shows the chromatograms obtained for mixtures of nitrite and nitrate without and with sulfate. This anion did not interfere due to it does not absorb at 220 nm.

### 3.2. Figures of Merit

Linearity, LODs, LOQs and precision were studied for IT-SPME-IE-CapLC-DAD. Suitable linearity was obtained for nitrite and nitrate as Table 3 shows. The obtained LOQs permit the quantification of a wide variety of environmental and drinking waters. Figure 3 shows the chromatogram obtained for the LOQs. Precision was evaluated from % RSD for LOQs and by injecting four replicates of the 0.1 and 1.0 mg/L standard solutions and satisfactory values were obtained (see Table 3). Precision of the retention times expressed as % RSD was <1%, n = 20 and 100 µL per run of wastes were generated.

Several organic acids assayed as possible interferents (Figure 4), with pKa between 2.93 and 5.4 [27–31], gave retention times different to those obtained by nitrite and nitrate (see Figure 2B). Retention time for nitrite (5.8 min) was between those presented by salycilic (5.3 min) and phathalic (6.0 min); this latter was the closest one but at the working pH value of the mobile phase, phathalic acid is protonated. On the other hand, its UV absorption spectra are easily distinguishable from that provided by nitrite.

**Table 3.** Analytical parameters obtained with the proposed method IT-SPME-IE-CapLC-DAD.

| Anion | Linearity [1] : Y = $b_1$x + $b_0$ | | | LOD (µg/L) | LOQ (µg/L); %RSD (n = 3) | % RSD |
|---|---|---|---|---|---|---|
| | $b_1 \pm s_{b1}$ | $b_0 \pm s_{b0}$ | $R^2$ | | | |
| nitrite | 1011 ± 45 | 64 ± 6 | 0.993 | 0.9 | 3; 6 | 1.8 [2]; 3 [3] |
| nitrate | 1669 ± 100 | 115 ± 40 | 0.996 | 9 | 30; 5 | 0.9 [2]; 2 [3] |

Notes: [1] mg/L n = 5, Established at a concentration of [2] 0.1 and [3] 1 mg/L n = 4.

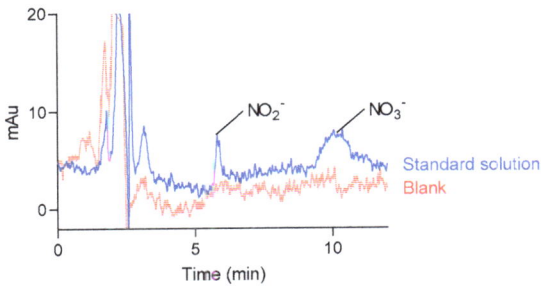

**Figure 3.** Limits of quantification for nitrite and nitrate (3 and 30 µg/L, respectively) by IT–SPME–IE–CapLC–DAD.

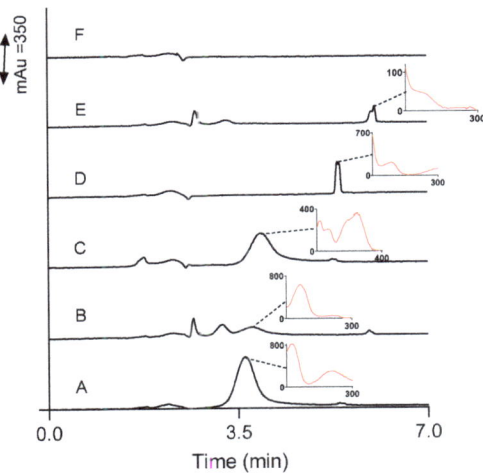

**Figure 4.** Chromatograms and spectra of organic acids: gallic ($pK_a$ = 4.2, trace A), benzoic ($pK_a$ = 4.1, trace B), chlorogenic ($pK_a$ = 3.6, trace C), salicylic ($pK_a$ = 2.9, trace D), phthalic ($pK_a$ = 5.4, trace E) and sulfate (trace F) using the optimized conditions for nitrite and nitrate determination by IT-SPME-IE-CapLC-DAD. Phthalic and salicylic acids were assayed at 1 mg/L, chlorogenic, benzoic and gallic acids at 3.5, 0.5 and 1.7 mg/L, respectively.

### 3.3. Analysis of Waters

To evaluate the applicability of the developed method, several real water samples including natural and drinking water and samples from a potable water treatment collected during the nitrate removal process (inlet, mid-process, and outlet water) were analysed. Representative chromatograms obtained are given in Figure 5.

The chromatograms obtained for samples, shown in Figure 5, maintained the shape of the peaks, the resolution, and the retention times compared to the standard solutions of nitrite and nitrate (see Figures 2 and 3). The inserts of Figure 5 contained the spectra obtained at the maximum of the chromatographic peak, which permit the identification of the analytes. Spiked samples provided % recoveries near to 100 ± 5% and then, the matrix effect was not present. Table 4 shows the obtained values for nitrite and nitrate in the analysed waters. A study of confirmation was carried out by analysing samples with the Griess method for nitrite and UV spectrophotometry for nitrate, and the results are also included in Table 4.

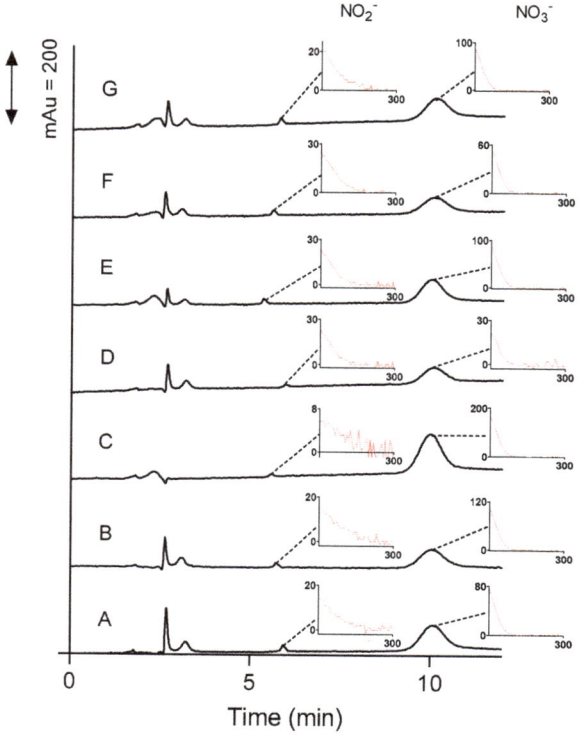

**Figure 5.** Chromatograms and spectra obtained at maximum of each chromatographic peak for several waters by IT−SPME−IE−CapLC−DAD. Natural water: well (trace G), river (trace F), and lake (trace E) diluted 1/40, 1/50 and 1/500, respectively, drinking water: tap (trace D) diluted 1/50 and undiluted bottled (trace C) and water treatment plant: inlet (trace A) and outlet (trace B), both diluted 1/50. The retention times for nitrite and nitrate are 5.8 and 10.5 min, respectively.

In reference to nitrite determination by the Griess method, a river water sample presented interference and its concentration could not be obtained. On the other hand, only samples containing nitrite concentrations higher than 0.03 mg/L could be quantified by the Griess method. However, the proposed IT-SPME-EI-CapLC-DAD provided results for all samples, as can be seen in Table 4.

In the case of nitrate, the proposed method provided quantitative results for all samples, which presented concentrations between 1.56 and 93 mg/L. Biased results were obtained for the sea water assayed from UV-spectrophotometry at 220 nm and this method did not permit quantifying nitrate below 9 mg/L. Table 4 shows that when the Griess and UV spectrophotometry methods could be applied, the results were statistically similar to those provided by the proposed method.

Table 4. Nitrite and nitrate concentration in real water samples (natural water, drinking water and water treatment plant) (n = 4). For more explanations see text.

| Sample | | Nitrite (mg/L) | | Nitrate (mg/L) | |
|---|---|---|---|---|---|
| | | Cap-ionLC | Griess Method [1] | Cap-ionLC | UV Spectrophotometry [2] |
| Natural water | Well | 0.4 ± 0.1 | - | 54.4 ± 0.6 | 55.1 ± 0.7 |
| | | 0.20 ± 0.05 | - | 19.3 ± 0.6 | 21.5 ± 0.7 |
| | River | 0.20 ± 0.05 | Not applicable [3] | 53 ± 1 | 51 ± 1 |
| | | 0.30 ± 0.09 | - | 36.5 ± 0.5 | 31 ± 1 |
| | | 0.40 ± 0.09 | 0.4 ± 0.1 | 93 ± 1 | 102 ± 1 |
| | | 0.30 ± 0.09 | - | 40.3 ± 0.6 | 41.3 ± 0.8 |
| | Channel | 0.30 ± 0.03 | - | 30.3 ± 0.4 | 30.5 ± 0.3 |
| | lake | 0.40 ± 0.03 | - | 13.3 ± 0.5 | 14 ± 0.7 |
| | sea | 0.30 ± 0.03 | - | 4.3 ± 0.3 | Not applicable [3] |
| Drinking water | tap | 0.05 ± 0.01 | <LOQ | 12.9 ± 0.5 | 12.6 ± 0.4 |
| | | 0.09 ± 0.01 | - | 17.4 ± 0.9 | 17 ± 1 |
| | | 0.10 ± 0.01 | - | 19.5 ± 0.1 | 18.0 ± 0.2 |
| | bottled | 0.010 ± 0.003 | <LOD | 2.23 ± 0.01 | <LOD |
| | | 0.030 ± 0.005 | ≈LOD | 1.56 ± 0.06 | <LOD |
| | | <LOD | - | 2.19 ± 0.03 | <LOD |
| Water treatment plant | inlet | 0.9 ± 0.1 | 1.0 ± 0.2 | 55.3 ± 0.6 | 56.0 ± 0.7 |
| | half process | 0.7 ± 0.1 | - | 37.9 ± 0.6 | 37.1 ± 0.4 |
| | outlet water | 0.10 ± 0.01 | - | 14.7 ± 0.6 | 13 ± 1 |

Notes: [1] PDMS Griess sensor [24], LOD 0.01 mg/L, [2] direct UV spectrophotometry at 220 nm, LOD 3 mg/L. [3] Not applicable due to interference.

## 4. Conclusions

In the present work, a novel method for the quantification of nitrite and nitrate in a variety of waters with variable concentrations was proposed. The ion exchange mechanism provided better results than ion-pair formation with respect to analysis time, selectivity and trueness. The IT-SPME-IE-CapLC-DAD analytical procedure was applied for water analysis, decreasing the amount of waste generation by minimizing the amount of solvents employed, 100 µL/run instead of 10 mL/run for conventional liquid chromatography, which uses flow rates around 1 mL/min and considering a retention time for nitrate of around 10 min. A goal of this work was also to demonstrate that minimizing sample treatment by online IT-SPME by using a fused silica capillary, and miniaturizing the system can remarkably improve the greenness and sustainability of an analytical method, maintaining or even improving its figures of merit as detection limits and applicability to different water matrices with different levels of nitrite and nitrate, between 0.01–0.9 mg/L and 1.56–93 mg/L, respectively. The proposed IT-SPME-IE-CapLC-DAD approach proposed here for the first time is a good choice for the direct analysis of environmental waters. The quantitative performance of the proposed method is suitable in terms of linearity and precision, LODs, and selectivity, besides the absence of the matrix effect for both anions. The precision of the retention times is also remarkable for standards and samples.

**Author Contributions:** Conceptualization, P.C.-F.; methodology, H.R.R.-J., N.J.-M. and P.C.-F.; validation, H.R.R.-J.; investigation, H.R.R.-J., N.J.-M. and P.C.-F.; writing—original draft preparation, H.R.R.-J.; writing—review and editing, P.C.-F.; supervision, P.C.-F.; funding acquisition, P.C.-F. All authors have read and agreed to the published version of the manuscript.

**Funding:** This research was funded by EU (FEDER) and MCI-AEI of Spain (PID2021-124554NB-I00), Generalitat Valenciana (PROMETEO, program 2020/078) and Agencia Valenciana de Innovación (INNEST/2021/15), EU (EASME LIFE and CIP ECO-Innovation) LIBERNITRATE LIFE 16 ENV/ES/000419. H. R. Robles-Jimarez expresses his gratitude to UV for his pre-doc contract.

**Data Availability Statement:** Data are included in the paper.

**Conflicts of Interest:** The authors declare no conflict of interest.

## References

1. Karlović, I.; Posavec, K.; Larva, O.; Marković, T. Numerical groundwater flow and nitrate transport assessment in alluvial aquifer of Varaždin region, NW Croatia. *J. Hydrol. Reg. Stud.* **2022**, *41*, 101084. [CrossRef]
2. Beeckman, F.; Motte, H.; Beeckman, T. Nitrification in agricultural soils: Impact, actors and mitigation. *Curr. Opin. Biotechnol.* **2018**, *50*, 166–173. [CrossRef]
3. Internation Agency for Research on Cancer (IARC). World health organization international agency for research on cancer. *Iarc. Monogr. Eval. Carcinog. Risks Hum.* **2010**, *94*, 1–464.
4. Picetti, R.; Deeney, M.; Pastorino, S.; Miller, M.R.; Shah, A.; Leon, D.A.; Dangour, A.D.; Green, R. Nitrate and nitrite contamination in drinking water and cancer risk: A systematic review with meta-analysis. *Environ. Res.* **2022**, *210*, 112988. [CrossRef]
5. European Parliament. Directive 2015/1787/EU: Amending Council Directive 98/83/EC of 3 November 1998 on the quality of water intended for human consumption. *Off. J. Eur. Union* **2015**, *L260*, 6–17.
6. European Parliament. Directiva 2006/118/CE: On the protection of groundwater against pollution and deterioration. *Off. J. Eur. Union* **2006**, *L372*, 19–31.
7. European Parliament. Directive 2013/39/EC: Amending Directives 2000/60/EC and 2008/105/EC as regards priority substances in the field of water policy. *Off. J. Eur. Union* **2013**, *L226*, 1–17.
8. Wang, Q.-H.; Yu, L.-J.; Liu, Y.; Lin, L.; Lu, R.-G.; Zhu, J.-P.; He, L.; Lu, Z.-L. Methods for the detection and determination of nitrite and nitrate: A review. *Talanta* **2017**, *165*, 709–720. [CrossRef] [PubMed]
9. Singh, P.; Singh, M.K.; Beg, Y.R.; Nishad, G.R. A review on spectroscopic methods for determination of nitrite and nitrate in environmental samples. *Talanta* **2018**, *191*, 364–381. [CrossRef] [PubMed]
10. APHA–AWWA–WEF. *Standard Methods for the Examination of Water and Wastewate*, 21st ed.; American Public Health Association: Washington, DC, USA, 2005.
11. Moldoveanu, S.C.; David, V. *Retention Mechanisms in Different HPLC Types in Essentials in Modern HPLC Separations*; Elsevier: Amsterdam, The Netherlands, 2013. [CrossRef]
12. Soleimani, M.; Yamini, Y.; Rad, F.M. A Simple and High Resolution Ion-Pair HPLC Method for Separation and Simultaneous Determination of Nitrate and Thiocyanate in Different Water Samples. *J. Chromatogr. Sci.* **2012**, *50*, 826–830. [CrossRef]
13. Khan, S.S.; Riaz, M. Determination of UV active inorganic anions in potable and high salinity water by ion pair reversed phase liquid chromatography. *Talanta* **2014**, *122*, 209–213. [CrossRef] [PubMed]
14. Horioka, Y.; Kusumoto, R.; Yamane, K.; Nomura, R.; Hirokawa, T.; Ito, K. Determination of Inorganic Anions in Seawater Samples by Ion Chromatography with Ultraviolet Detection Using Monolithic Octadecylsilyl Columns Coated with Dodecylammonium Cation. *Anal. Sci.* **2016**, *32*, 1123–1128. [CrossRef] [PubMed]
15. Sedyohutomo, A.; Suzuki, H.; Fujimoto, C. The Utilization of Triacontyl-Bonded Silica Coated with Imidazolium Ions for Capillary Ion Chromatographic Determination of Inorganic Anions. *Chromatography* **2021**, *42*, 119–126. [CrossRef]
16. Fitzhenry, C.; Jowett, L.; Roche, P.; Harrington, K.; Moore, B.; Paull, B.; Murray, E. Portable analyser using two-dimensional ion chromatography with ultra-violet light-emitting diode-based absorbance detection for nitrate monitoring within both saline and freshwaters. *J. Chromatogr. A* **2021**, *1652*, 462368. [CrossRef]
17. Moliner-Martinez, Y.; Herráez-Hernández, R.; Verdú-Andrés, J.; Molins-Legua, C.; Campíns-Falcó, P. Recent advances of in-tube solid-phase microextraction. *TrAC Trends Anal. Chem.* **2015**, *71*, 205–213. [CrossRef]
18. Moliner-Martinez, Y.; Ballester-Caudet, A.; Verdú-Andrés, J.; Herráez-Hernández, R.; Molins-Legua, C.; Campíns-Falcó, P. In-tube solid-phase microextraction in Solid-Phase Extraction. In *Handbooks in Separation Science*; Poole, C.F., Ed.; Elsevier: Amsterdam, The Netherlands, 2020; Chapter 14; pp. 387–427. [CrossRef]
19. Fowkes, M.R. Planetary forecast: The roots of sustainability in the radical art of the 1970s. *Third Text* **2009**, *23*, 669–674. [CrossRef]
20. Gałuszka, A.; Migaszewski, Z.; Namiesnik, J. The 12 principles of green analytical chemistry and the SIGNIFICANCE mnemonic of green analytical practices. *TrAC Trends Anal. Chem.* **2013**, *50*, 78–84. [CrossRef]
21. Horváth, I.T. Introduction: Sustainable Chemistry. *Chem. Rev.* **2018**, *118*, 369–371. [CrossRef]
22. Ballester-Caudet, A.; Campíns-Falcó, P.; Pérez, B.; Sancho, R.; Lorente, M.; Sastre, G.; González, C. A new tool for evaluating and/or selecting analytical methods: Summarizing the information in a hexagon. *TrAC Trends Anal. Chem.* **2019**, *118*, 538–547. [CrossRef]
23. Ballester-Caudet, A.; Navarro-Utiel, R.; Campos-Hernández, I.; Campíns-Falcó, P. Evaluation of the sample treatment influence in green and sustainable assessment of liquid chromatography methods by the HEXAGON tool: Sulfonate-based dyes determination in meat samples. *Green Anal. Chem.* **2022**, *3*, 100024. [CrossRef]
24. Hakobyan, L.; Monforte-Gómez, B.; Moliner-Martínez, Y.; Molins-Legua, C.; Campíns-Falcó, P. Improving Sustainability of the Griess Reaction by Reagent Stabilization on PDMS Membranes and ZnNPs as Reductor of Nitrates: Application to Different Water Samples. *Polymers* **2022**, *14*, 464. [CrossRef] [PubMed]
25. Kadlecová, Z.; Kalíková, K.; Folprechtová, D.; Tesařová, E.; Gilar, M. Method for evaluation of ionic interactions in liquid chromatography. *J. Chromatogr. A* **2020**, *1625*, 461301. [CrossRef] [PubMed]

26. Sulpizi, M.; Gaigeot, M.-P.; Sprik, M. The Silica–Water Interface: How the Silanols Determine the Surface Acidity and Modulate the Water Properties. *J. Chem. Theory Comput.* **2012**, *8*, 1037–1047. [CrossRef]
27. Beltrán, J.; Sanli, N.; Fonrodona, G.; Barrón, D.; Özkan, G.; Barbosa, J. Spectrophotometric, potentiometric and chromatographic pKa values of polyphenolic acids in water and acetonitrile–water media. *Anal. Chim. Acta* **2003**, *484*, 253–264. [CrossRef]
28. Borgo, L. Evaluation of buffers toxicity in tobacco cells: Homopiperazine-1,4-bis (2-ethanesulfonic acid) is a suitable buffer for plant cells studies at low pH. *Plant Physiol. Biochem* **2017**, *115*, 119–125. [CrossRef]
29. Geiser, L.; Henchoz, Y.; Galland, A.; Carrupt, P.-A.; Veuthey, J.-L. Determination of pKa values by capillary zone electrophoresis with a dynamic coating procedure. *J. Sep. Sci.* **2005**, *28*, 2374–2380. [CrossRef]
30. Hollingsworth, C.A.; Seybold, P.G.; Hadad, C.M. Substituent effects on the electronic structure and pKa of benzoic acid. *Int. J. Quantum Chem.* **2002**, *90*, 1396–1403. [CrossRef]
31. Kabir, F.; Katayama, S.; Tanji, N.; Nakamura, S. Antimicrobial effects of chlorogenic acid and related compounds. *J. Korean Soc. Appl. Biol. Chem.* **2014**, *57*, 359–365. [CrossRef]

**Disclaimer/Publisher's Note:** The statements, opinions and data contained in all publications are solely those of the individual author(s) and contributor(s) and not of MDPI and/or the editor(s). MDPI and/or the editor(s) disclaim responsibility for any injury to people or property resulting from any ideas, methods, instructions or products referred to in the content.

*Review*

# Geochemical Indicators for Paleolimnological Studies of the Anthropogenic Influence on the Environment of the Russian Federation: A Review

Zakhar Slukovskii [1,2]

[1] Institute of the North Industrial Ecology Problems of Kola Science Center of RAS, 184209 Apatity, Russia
[2] Institute of Geology of Karelian Research Centre of RAS, 185910 Petrozavodsk, Russia

**Abstract:** Lake sediments are a reliable source of information about the past, including data of the origin of water bodies and their changes. Russia has more than 2 million lakes, so paleolimnological studies are relevant here. This review deals with the most significant studies of sequential accumulation of pollutants, including heavy metals in recent lake sediments in Russia. The key areas are northwestern regions of Russia (Murmansk Region, the Republic of Karelia, Arkhangelsk Region), the Urals (Chelyabinsk Region, the Republic of Bashkortostan), and Siberia. The review presents the data of pollutants accumulation, the sedimentation rate in lakes in the anthropogenic period, and the key sources of pollution of the environment in each of the mentioned regions. The article is divided into three parts (sections): industrial areas, urbanized areas, and background (pristine) areas so that readers might better understand the specifics of particular pollution and its impact on lake ecosystems. The impact of metallurgical plants, mining companies, boiler rooms, coal and mazut thermal power plants, transport, and other anthropogenic sources influencing geochemical characteristics of lakes located nearby or at a distance to these sources of pollution are considered. For instance, the direct influence of factories and transport was noted in the study of lake sediments in industrial regions and cities. In the background territories, the influence of long-range transport of pollutants was mainly noted. It was found that sedimentation rates are significantly lower in pristine areas, especially in the Frigid zone, compared to urbanized areas and industrial territories. In addition, the excess concentrations of heavy metals over the background are higher in the sediments of lakes that are directly affected by the source of pollution. At the end of the article, further prospects of the development of paleolimnological studies in Russia are discussed in the context of the continuing anthropogenic impact on the environment.

**Keywords:** freshwater ecosystems; lake sediments; human impact; heavy metals; Russia; Arctic

**Citation:** Slukovskii, Z. Geochemical Indicators for Paleolimnological Studies of the Anthropogenic Influence on the Environment of the Russian Federation: A Review. *Water* 2023, *15*, 420. https://doi.org/10.3390/w15030420

Academic Editors: Weiying Feng, Fang Yang and Jing Liu

Received: 23 December 2022
Revised: 11 January 2023
Accepted: 17 January 2023
Published: 19 January 2023

**Copyright:** © 2023 by the author. Licensee MDPI, Basel, Switzerland. This article is an open access article distributed under the terms and conditions of the Creative Commons Attribution (CC BY) license (https://creativecommons.org/licenses/by/4.0/).

## 1. Introduction

The anthropogenic impact on the environment for the last two or three centuries is an indisputable fact. One of the well-known manifestations of this process is chemical pollution, reflected in the increased concentrations of various elements and substances in the main environmental components—air, soil, surface and underground waters, sediments of water bodies, and living organisms. Heavy metals and metalloids are among the most dangerous environmental pollutants, as their compounds are quite stable, can exhibit toxic properties, migrate along trophic chains from abiotic components of ecosystems to biota, and accumulate in sediments, soils, tissues, and organs of organisms [1–4]. Paleo-archive methods allow for simultaneous analysis of the current state of the environment and the historical trends often under the influence of the anthropogenic factors. Environmental archives often examined for anthropogenic contamination include ice and tree cores, and peat and lake sediments cores [5–9]. Lake sediments can best perform the present and past environmental assessments of anthropogenic metal contamination, as lakes sediments act as a passive sampler of the environment, can be readily dated with radiometric methods, and are generally common in the vicinity of urban and industrial centers.

All over the world, researchers conduct paleolimnological reconstructions based on the detailed (layer-by-layer) study of sediment cores of the lakes, thus restoring the main stages of the anthropogenic influence on the studied water bodies and their surrounding areas [5,10–14]. Such works are certainly widely developed in Russia, where there are more than 2 million lakes with a surface area of ~350 thousand km$^2$ (excluding the Caspian sea). Paleolimnological studies and reconstructions are especially relevant for regions with large industrial histories (the Southern Ural, Murmansk Region, Western Siberia) [15–17]. Besides, the close location of the aquatic ecosystems to the direct sources of the anthropogenic emissions is important for such research. Therefore, paleolimnological studies are either impossible or barely conducted in the regions with a small number of lakes or in inaccessibility areas.

This review aims to highlight the main paleolimnological studies of the anthropogenic impact on the environment of the Russian Federation, published so far in Russian (in most cases) and English in scientific journals, books, and theses. This is extremely important, since, for example, this review of studies of natural archives [18] is not full without the data of Russian scientists. The focus was on the regions of Russia, where paleolimnological studies based on the analysis of the accumulation dynamics of heavy metals and metalloids in recent sediments have long been a part of the environmental monitoring system. These are the regions of Northwest Russia, the Southern Ural, and Siberia (Figure 1).

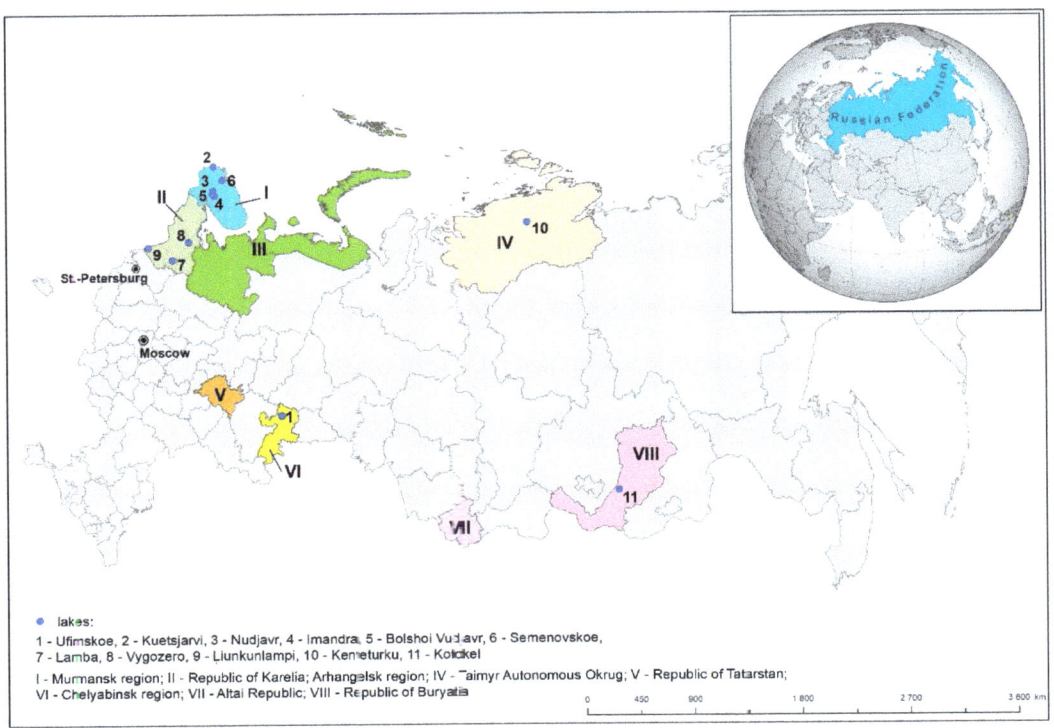

**Figure 1.** A map with the designation of lakes and key regions of paleolimnological studies. The main characteristics of water bodies from the map are in Table 1.

Table 1. Main parameters of water bodies are shown on the map in Figure 1. Note: n/d—there are no data.

| Water Bodies | Area | Coordinates | Square, km² | Depth, m | |
|---|---|---|---|---|---|
| | | | | Maximum | Average |
| Lake Ufimskoe | Chelyabinsk Region | 60.11862, 55.52231 | 0.89 | 3.5 | 1.1 |
| Lake Kuetsjarvi | | 30.16771, 69.43524 | 17.00 | 37.0 | n/d |
| Lake Nudjavr | | 32.88535, 67.92346 | 3.97 | 2.0 | 1.6 |
| Lake Imandra | Murmansk Region | 33.08029, 67.64688 | 876.00 | 67 | 13.3 |
| Lake Bolshoi Vudjavr | | 33.67456, 67.63246 | 3.49 | 38.6 | n/d |
| Lake Semenovskoe | | 33.09001, 68.99101 | 0.21 | 11.3 | 2.4 |
| Lake Lamba | | 34.24950, 61.80713 | 0.01 | 5.2 | 3.4 |
| Vygozero Reservoir | Republic of Karelia | 34.69694, 63.59750 | 1270 | 25 | 7.1 |
| Lake Liunkunlampi | | 29.87730, 61.49913 | 0.1 | 6.8 | 3.6 |
| Lake Kenteturku | Krasnoyarsk region | 96.43925, 73.46444 | 2.5 | 20 | 10 |
| Lake Kotokel | Republic of Buryatia | 108.15000, 52.81667 | 70 | 14 | 4.25 |

## 2. Materials and Methods

Publications, including articles, conference materials, books, and chapters in books published so far to the end of 2021, studied by the author, were used to prepare the review. The main criterion for using publications was the presence of the data on the studies of cores (up to 1 m) sediments of lakes and water bodies with the analysis of the layer-by-layer distribution of chemical elements (mainly heavy metals) and/or isotopes of $^{210}$Pb or $^{137}$Cs in these cores. Although the most studied materials were published in Russian language, they are still important for world science, as researchers have been using methods recognized in paleolimnology for studying geochemistry and the age of sediments of water bodies. This review will allow scientists from all over the world who do not speak Russian to become better acquainted with these studies, considering that Russia is a country with one of the largest number of lakes in the world, and thus has some of the largest numbers of limnological studies which should be known and recognizable. All the publications in Russian are marked in References as (in Russian).

Another criterion for choosing publications was dividing recent sediments cores by researchers into layers no more than 5 cm, with a few exceptions of 10 cm. Personal experience shows that larger layers do not allow for accurately assessing the impact of sources of anthropogenic emissions on the aquatic ecosystem. The best option is to divide cores into 1–2 cm layers, however, studies where cores were divided into 3–10 cm layers were also included in the review. Besides, the review focused on studying lake ecosystems, with rare exception being reservoir ecosystems. This choice resulted from the fact that the water bodies with relatively stagnant water are best suitable for paleolimnological reconstructions as sedimentary material does not mix, and thus accumulates more sequentially, which allows for accurately fixing various changes in the water body and its catchment area. River sediments were excluded from the review as sedimentary material in rivers accumulates in a dynamic environment constantly mixing, which can provide only a general picture of sediment geochemistry. Marine sediments were also excluded since the sedimentation rate in seas, oceans, and large marine water bodies at all is usually very low, which does not allow for fixing point changes in the geochemistry of recent sediments over the last 100–300 years.

Concentrations of chemical elements in the article are presented in mg/kg. If concentrations could be taken only from charts, graphs, or figures, then approximate concentrations were used. All figures in this review are made by the author, and are made either on the basis of the numerical data from publications or the charts from the same works. In this case, there is no copyright infringement as charts were not copied—they were taken from open access sources and then remade either to another format or using other software for illustrations. In special cases, the researchers gave permission to use their data.

## 3. Results and Discussion

### 3.1. Industrial Areas

#### 3.1.1. Ural Region

Chelyabinsk and Murmansk Regions are some of the most industrially developed regions of Russia. There are metallurgical plants for mining and processing copper and copper–nickel ores in both regions. There are also a large number of lakes subject to substantial pollution due to operations of these industrial enterprises in these regions [19–21]. Karabashmed (Karabashskiy Copper-Smelting Plant) (the city of Karabash, Figure 2), producing blister copper, has been operating in Chelyabinsk Region since 1910. Many paleolimnological studies assessing the dynamics of pollutants in water bodies of the Chelyabinsk Region have been conducted in the impact area of this plant. For instance, the analysis of dynamics of heavy metals and stable $^{210}$Pb isotopes behavior in the core of recent sediments of Lake Serebry located 4 km from Karabashmed showed increased concentrations of Cu (up to ~6000 mg/kg, while background level is about 50 mg/kg), Zn (up to ~6000 mg/kg, background is ~70 mg/kg), Pb (up to ~2000 mg/kg, background is ~20 mg/kg), and Mn (up to ~1000 mg/kg, background is ~410 mg/kg) in the upper layers compared to the lower ones [16,22,23]. The increase in concentrations of these metals started according to different references at a depth of 50–80 cm, likely corresponding to the start of the plant operations. The average sedimentary rate in Lake Serebry in the industrial period was 4.8 mm/year [22]. However, more recent data show that this value can be higher, up to ~9 mm/year (calculated based on data from [16]).

**Figure 2.** The view of Karabashskiy Copper-Smelting Plant (photo by the author).

Similar trends of heavy metals (Cu, Zn, Pb, Sb, Cd) can also be seen in sediments of other lakes located in the impact area of Karabashskiy Copper-Smelting Plant [19,20]. For instance, this is well-demonstrated in the example of Lake Ufimskoe located 7 km from the plant (Figure 3). The uppermost layers of lake sediments are enriched with Cu (up to 2341 mg/kg while the background level is ~120 mg/kg), Zn (up to 1256 mg/kg, background is ~54), Pb (up to 1039 mg/kg, background is ~8), Sb (up to 21 mg/kg, background is ~0.3), and Cd (up to 13 mg/kg, background is ~0.4) [19]. These metals are closely related to the copper-smelting plant operations. Other elements (e.g., V, Co, Li, rare earth elements, etc.) did not have a similar tendency towards an increase in the upper layers compared to the lower ones, which indicates that their origin in sediments is not related to the anthropogenic impact on the water body [19].

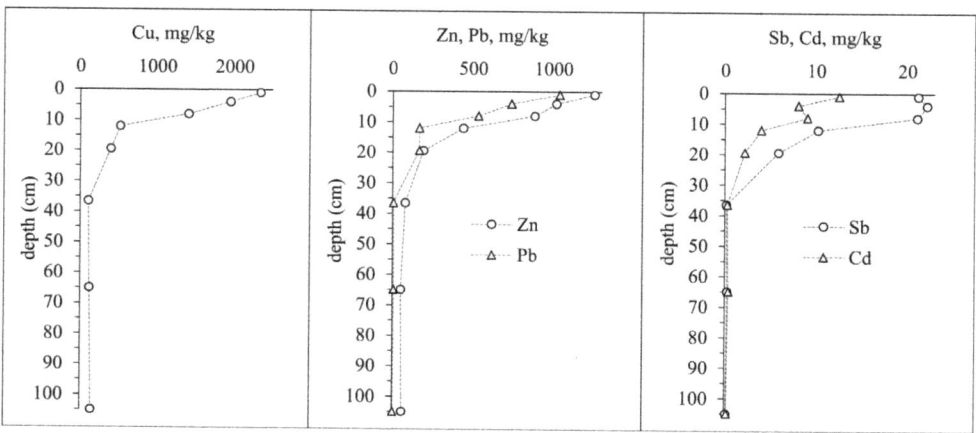

**Figure 3.** The vertical distribution of heavy metals in sediments of Lake Ufimskoe (Chelyabinsk Region) [19].

Scientists from the Institute of Mineralogy, Ural Branch of the Russian Academy of Sciences, revealed the similar dynamics of accumulation of heavy metals in Lake Syrytkul (~30 km from the plant) and Lake Turgoyak (~40 km from the plant) [22,23]. However, total values of Cu, Zn, and Pb concentrations in the upper layers of sediments of these lakes were lower than in Lake Serebry and Lake Ufimskoe. It was noted that the concentration of Cu reached 800 mg/kg, Zn—260 mg/kg, Pb—200 mg/kg in sediments of Lake Turgoyak, where the sedimentation rate was 1.7 mm/year [22]. Thus, there is a tendency towards decreasing concentrations of heavy metals in lake sediments with increasing distance to Karabashskiy Copper-Smelting Plant. In the north of Chelyabinsk Region, 100 km from the city of Karabash, Cu concentrations in recent sediments (0–16 cm) of Lake Itkul varied from 58 to 91 mg/kg, Zn from 94 to 228 mg/kg, and Pb from 20 to 64 mg/kg [19]. However, researchers include these three elements and Cd, Bi, Sb, Co, and Te in the anthropogenic geochemical association of studied sediments of Lake Itkul, as there is a stable tendency towards their increased concentrations in upper layers compared to the background level.

Furthermore, in the Urals (the Republic of Bashkortostan), Cu, Zn, Co, and Ni were also studied in recent sediments of Lake Bolshye Uchaly subject to the Uchaly geotechnical system (the city of Uchaly) [24]. Paleolimnological studies indicated that due to massive quarry blasting in the 1970–1980s and aerial dust from the processing plant, upper layers of sediments aged 40 years of Lake Bolshye Uchaly were enriched with Zn (up to ~6000 mg/kg, while the background level is ~100 mg/kg), Cu (up to ~600 mg/kg, background is ~200 mg/kg), Ni (up to ~45 mg/kg, background is ~27 mg/kg), Co (up to ~15 mg/kg, background is ~10 mg/kg), and Cd (up to ~4 mg/kg). Scientists associate these processes not only with the mine and the concentrating plant operations but also with transport emissions, which is reflected in increased concentrations of Pb (up to ~70 mg/kg, while the background level of Pb in lake sediments of Ural region is 21 mg/kg) in sediments of this water body [24].

### 3.1.2. Murmansk Region

The main sources of pollution in Murmansk Region (Figure 4) are two plants of Kola Mining and Metallurgical Company (Kola MMC), located near the city of Monchegorsk–"Severonickel" combine (the central part of the region) and the urban-type settlement of Nikel–"Pechenganickel" combine (the northwestern part of the region, near the Norway–Russia border) [20]. As the company deals with the mining and processing of copper–nickel ore, the key pollutants of lakes nearby are heavy metals Ni and Cu. Both combines started operating in the 1930s, which caused a significant anthropogenic load on terrestrial and aquatic ecosystems nearby [25–27].

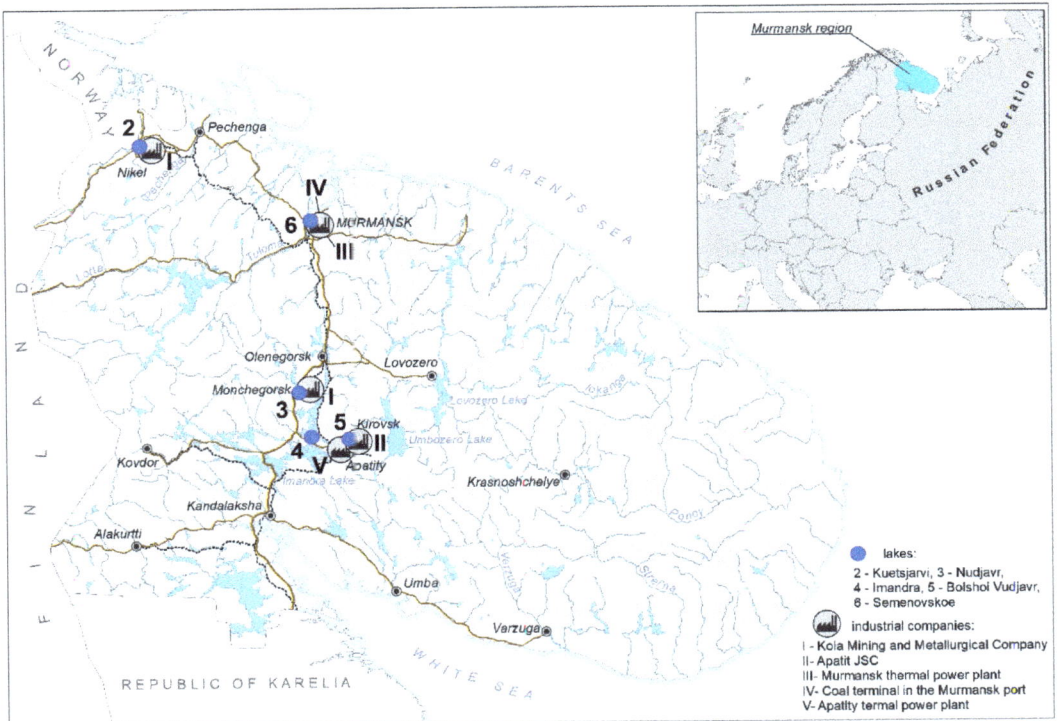

**Figure 4.** A map of Murmansk region with key lakes, cities, and industrial companies that are noted in the article.

In the area of Nikel, paleolimnological studies of the anthropogenic load on the aquatic ecosystems were mostly focused on lakes of the Pasvik river system. The largest lake, on the banks of which the plant operates (Figure 5), is Lake Kuetsjarvi. Researchers in Institute of the North Industrial Ecology Problems of Kola Science Center of Russian Academy of Sciences have been conducting environmental monitoring of this water body for about 30 years [27]. Studies showed that the upper 10–15 cm of sediments of Lake Kuetsjarvi were polluted with heavy metals [28,29]. The sedimentation rate in the lake varied from 1.5 to 3 mm/year depending on the study area [29]. The increases in concentrations of Ni (up to 4892 mg/kg, while background level is 32 mg/kg), Cu (up to 1496 mg/kg, background is 40 mg/kg), Zn (up to 301 mg/kg, background is 80 mg/kg), Co (up to 184 mg/kg, background is 16 mg/kg), Cd (up to 3.14 mg/kg, background is 0.10 mg/kg), Pb (up to 45.7 mg/kg, background is 6.6 mg/kg), As (up to 59.3 mg/kg, background is 2.6 mg/kg), and Hg (0.57 mg/kg, background is 0.05 mg/kg) were noted at all studied sites of Lake Kuetsjarvi (Figure 6), which is related to the start of the metallurgical plant operations in the 1930s. In 2020, the melting shop stopped working, which will probably lead to a decrease in the anthropogenic load on the lake. However, due to the pollution of soils around the water body with heavy metals, pollutants will continue to enter the Kuetsjarvi.

**Figure 5.** The view of Kola Mining and Metallurgical Company, "Pechenganickel" combine (photo by the author). Lake Kuetsjarvi is in the foreground.

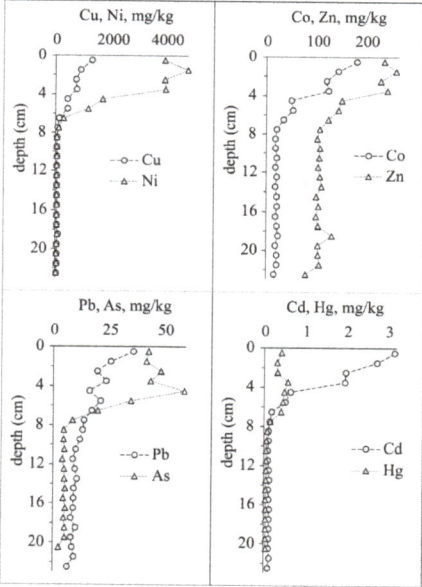

**Figure 6.** The vertical distribution of heavy metals in sediments of Lake Kuetsjarvi (Murmansk Region) [28,29].

Similar tendencies of accumulation of heavy metals in upper layers of sediments were also noted in other lakes located in the northwestern part of Murmansk Region and in the lakes of the north of Finland and Norway, also subject to the influence of Kola MMC ("Pechenganickel" combine). For instance, increased concentrations of Ni (up to

373 mg/kg, while background level is 45 mg/kg), Cu (up to 185 mg/kg, background is 38 mg/kg) and Co (up to 35 mg/kg, background is 21 mg/kg) were found in the upper layer of sediments of Lake Bjørnevatn located ~10 km to the north from the metallurgical plant [30,31]. Only the uppermost 4 cm of sediments were polluted with both heavy metals, which indicated the low sedimentation rate in this water body (~1 mm/year). In total, based on the data on the distribution of radionuclides $^{137}$Cs and $^{210}$Pb in sediment cores in the area of borders of Russia, Finland, and Norway, the average sedimentation rates varied from 0.65 to 3 mm/year [30,32]. In the sediment core of Lake Rabbvatnet (Norway) located ~40 km to the north from the metallurgical plant Ni concentrations reached ~250 mg/kg, Cu—~300 mg/kg, and Co—~12 mg/kg despite the distance [32]. As in other studied lakes, the increases in the content of mentioned heavy metals in sediments of this lake were found in layers dating to the 1920–1930s, and maximum concentrations were fixed in the 1970–1980s due to the most intensive work of the plant and the highest atmospheric emissions of pollutants. The impact of Kola MMC was also noted in the lakes of the north of Finland [33]. For instance, a slight increase in the concentrations of Ni (from ~14 to ~20 mg/kg) and Cu (from ~18 to ~23 mg/kg) was fixed in the uppermost 2–3 cm of sediments of Lake Vassikajarvi, despite the fact that this water body is located ~150 km from the direct source of pollution. Therefore, similarly to research in Chelyabinsk Region, limnological studies in the northwest of Murmansk Region and in the border area showed a tendency towards decreasing concentrations of main pollutants from the metallurgical plant in the upper layers of lake sediments with increasing distance from the industrial enterprise. The negative impact of emissions from the second combine of Kola MMC located near the city of Monchegorsk was also well studied on the example of lakes, including Lake Imandra, which is the largest water body in Murmansk Region [34,35]. The metallurgical plant in Monchegorsk ("Severonickel" combine) started operating in the 1930s refining copper–nickel ore. By mass, these elements (Ni and Cu) are the main pollutants of the local environment. One of the most polluted water bodies of this region is Lake Nudjavr, receiving waste and mine water from the combine [21,36]. Figure 7 illustrates that due to the impact of the metallurgical plant there was an increase in concentrations of Ni from 191 to 129,516 mg/kg and Cu—from 34 to 22,965 mg/kg in sediments (0–13 cm) [21]. The increases in content of Co (up to 1498 mg/kg, while background level is 12 mg/kg), Zn (up to 376 mg/kg, background is 19 mg/kg), Cd (up to 13.8 mg/kg, background is 0.1 mg/kg), and Pb (up to 127.7 mg/kg, background is 1.4 mg/kg) were also noted. Moreover, there were some of the highest concentrations of chalcophile elements (Cd, Pb) of all lakes of Murmansk Region in this water body, which is related to the fact that pollutants enter Lake Nudjavr not only by air and through polluted soil, but also directly with wastewater from Kola MMC

Studies of other lakes located in the impact zone of atmospheric emissions from the metallurgical plant [34] showed that there were similar dynamics of increased concentrations of heavy metals in all lakes, despite the different distances (from 7.5 to 12 km) from the source of pollution. Depending on the water body, the increases in the content of pollutants were found at depths of 10–15 cm, which indicated that the sedimentation rate in these lakes was ~2.3 mm/year. The concentrations of heavy metals in these lakes were by an order or even several orders of magnitude lower than in sediments of Lake Nudjavr and varied from ~500 to ~2200 mg/kg for Ni, from ~150 to ~1100 mg/kg for Cu, and from ~25 to ~115 mg/kg for Co [34]. The lowest concentrations of mentioned metals were found in Lake Pagel, located 12 km from the metallurgical plant.

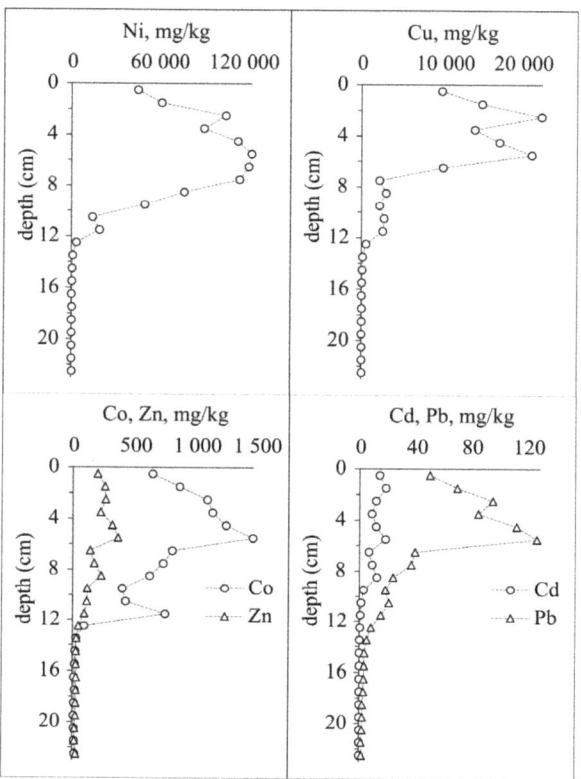

**Figure 7.** The vertical distribution of heavy metals in sediments of Lake Nudjavr (Murmansk Region) [21].

Lake Imandra, on the bank of which the city of Monchegorsk is located, is also subject to the impact of Kola MMC. The most polluted area is Monche Bay, the part of the lake near the city and the metallurgical plant [35]. Here, the increases in concentrations of heavy metals (Ni, Cu, Co, Zn и Pb) were fixed at a depth of 10 cm, which corresponds to the start of operating of the combine in the late 1930s. The maximum contents of almost all mentioned pollutants were found in the uppermost layer of sediments (0–1 cm): ~16000 mg/kg for Ni, ~1400 mg/kg for Cu, ~315 mg/kg for Co, ~260 mg/kg for Zn, and ~75 mg/kg for Pb (Figure 8). Paleolimnological studies revealed that there are similar dynamics of behavior of main pollutants from the metallurgical plant in another part of Lake Imandra, Kunchast Bay, located ~100 km from Kola MMC [35], which may be related to both atmospheric and aquatic transport of substances in the largest water body of Murmansk Region. However, total values of heavy metals concentrations in sediments of Lake Imandra in the area of Kunchast Bay were lower than in the area of Monche Bay. Thus, the maximum content of Ni in Kunchast Bay sediments was ~300 mg/kg, Cu—~120 mg/kg, Co—~25 mg/kg, Zn—~130 mg/kg, Pb—~36 mg/kg. Moreover, the increase in concentrations of pollutants began at a depth of 5 cm. Therefore, according to the knowledge of the timing of smelting/mining operations in the studied region, the sedimentation rate in these areas of Lake Imandra varied from ~0.8 to 1.6 mm/year.

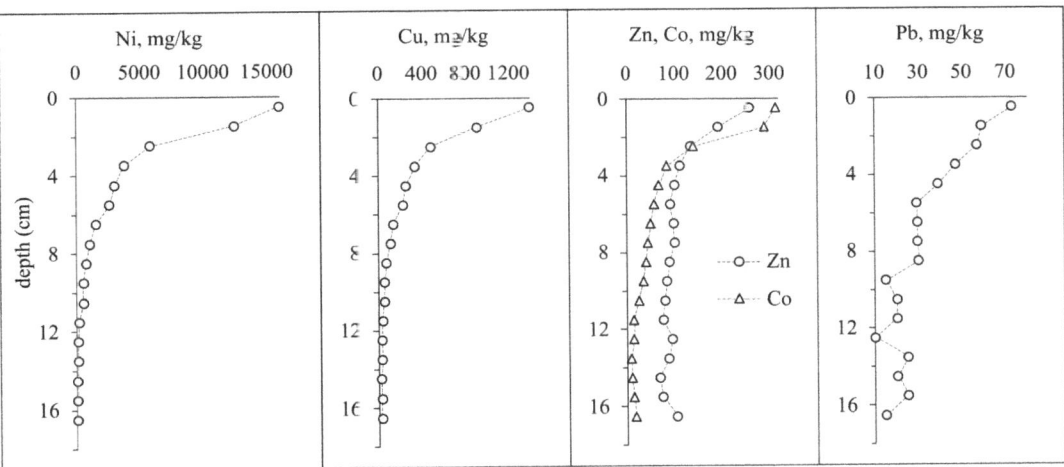

**Figure 8.** The vertical distribution of heavy metals in sediments of Lake Imandra (Murmansk Region) [35].

Another large industrial enterprise of Murmansk Region is Apatit JSC in Kirovsk, mining apatite-nepheline ore from the Khibiny deposit. Studies reported that wastewater and dust emissions from Apatit JSC played a significant role in the pollution of lakes located near this enterprise and its mines. The paleolimnological studies of Lake Bolshoi Vudjavr and Lake Imandra are the most illustrative [37–39]. Lake Bolshoi Vudjavr is the largest water body of the Khibiny Massif. This water body is mostly influenced by wastewater from mines. There were the increases in concentrations of P (up to ~15,000 mg/kg, while background level is less ~1000 mg/kg), Ca (up to ~76,500 mg/kg, background is ~1500 mg/kg), Sr (up to ~2900 mg/kg, background is ~400 mg/kg), Pb (up to 45.9 mg/kg, background is 9.7 mg/kg), and Cu (up to 225 mg/kg, background is 55 mg/kg) in upper layers of sediments of Lake Bolshoi Vudjavr [37,39–41], which is related to the composition of apatite-nepheline ore and also the influence of the city and the long-range transport of pollutants, including those from the metallurgical combine in Monchegorsk located ~45 km from this lake [42]. Recent studies of sediments of Lake Bolshoi Vudjavr have confirmed previously received data, broadened the range of identified elements, and specified the sedimentation rate in the lake [39,41]. Based on the data on the vertical distribution of the $^{210}$Pb isotope, the sedimentation rate in the water body was 2.3 mm/year. Figure 9 illustrates that the increase in concentrations of heavy metals started in the early 1930s when the city of Kirovsk and the mining and concentrating company were founded. The majority of pollutants (Pb, Cu, Zn) enter the water body with wastewater from mines. However, V is probably related to the operations of boiler room, functioning until 2013, and has used heavy residual fuel oil (mazut). Additionally, increased contents of Sb and W in the sediments are related to the operations of the thermal power plant located ~10 km from the lake using coal as fuel [41].

Besides the anthropogenic impact on Lake Bolshoi Vudjavr, paleolimnological studies determined the natural geochemical anomaly of Mo in sediments of the studied lake [39,43]. It was shown that the sediment cores were enriched with Mo both in upper layers (up to 9.9 mg/kg) due to the influence of mine waters and lower layers (up to 15.1 mg/kg) due to the influence of underlying rocks with increased concentrations of this metal. Previously, the increased concentrations of Mo were found in rivers, streams, and industrial wastewater entering Lake Bolshoi Vudjavr [44].

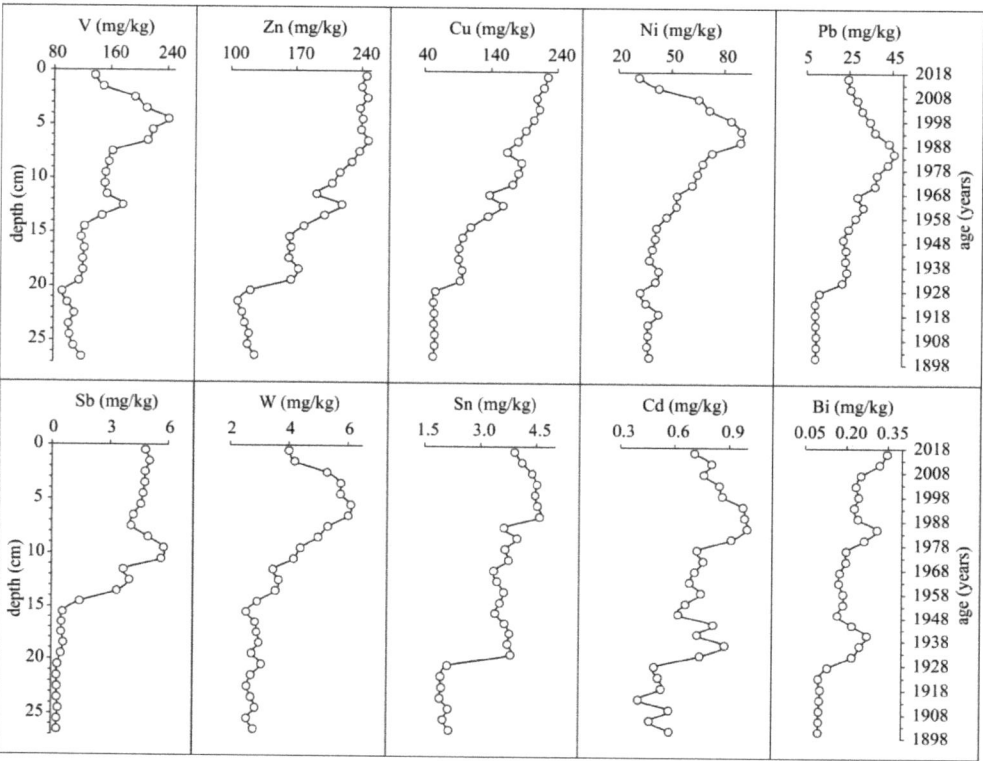

**Figure 9.** The vertical distribution of heavy metals in sediments of Lake Bolshoi Vudjavr (Murmansk Region) [39,41].

The impact of the Apatit JSC operations was also revealed in Lake Imandra, the largest lake of Murmansk Region. This was reflected in the upper layers of sediments in the increased content of P, Ca, Sr, and rare earth elements [38], enriching rocks in the Khibiny Massif [45]. Paleolimnological studies determined the increase in concentrations of rare earth elements in sediments of Lake Imandra at a depth of 10 cm, which corresponds to the start of operating of the ore-processing plant. The highest concentrations of rare earth elements (up to ~240 mg/kg for La (background is 56.5 mg/kg) and up to ~400 mg/kg for Ce (background is 80.6 mg/kg)) in studied sediments date back to the 1970s, the period of the most active ore production (Figure 10) [38]. Even higher concentrations of rare-earth elements due to the activities of JSC Apatit were found in the upper layers of the sediments of Lake Bolshoi Vudjavr [39]. For instance, the detailed analysis of the sediment core of this lake revealed a tendency towards increased concentrations of La (up to 535 mg/kg, while minimum in the core is 84 mg/kg), Ce (up to 802 mg/kg, while minimum in the core is 128 mg/kg), Sm (up to 44 mg/kg, while minimum in the core is 7.3 mg/kg), and Eu (up to 13 mg/kg, while minimum in the core is 2). Obviously, due to the close proximity of Lake Bolshoy Vudjavr to the plant, the concentration of rare earth elements in the sediments of this lake is significantly higher than in Lake Imandra.

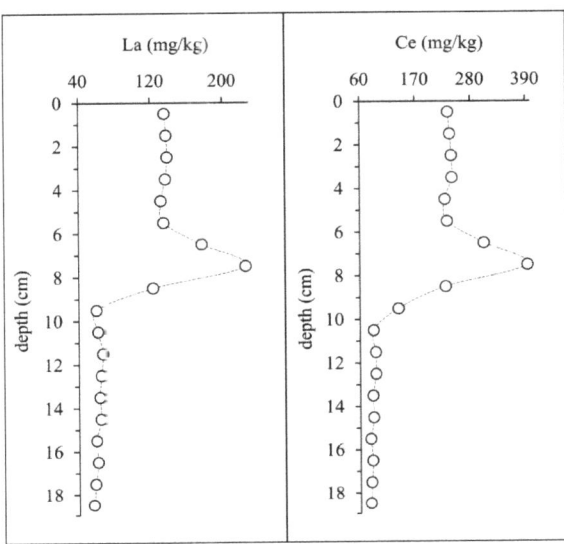

**Figure 10.** The vertical distribution of La and Ce in sediments of Lake Imandra (Murmansk Region) [38].

The studies of lakes of Murmansk Region, including Lake Imandra, located in the impact zone of Olcon JSC, mining, and processing iron-bearing ores, demonstrated the increase of the contents of Fe (up to 20%, while background level is ~3%) and Mn (up to 4%, background is ~1%) in the upper layers of studied sediments [46]. Moreover, the pollution of these lakes by Kola MMC was fixed using marker elements Ni, Cu, and Co with their increased concentrations in 0–10 cm layers similar to other studies. The average sediment rate in studied lakes was 1–2 mm/year, which is close to the average sediment rates in lakes of Murmansk Region in the industrial period [47].

3.1.3. Conclusions of Section 3.1

Anthropogenic influence is considered on the example of the Murmansk region and the Ural region (mainly the Chelyabinsk region). The impact on lake ecosystems from metallurgical enterprises and the mining industry is shown. Lake sediments formed during the 20th and 21st centuries are characterized by a significant level of enrichment in heavy metals (Ni, Cu, Zn, Co, Mo, Pb, Cd) and other elements (for instance, rare earth metals).

*3.2. Urbanized Areas*

A great number of potential sources of anthropogenic pollution are often concentrated in cities. These sources are industrial enterprises, all means of transport, road, and construction dust, and household waste [48–52]. Moreover, the long-range transport of pollutants influences the city areas similar to other (non-urban) areas. The targeted detailed paleolimnological studies of urban areas in Russia were conducted only by the author and his colleagues from the Institute of the North Industrial Ecology Problems of Kola Science Center of RAS and the Institute of Geology, Karelian Research Centre of RAS in Murmansk Region and the Republic of Karelia. It should be noted that these studies are still ongoing.

According to different monitoring services, the Republic of Karelia is one of the clean regions of Russia. There, the anthropogenic pollution of the aquatic environment is mainly related to urban areas and rarely to industrial areas [53]. The majority of paleolimnological studies of the anthropogenic impact on lakes were conducted in Petrozavodsk, the largest city of Karelia [54,55]. For instance, the detailed analysis of the sediment core of Lake Lamba located in the northern part of the city district revealed a tendency towards increased concentrations of heavy metals, including Pb (up to 137 mg/kg, while background is 4 mg/kg), Cd (up to 1.2 mg/kg, background is 0.2 mg/kg), Ni (up to 607 mg/kg, background is 22 mg/kg), V (up to 4785 mg/kg, background is 17 mg/kg), Cr (up to 179 mg/kg, background is 10 mg/kg), Cu (up to 1189 mg/kg, background is 45 mg/kg), Zn (up to 963 mg/kg, background is 136 mg/kg), etc. (Figure 11). The analysis of concentrations of mentioned elements in lower (Holocene) layers of sediments showed that they were similar to the background, or even lower [56].

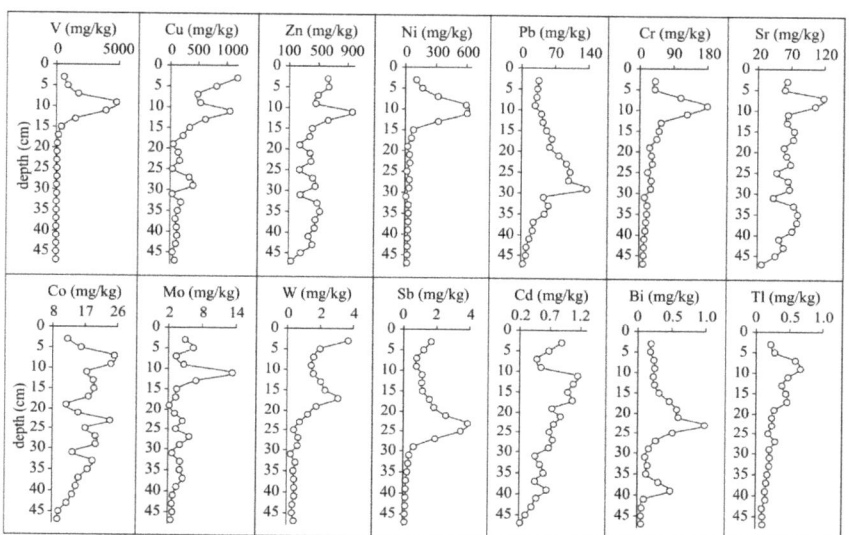

Figure 11. The vertical distribution of heavy metals in sediments of Lake Lamba (the Republic of Karelia) [55].

The analysis revealed that the lake was exposed to the multifactorial outside load. For example, the increased levels of V and Ni are related to the operations of the thermal power plant (Figure 12), which had been using mazut from 1978 until 2000 and then started using gas. Studies show that mazut boiler rooms and thermal power plants always induce increased concentrations of V and Ni in the environment [57,58]. The transition of the Petrozavodsk thermal power plant from mazut to gas resulted in a sharp decrease in concentrations of both heavy metals in the uppermost layers of sediments of Lake Lamba (Figure 11). The increase in concentrations of Zn, Cu, W, and Mo in sediments of the water body is associated with the operations of engineering and instrument-making plants [59], and Pb is related to transport, which had used leaded gasoline with tetraethyllead in Russia until 2002 [13,18,60,61]. Similar behavior of mentioned heavy metals was observed in the paleolimnological study of Lake Chetyrekhverstnoe, also located in the Petrozavodsk city area [54]. The exception was V and Ni behavior. Concentrations of these metals were significantly lower in sediments of Lake Chetyrekhverstnoe compared to Lake Lamba, as the first lake (the Chetyrekhverstnoe) is located 11 km from the thermal power plant and the other is 500 m from the plant. It is known that the range of transfer of particles from mazut thermal power plants and boiler rooms usually does not exceed ~15 km [62].

**Figure 12.** The view of the Petrozavodsk thermal power plant (the Republic of Karelia).

Based on the paleolimnological studies, in other cities of the Republic of Karelia (Medvezhyegorsk, Suoyarvi, Sortavala) the main pollutants are Pb, related to the transport activities and the long-range transport of pollutants [63], Sb and Cd, entering lakes due to fuel combustion all around the world [64], and rarely Cu, Zn, and Sn, which may be related to both transport and dust pollution of urban areas [65–67]. The main geochemical markers allowed for determining that the sedimentation rate in urban lakes of Karelia varied from 2 to 5 mm/year [66].

The impact of urbanized areas in Karelia was also shown in the analysis of geochemistry of sediments of Vygozero Reservoir located in the center of this region [68,69]. The increase in concentrations of V (up to 171 mg/kg, while background for Vygozero is 48 mg/kg), Ni (up to 46 mg/kg, background for Vygozero is 26 mg/kg), Pb (up to 26 mg/kg, background for Vygozero is 6.4 mg/kg), Cd (1.3 mg/kg, background for Vygozero is 0.7 mg/kg), and Bi (up to 0.28 mg/kg, background for Vygozero is 0.09 mg/kg) at depths of 16–24 cm was fixed in the sediment core sampled ~5 km from the city of Segezha (Figure 13). Considering that recent sediments of Vygozero Reservoir have formed in the last 80 years, the age of the studied core was no more than 30–40 years. Therefore, the newest anthropogenic processes relevant to the environment of Segezha were fixed in this case. For instance, the increased level of V and Ni accumulation is associated with emissions from the Mazut thermal power plant operating since the 2000s [69]. The increase in concentrations of Pb, Cd, and Bi in the uppermost layers of sediments of Vygozero Reservoir is evidence of the perpetual entering of pollutants into the area of the North of Russia due to the long-range atmospheric transport [70,71].

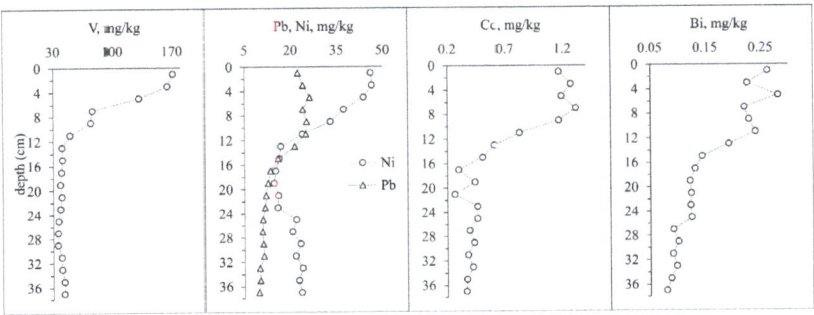

**Figure 13.** The vertical distribution of heavy metals in sediments of Vygozero Reservoir (Republic of Karelia) [69].

Murmansk Region is one of the most urbanized regions of Russia. However, paleolimnological studies in the city areas of this industrial region have not been conducted until recently, since they were focused mainly on the impact of the metallurgical plants and mining enterprises. Similar to Karelia, the urban lakes of Murmansk Region are subject to pollution from the energy industry, transport, and also the metallurgical plants mentioned before [52,72,73].

In the city of Murmansk, which is the capital of Murmansk Region, the analysis of the recent sediment core of Lake Semenovskoe (Figure 14) showed the increases in concentrations of heavy metals starting from a depth of 32 cm for Pb (up to 125 mg/kg, while background is ~4 mg/kg), 28 cm for Zn (up to 694 mg/kg, background is ~76 mg/kg), Co (up to 38 mg/kg, background is ~5 mg/kg), Ni (up to 263 mg/kg, background is ~27 mg/kg), Cd (up to 3.2 mg/kg, background is ~0.3 mg/kg), and Sb (up to 4.1 mg/kg, background is ~0.08 mg/kg), and 16 cm for V (up to 904 mg/kg, background is ~70 mg/kg). Studies [74,75] revealed that the Mazut thermal power plant and boiler rooms play a significant part in the pollution of Murmansk lakes with V and Ni, because mazut has been used at this enterprise since the 1960s as the main fuel [76]. Based on the dynamics of behavior of the two mentioned pollutants, it was determined that the average sedimentation rate in the lake in the industrial period was ~3 mm/year. Other metals are related to dust emissions from the coal terminal in the Murmansk port (Zn, Co, Pb, Cu, Cd, Sb), transport using leaded fuel (Pb) [5], the incineration plant, and also the influence of the long-range transport of pollution from the local plants and the plants located in other regions of Russia and other countries [64,77]. It should be noted that all studied lakes of Murmansk are characterized by similar dynamics of behavior of mentioned heavy metals [72].

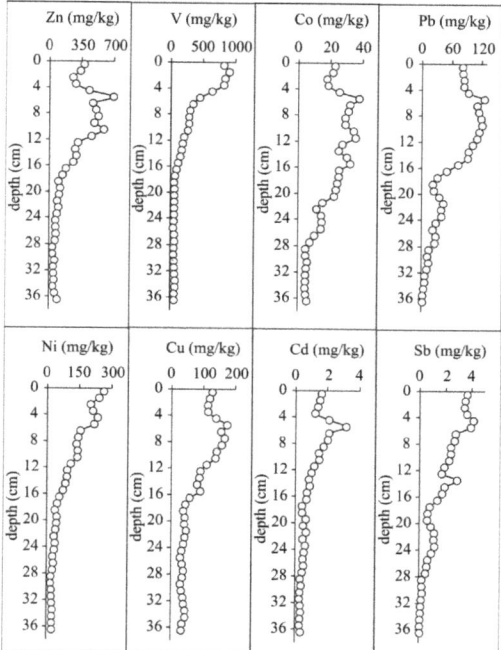

**Figure 14.** The vertical distribution of heavy metals in sediments of Lake Semenovskoe (Murmansk Region) [52].

In addition, studies of the urban lakes of Murmansk showed that rare earth elements can also be indicators of technogenic impact on water bodies [78]. In the course of the work, the general dynamics of the accumulation of rare earth elements and «classical» heavy metals in the upper layers of sediments of polluted lakes were established. Basically, rare earth elements enter aquatic ecosystems as a result of dust emissions (from transport, enterprises, wear of buildings and roads, destruction of soil cover and rocks) [49,79,80]. Similar patterns have not been established in the remote territories of the Murmansk region, since the described processes have a minimal manifestation there.

In Monchegorsk, the other city of Murmansk Region mentioned before, the main anthropogenic load on Lake Komsomolskoe comes from Kola MMC emissions [81]. The stable dynamics of increased concentrations of a wide range of heavy metals such as Ni (up to 2140 mg/kg, while the background is 89 mg/kg), Cu (up to 2607 mg/kg, background is 68 mg/kg), Cr (up to 335 mg/kg, background is 54 mg/kg), Zn (up to 335 mg/kg, background is 41 mg/kg), Co (up to 129 mg/kg, background is 4 mg/kg), V (up to 140 mg/kg, background is 35 mg/kg), Pb (up to 100 mg/kg, background is 8 mg/kg), Cd (up to 2.5 mg/kg, background is 0.4 mg/kg), Sb (up to 3.3 mg/kg, background is 0.2 mg/kg), etc., were fixed in recent sediments of this lake [82]. Similar tendencies of heavy metals behavior, shown earlier on the example of other lakes in the impact area of the plant, remain there, mainly because Lake Komsomolskoe is located 4 km from the metallurgical plant [34,35]. Besides, the impact of the Mazut thermal power plant located on the premises of the metallurgical plant was noted for the first time using marker element V. The average sedimentation rate, calculated using $^{210}$Pb isotope activity in this urban lake, was 2.7 mm/year [82]. The comparison of the age of sediments and the dynamics of behavior of heavy metals showed that the increase in main pollutants content began in the late 1930s when the plant near Monchegorsk started operating.

Other urbanized areas of Russia are poorly studied from the paleolimnological point of view. Unfortunately, despite the great activity of lake researchers in the Republic of Tatarstan and a large number of publications on the content of heavy metals in recent lake sediments [83–85], there are almost no studies with the detailed analysis (layers from 0 to 10 cm) of the vertical distribution of pollutants in sediment cores. There is only one example of such a study of the urban water body in the Republic of Tatarstan. In particular, the studies of the geochemistry of the sediment core 110 cm long of Lake Verkhny Kaban located in the city of Kazan revealed the anthropogenic impact on the lake by the vertical distribution of Pb (up to 45 mg/kg, minimum for the sediment core is 6.1 mg/kg), Cd (up to 4.7 mg/kg, minimum for the sediment core is 0.01 mg/kg), Cu (up to 176 mg/kg, minimum for the sediment core is 0.2 mg/kg), and Zn (up to 480 mg/kg, minimum for the sediment core is 1.4 mg/kg) [86]. The highest concentrations of mentioned heavy metals were found in the upper layers of sediments accumulated in the area of the discharge channel of the thermal power plant. Moreover, other industrial enterprises of Kazan use this channel for untreated water disposal.

Conclusions of Section 3.2

According to published data, paleolimnological studies of the anthropogenic impact of cities on the environment were carried out on the example of urbanized areas of the Republic of Karelia and the Murmansk region (north-west Russia). The pollutants of water ecosystems are industrial enterprises, thermal power plants, boiler houses, waste processing plants, and transport (primarily cars). In recent sediments of lakes, background excesses for V, Ni, Pb, Zn, Cu, Cd, and others have been established. In addition, it was found that lithophile elements that enter the environment with dust from the destruction of soil, road surfaces, and buildings can also be indicators of urban impact on water bodies.

*3.3. Background (Pristine) Areas*

The important part of paleolimnological studies in Russia is the study of lakes in the regions remote from the anthropogenic sources of pollution. To a certain extent, these regions can be considered as a background. First of all, this concerns the Arctic zone of the Russian Federation. The studies of the levels of heavy metals accumulation in sediments of lakes in such areas are of interest, mainly in terms of the study of the long-range transport of pollutants [3,9,18,70,87].

Udachin V.N. and his colleagues conducted studies of the arctic lake Kenteturku located in the center of the Taimyr Peninsula [88]. Researchers sampled the sediment core 30 cm long and divided it into 1 cm layers. It was found that the lake is still practically pristine. There was no significant exceedance of heavy metals concentrations in the upper layers of studied sediments, except for Pb in the 1–2 cm layer (Figure 15).

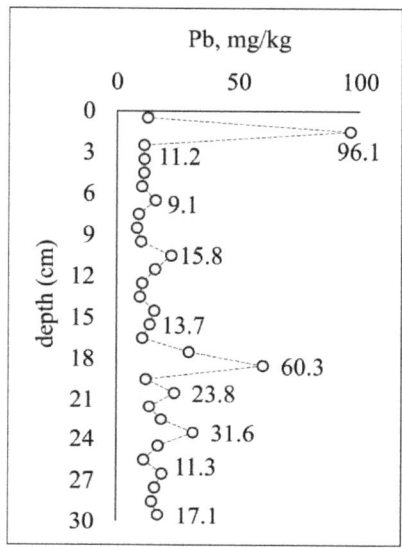

**Figure 15.** The vertical distribution of Pb in sediments of Lake Kenteturku (the Arctic) [88].

First of all, it is interesting to note that there were no abnormal peaks of Ni and Cu concentrations, considering that the Norilsk industrial hub is located 550 km from the lake [89]. Emissions from the metallurgical plant are likely to not reach the studied area and Lake Kenteturku. The peak concentration of Pb (96 mg/kg), in contrast to the median content (14 mg/kg) throughout the sediment core, seems to be a measurement error on the one hand. However, on the other hand, taking into account possible extremely low sedimentation rates in the lake, the sharp increase in Pb concentrations may indicate the influence of the long-range transport of pollutants, which is typical for the recent lake sediments in the Northern Hemisphere [14,34].

The studies of the geochemistry of sediments of other arctic lakes located in the Yamal and the Gyda Peninsulas also did not show the significant dynamics of the majority of elements except for Hg [90]. In sediments of Lake Langtibeito, the concentration of this metal slightly increased to ~0.08 mg/kg, starting from a depth of ~10 cm. The sedimentation rate in lakes of mentioned areas was from 1.7 to 2.0 mm/year based on the $^{210}$Pb activity.

In the Murmansk Region, which also belongs to the Arctic zone, the studies of lakes of the background areas showed a tendency towards an increase in traditional pollutants from among chalcophile elements (Pb, Cd, Hg, As) and local pollutants Ni and Cu from Kola MMC emissions in the uppermost layers of sediments [34,91]. The increased concentrations of Ni (from 12 up to 111 mg/kg) and Pb (from 8 up to 36 mg/kg) in the uppermost layers (5–6 cm) of sediments were even found in the lakes located in the mountainous areas, which can act as a barrier for pollutants distribution [92]. The sedimentation rate in such lakes can be assessed as ~1 mm/year or less based on the marker pollutants. In Lake Umbozero, the second largest lake of Murmansk Region, concentrations of heavy metals in sediments increased from a depth of ~10 cm (typical for Pb, Cd, As) and ~5 cm (typical for Ni and

Cu) [14,93]. In total, both the largest lakes of Murmansk Region are characterized by similar patterns of accumulation of heavy metals, which are the main pollutants in the region.

In recent years, when studying the lakes of Murmansk Region, a range of pollutants also indicating the long-range atmospheric transport has been extended due to the use of new methods for the analysis of microelements in sediments (ICP-MS) [94]. In lake sediments of pristine areas of the northern part of Murmansk Region (the area of the Rybachy Peninsula), the increased concentrations of Ni (up to 127 mg/kg, background is 25 mg/kg), Cu (up to 370 mg/kg, background is 35 mg/kg), Co (up to 40 mg/kg, background is 4.2 mg/kg), Pb (up to 82 mg/kg, background is 10 mg/kg), As (up to 6.3 mg/kg, background is 1.8 mg/kg), Sn (up to 32 mg/kg, background is 0.8 mg/kg), Bi (up to 1 mg/kg, background is 0.06 mg/kg), Sb (up to 0.5 mg/kg, background is 0.08 mg/kg), and Tl (up to 0.11 mg/kg, background is 0.09 mg/kg) were found. Paleolimnological studies of lakes in the south of Murmansk Region and the north of the Republic of Karelia showed that the range of transport of emissions from Kola MMC enterprises reached ~250 km [82]. The sedimentation rate in lakes of pristine taiga landscapes of Northwest Russia can be ~0.6 mm/year based on the $^{210}$Pb isotope [82].

Similar tendencies of the behavior of the above-mentioned heavy metals in lake sediments are found in two regions located to the south of Murmansk Region. These are the Republic of Karelia [54,95] and Arkhangelsk Region [96]. In the study of the geochemical analysis of recent sediments of Lake Maselgskoe (the south-west of Arkhangelsk Region), it was determined that the upper layers of sediments were enriched with Pb, Cd, Sb, Bi, and W. Particularly, the increase in concentrations of Pb (up to ~50 mg/kg) started at a depth of ~30 cm. The sedimentation rate was 4.1 mm/year based on the nonequilibrium $^{210}$Pb [96]. Other studied elements (for instance, Sc and Zn) do not tend to increase in the upper layers of sediments of Lake Maselgskoe, since they are not the agents of the long-range atmospheric transport.

The sedimentation rate in recent sediments of Lake Ukonlampi located in the south-eastern part of Karelia (near the Finnish–Russian border) was 1.25 mm/year based on the $^{210}$Pb activity [97,98]. The similar tendency towards increased concentrations of Pb (up to 91.1 mg/kg, background is 3.8 mg/kg), Cd (up to 2.69 mg/kg, background is 0.39 mg/kg), Sb (up to 1.97 mg/kg, background is 0.10 mg/kg), Sn (up to 5.34 mg/kg, background is 0.46 mg/kg), Tl (up to 0.84 mg/kg, background is 0.06 mg/kg), Bi (up to 4.06 mg/kg, background is 0.08 mg/kg), Cu (up to 51.2 mg/kg, background is 12 mg/kg), and Zn (up to 263.8 mg/kg, background is 40 mg/kg) was found in this lake and two water bodies nearby (Figure 16). It should be noted that pollution of these background water bodies might be associated not only with the global pollution of the Northern Hemisphere but also with the proximity of this region to industrial enterprises of Finland in Imatra and Kotka [99]. This explains the increased content of Zn and Cu, which usually are not categorized as indicators of the atmospheric transport in pollution of the North background regions. In total, the analysis of recent sediment cores of 30 small lakes of the south of Karelia and Vygozero Reservoir showed that the main pollutants in the region are due to the long-range atmospheric transport of Pb, Sb, Cd, Bi, and Tl [67,69]. The close correlation between concentrations of these metals (for instance, Pb and Sb, Figure 17) confirmed the unity of their entering to the aquatic ecosystem and accumulation in lake sediments.

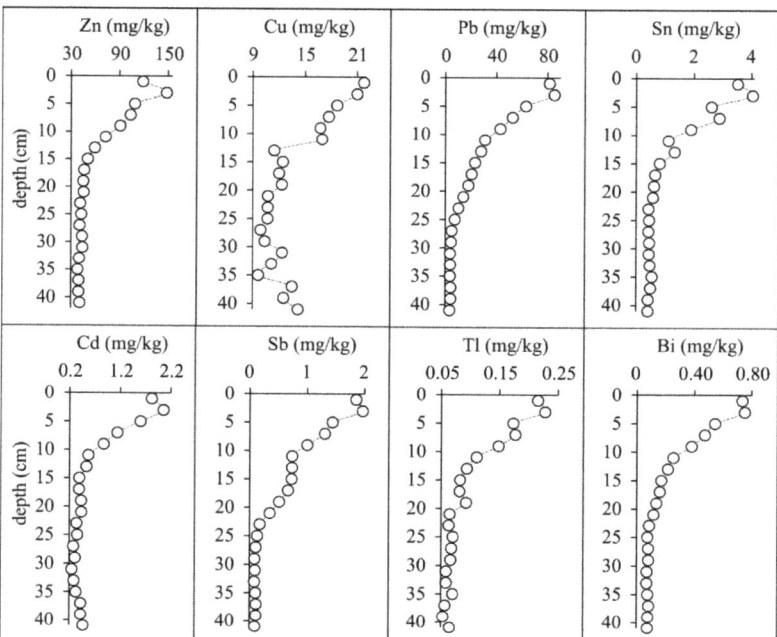

**Figure 16.** The vertical distribution of Pb in sediments of Lake Liunkunlampi (the Republic of Karelia) [98].

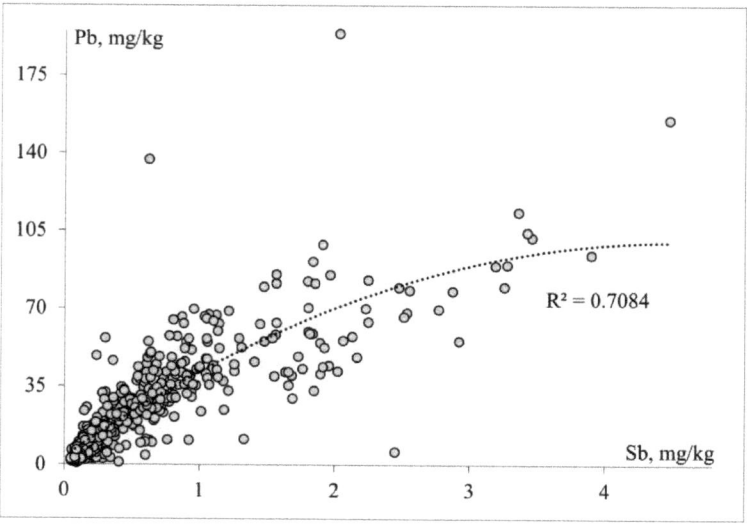

**Figure 17.** The correlation between Pb and Sb in lake sediments of Karelia (author data).

Similar patterns are observed in Lake Onega, the largest reservoir in the Republic of Karelia and the second largest in Europe [95]. In technogenesis, the rate of sedimentation in Lake Onega is not high, 1 mm/year, which is estimated from the activity of $^{137}$Cs and $^{210}$Pb isotopes [100]. This value is close to the average sedimentation rates in lakes in remote (background) areas. It has been established that the content of Pb, Cd and Sb increases in the upper (up to 10 cm) sediment layers of Lake Onega. The authors attribute this dynamic to the technogenic impact on the lake, primarily due to the long-range transport of pollutants. In particular, an increase in the concentration of Pb up to ~40 mg/kg (with a background level is ~10) and Cd up to ~1 mg/kg (background is ~0.2) was found.

Siberia is a region of Russia where limnological geochemical studies of freshwater sediments are well-developed. The focus of Siberian paleolimnology has been on the analysis of natural variations of chemical elements in sediment cores. At the same time, there are studies on the determination of the anthropogenic impact on lake ecosystems and the environment.

The study of the sediment core of Lake Manzherok in the Altai Republic (Siberia) showed the difference in the accumulation of heavy metals such as Pb (up to ~13 mg/kg, minimum for the core is ~4.5), Cd (up to ~1.7 mg/kg, minimum for the core is ~0.15), and As (up to ~35 mg/kg, minimum for the core is ~4) in the uppermost layers of sediments (0–20 cm) compared to lithophile elements [15]. Paleolimnologists suggest that this behavior of the mentioned heavy metals can be explained by the anthropogenic impact on the studied lake and its catchment area. In another Siberian lake (the Kolyvanovskoe) located in the southwest of Altai Krai, similar behavior of Pb, Cd, and Hg was observed in recent sediments [101]. The concentrations of mentioned heavy metals increased from the lower to the upper layers in the 50-cm core dated using $^{137}$Cs and $^{210}$Pb isotopes. The age of studied sediments of Lake Kolyvanovskoe showed that the increase in the concentrations of Pb (up to ~25 mg/kg, background is ~9), Cd (up to ~0.3 mg/kg, background is ~0.05), and Hg (up to ~0.3 mg/kg, background is ~0.02) started in the period from the end of the 19 century to the present. Other trace elements such as Cu, Co, Zn, and Ni do not have similar accumulation dynamics in the studied sediments of Lake Kolyvanovskoe.

The extensive studies of Siberian lakes demonstrate common patterns of increased concentrations of Pb, Cd, Hg, and Sb in sediments dated back to the last three centuries using $^{137}$Cs and $^{210}$Pb isotopes [17,102,103]. The majority of scientists admit the significant anthropogenic impact on the formation of geochemical anomalies of mentioned elements [101,104]. The high concentrations of some heavy metals were found in recent sediments of small lakes: 3345 mg/kg of Pb in sediments of Lake Bolshye Rakity, adjacent to the city of Rubtsovsk, 112 mg/kg of Sb, and 4.2 mg/kg of Cd in sediments of Lake Yakov (Tomsk Region) [17]. However, such concentrations of heavy metals are not common for small lakes of Siberia, even in cases of the anthropogenic impact on studied lakes. For instance, in the uppermost layers of sediments of Lake Kotokel (Pribaykalsky District), the concentrations of Pb reached 11.5 mg/kg (minimum for the core is 6.4) and Cd—0.4 mg/kg (minimum for the core is 0.13) (Figure 18) [104]. In the uppermost layers of sediments of Lake Shchuchie (Tomsk region), Cd concentrations reached 0.83 mg/kg (minimum for the core is 0.08). Despite the historical dynamics of the anthropogenic input of heavy metals into the water bodies, median background levels of Pb (20 mg/kg), and Cd (0.14 mg/kg) for Siberia, in the lake, sediments are not often exceeded [17].

Conclusions of Section 3.3

The geography of paleolimnological studies of the anthropogenic impact on the pristine areas of Russia is more extensive than in the previous sections of the article. Studies of lake sediments were carried out here in the Arctic, in the taiga zone of Karelia, and in different regions of Siberia. Practically everywhere, the influence of the long-range transport of heavy metals (Pb, Sb, Cd, Bi, Tl) associated with the combustion of fossil fuels at the enterprises of North America, Europe, and Asia is manifested. According to isotopic dating, low sedimentation rates are noted in the lakes of the background areas compared to industrial and urban areas.

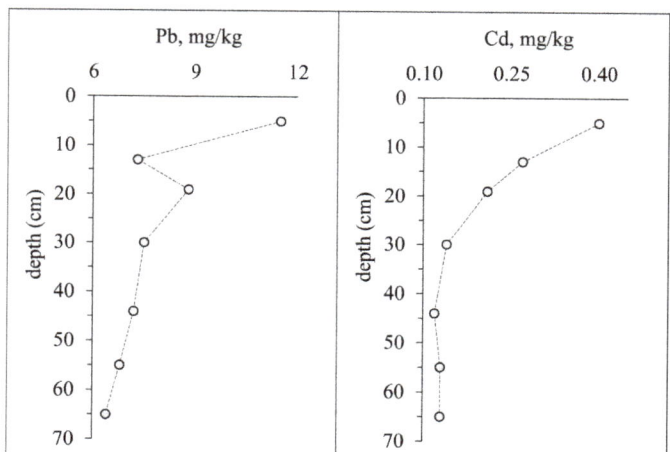

**Figure 18.** The vertical distribution of Pb and Cd in sediments of Lake Kotokel (Siberia) [104].

*3.4. Comparison of Element Concentrations*

In order to compare the values of the content of chemical elements in the sediments of lakes from different regions of Russia, it was necessary to carry out a normalization procedure. For this, the average content of chemical elements in the upper part of the Earth's crust [105] was used, by which the concentrations of elements in the sediments of the lakes were divided (the uppermost layers of the lake cores were taken). After that, the obtained data (enrichment factors) were logarithmic so that they could be placed on one chart (Figure 19).

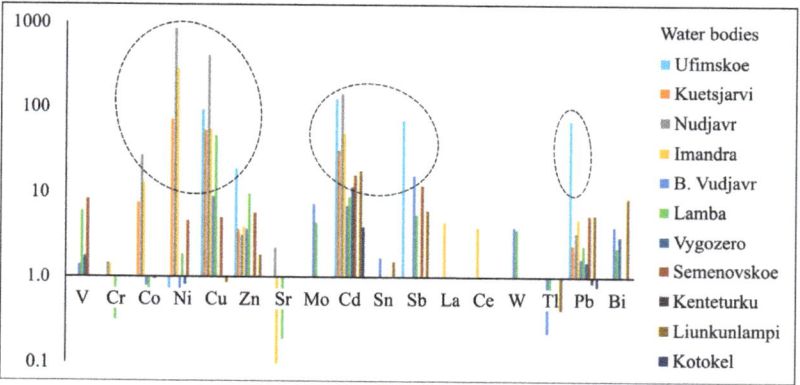

**Figure 19.** Enrichment factors of chemical elements from lake sediments described in this review (Figure 1).

It can be seen that the sediments of industrial lakes (Ufimskoe, Kuetsjavri, Nudjavr) are the most polluted. The highest enrichment factors are noted for Ni, Cu, Cd, Pb, and Sb. On the other hand, even in the lakes of the background areas, there are excesses of heavy metals, which are associated with the long-range atmospheric transport of pollutants (Cd, Pb, Bi, Sb). Vanadium is a specific pollutant in the urban lakes of the North-West of Russia, which, as noted, is associated with emissions from mazut thermal power plants and boiler houses.

## 4. General Conclusions and Perspectives

### 4.1. Conclusions

The analysis of the large number of paleolimnological studies of the recent anthropogenic impact on lakes of Russia showed that, despite the large distances between regions, likely different geology, and other factors influencing the sedimentation, there were a lot of similarities in the accumulation dynamics of pollutants in lake sediments. For instance, the specifics of metallurgical and mining plants are well fixed both in Chelyabinsk Region (the Southern Ural) and Murmansk Region (Northwest Russia). The increases in Cu, Zn, and Pb concentrations in the upper layers of sediments were observed in lakes near the metallurgical plant of the city of Karabash [16]. Moreover, this enterprise influences even lakes located at a distance of 100 km from emissions. A similar situation can be observed in Murmansk Region [21,27], where Cu and Ni are the key pollutants. Sediments most polluted with these metals were found in Lake Kuetsjarvi (the area of the Norway–Russia border) and Lake Nudjavr (the central part of the region). In Murmansk Region, paleolimnological studies of the impact of the mining enterprises determined the increased concentrations of P, Ca, Sr, and rare earth elements in recent sediments, as all these elements are included in produced ore entering water bodies with mine waters and dust [39].

The similarity in the impact of Mazut boiler rooms and thermal power plants on sediments of lakes of Petrozavodsk, Segezha, Murmansk, and Monchegorsk was found in urban areas of Karelia and Murmansk Region. It was shown that all water bodies were enriched with V and Ni, included in ash from mazut burning [54]. Moreover, there were increased concentrations of Pb in sediments of these cities due to the active use of leaded fuel in cars all over the world [61]. In Russia, the fuel containing Pb was banned in 2002. Besides, the impact of engineering and instrument-making companies was fixed in sediments of Petrozavodsk lakes, and the impact of the coal port, the incineration plant, and metallurgical industry was observed in lakes of Murmansk Region [72,74].

The special analysis of areas not subject to the direct anthropogenic impact showed that lake sediments in the Arctic zone of Russia, Karelia, Arkhangelsk Region, and Siberia were still influenced by the long-range transport of pollutants. Mostly, it is related to the burning of fossil fuel (coal), therefore Pb, Cd, Sb, Hg, Bi, and Tl, included in coal as additives, are the main geochemical agents of this process [3,87]. The clearest patterns of the increase in concentrations of these heavy metals are found in Northwest Russia, possibly due to its proximity to Europe [97]. The studies of a great number of lakes of the Republic of Karelia showed that Pb is closely associated with Sb in sediments of this region, which indicates the similar pattern of the input and accumulation of metals in sediments of lakes in pristine areas [67].

Studies also demonstrated that sedimentation rates estimated using $^{210}$Pb and $^{137}$Cs isotopes varied to a large extent in modern times. The lowest sedimentation rates (less than 1 mm/year) were fixed in small lakes of background areas or in large water bodies of the Russian North [92]. The highest sedimentation rates (from 3 to 5 mm/year) were found in lakes of urban or industrial areas [22]. In regions subject to the increased level of the anthropogenic load, the more intensive weathering, together with the atmospheric inputs of pollutants to water, possibly lead to the larger amount of matter accumulated in lakes.

### 4.2. Perspectives

The review of all known paleolimnological research aimed at the influence of the modern anthropogenic load on the environment of Russia showed that all these studies were concentrated in three regions—the Northwest, the Urals, and Siberia. In the author's opinion, it is related not only to the fact that there are a lot of lakes and several large anthropogenic objects in these regions, but also to the lack of the necessary equipment for the detailed sediment core sampling and the analysis of a wide range of chemical elements including heavy metals, and the lack of human resources for conducting paleolimnological studies in other regions. For instance, despite the fact that a lot of studies of geochemistry of lake sediments are conducted in order to analyze the anthropogenic impact on water ecosystems in the Republic of Tatarstan [86], there are almost no works with the detailed analysis of sediment cores of urban and remote lakes. Unfortunately, there are no such works in other regions of Russia, where they can be highly in demand. These studies can be relevant for Moscow, Saint Petersburg, large cities of Siberia, and the Russian Far East. Recently, the author conducted detailed studies of the sediment cores of lakes of Arkhangelsk, which will be published soon. However, this is obviously not enough, considering that there are a lot of significant regions of Russia where paleolimnoligical studies of the anthropogenic load on the environment have not yet been conducted. Hopefully, such works will be done in future involving international cooperation, since the equipment for sampling recent sediments with an option of the detailed dividing cores into layers is mainly produced abroad (for instance, in Finland and Norway), and lake research is almost always included in European scientific projects on the environmental quality assessment.

**Funding:** This research is supported by the State order of Laboratory of Geoecology and Environmental Management of the Arctic of INEP KSC RAS No. 1021111018324-1 and the State order of Institute of Geology of Karelian Research Centre of RAS No. 1022040500826-4.

**Data Availability Statement:** The data that support the findings of this study are available on request from the corresponding author. The data are not publicly available due to privacy restrictions.

**Acknowledgments:** The author sincerely thank the colleagues V.A. Dauvalter for providing initial paleolimnological data and O.V. Petrova for creating of maps with the designation of lakes and key regions of paleolimnological studies.

**Conflicts of Interest:** There are no any competing interest as defined by Springer, or other interests that might be perceived to influence the results and/or discussion reported in this paper.

## References

1. Moore, J.W.; Ramamoorthy, S. *Heavy Metals in Natural Waters. Applied Monitoring and Impact Assessment*; Springer: New York, NY, USA, 1987.
2. Nriagu, J.O. Global metal pollution. *Environment* **1990**, *32*, 7–33.
3. Pacyna, J.M.; Pacyna, E.G. An assessment of global and regional emissions of trace metals to the atmosphere from anthropogenic sources worldwide. *Environ. Rev.* **2001**, *9*, 269–298. [CrossRef]
4. Pyle, G.G.; Rajotte, J.W.; Couture, P. Effects of industrial metals on wild fish populations along a metal contamination gradient. *Ecotoxicol. Environ. Saf.* **2005**, *61*, 287–312. [CrossRef]
5. Escobar, J.; Whitmore, T.J.; Kamenov, G.D.; Riedinger-Whitmore, M.A. Isotope record of anthropogenic lead pollution in lake sediments of Florida, USA. *J. Paleolimnol.* **2013**, *49*, 237–252. [CrossRef]
6. Frechen, M.; Sierralta, V.; Oezen, D.; Urban, B. Uranium-series dating of peat from central and Northern Europe. In *Developments in Quaternary Sciences*; Elsevier: Amsterdam, The Netherlands, 2007; Volume 7, pp. 93–117. [CrossRef]
7. McConnell, J.R.; Gregg, W.L.; Manuel, A.H. A 250-year high-resolution record of Pb flux and crustal enrichment in Central Greenland. *Geophys. Res. Lett.* **2002**, *29*, 2130. [CrossRef]
8. Pratte, S.; Bao, K.; Shen, J.; Mackenzie, L.; Klamt, A.M.; Wang, G.; Xing, W. Recent atmospheric metal deposition in peatlands of Northeast China: A review. *Sci. Total Environ.* **2018**, *626*, 1284–1294. [CrossRef]
9. Sarkar, S.; Ahmed, T.; Swami, K.; Judd, C.D.; Bari, A.; Dutkiewicz, V.A.; Husain, L. History of atmospheric deposition of trace elements in lake sediments, ~1880 to 2007. *J. Geophys. Res. Atmos.* **2015**, *120*, 5658–5669. [CrossRef]
10. Cooke, C.A.; Abbott, M.B. A paleolimnological perspective on industrial-era metal pollution in the central Andes, Peru. *Sci. Total Environ.* **2008**, *393*, 262–272. [CrossRef]
11. Förstner, U.; Heise, S.; Schwartz, R.; Westrich, B.; Ahlf, W. Historical contaminated sediments and soils at the river basin scale. Examples from the Elbe River catchment area. *J. Soils Sediments* **2004**, *4*, 247–260. [CrossRef]
12. Håkanson, L. Sediment sampling in different aquatic environments: Statistical aspects. *Water Resour. Res.* **1984**, *20*, 41–46. [CrossRef]
13. Hosono, T.; Alvarez, K.; Kuwae, M. Lead isotope ratios in six lake sediment cores from Japan Archipelago: Historical record of trans-boundary pollution sources. *Sci. Total Environ.* **2016**, *559*, 24–37. [CrossRef]
14. Jernström, J.; Lehto, J.; Dauvalter, V.A.; Hatakka, A.; Leskinen, A.; Paatero, J. Heavy metals in bottom sediments of Lake Umbozero in Murmansk Region, Russia. *Environ. Monit. Assess.* **2010**, *161*, 93–105. [CrossRef]
15. Blyakharchuk, T.; Udachin, V.; Li, H.; Kang, S. AMS 14 C Dating problem and high-resolution geochemical record in Manzherok Lake sediment core from Siberia: Climatic and environmental reconstruction for Northwest Altai over the past 1,500 years. *Front. Earth Sci.* **2020**, *8*, 206. [CrossRef]

16. Spiro, B.; Udachin, V.; Williamson, B.J.; Purvis, O.W.; Tessalina, S.G.; Weiss, D.J. Lacustrine sediments and lichen transplants: Two contrasting and complimentary environmental archives of natural and anthropogenic lead in the South Urals, Russia. *Aquat. Sci.* **2013**, *75*, 185–198. [CrossRef]
17. Strakhovenko, V.D. Geochemistry of Sediments of Small Continental Lakes in Siberia. Ph.D. Thesis, Novosibirsk State University, Novosibirsk, Russia, 2011; 24p. (In Russian).
18. Marx, S.K.; Rashid, S.; Stromsoe, N. Global-scale patterns in anthropogenic Pb contamination reconstructed from natural archives. *Environ. Pollut.* **2016**, *213*, 283–298. [CrossRef] [PubMed]
19. Maslennikova, A.V.; Udachin, V.N.; Deryagin, V.V. *Paleoecology and Geochemistry of Lacustrine Sedimentation of Ural*; Editorial and Publishing Department of the Ural Branch of the RAS: Yekaterinburg, Russia, 2014. (In Russian)
20. Moiseenko, T.I.; Kudryavtseva, L.P.; Rodyushkin, I.V.; Dauvalter, V.A.; Lukin, A.A.; Kashulin, N.A. Airborne contamination by heavy metals and aluminum in the freshwater ecosystems of the Kola Subarctic Region (Russia). *Sci. Total Environ.* **1995**, *160–161*, 715–727. [CrossRef]
21. Slukovskii, Z.I.; Dauvalter, V.A. Morphology and composition of technogenic particles in bottom sediments of the Lake Nudyavr, Murmansk region. *Proc. Russ. Mineral. Soc.* **2019**, *3*, 202–217. (In Russian) [CrossRef]
22. Udachin, V.N.; Deryagin, V.V.; Kitagava, R.; Aminov, P.G. Isotope geochemistry of sediments of lakes Southern Ural for an estimation of scales mining of technogenesis. *UT Res. J. Nat. Resource Use Ecol.* **2009**, *3*, 144–149. (In Russian)
23. Maslennikova, A.V.; Udachin, V.N.; Aminov, P.G. Lateglacial and Holocene environmental changes in the Southern Urals reflected in palynological, geochemical and diatom records from the Lake Syrytkul sediments. *Quatern. Int.* **2016**, *420*, 65–75. [CrossRef]
24. Shafigullina, G.T.; Udachin, V.N. Content of heavy metals in bottom sediments of the Uchalinskaya geotechnical system. *Explor. Prot. Miner. Resour.* **2009**, *1*, 60–66. (In Russian)
25. Denisov, D.; Terentjev, P.; Valkova, S.; Kudryavtzeva, L. Small Lakes Ecosystems under the Impact of Non-Ferrous Metallurgy (Russia, Murmansk Region). *Environments* **2020**, *7*, 29. [CrossRef]
26. Lyanguzova, I.V.; Goldvirt, D.K.; Fadeeva, I.K. Spatiotemporal dynamics of the pollution of Al–Fe-humus podzols in the impact zone of a nonferrous metallurgical plant. *Eurasian Soil Sci.* **2016**, *49*, 1189–1203. [CrossRef]
27. Zubova, E.M.; Kashulin, N.A.; Dauvalter, V.A.; Denisov, D.B.; Valkova, S.A.; Vandysh, O.I.; Slukovskii, Z.I.; Terentyev, P.M.; Cherepanov, A.A. Long-Term Environmental Monitoring in an Arctic Lake Polluted by Metals under Climate Change. *Environments* **2020**, *7*, 34. [CrossRef]
28. Dauvalter, V. Impact of mining and refining on the distribution and accumulation of nickel and other heavy metals in sediments of subarctic lake Kuetsjärvi, Murmansk region, Russia. *J. Environ. Monit.* **2003**, *5*, 210–215. [CrossRef]
29. Dauvalter, V.A.; Kashulin, N.A.; Sandimirov, S.S.; Terentyev, P.M.; Denisov, D.B.; Amudsen, P.-A. Chemical composition of lake sediments along a pollution gradient in a subaectic watercourse. *J. Environ. Sci. Health Part A* **2011**, *4529*, 1020–1033. [CrossRef]
30. Dauvalter, V.A. Heavy metals in bottom sediments of system of Lake Inari and Pasvik River. *Water Resour.* **1998**, *25*, 494–500. (In Russian)
31. Dauvalter, V.A.; Rognerud, S. Heavy metal pollution in sediments of the Pasvik River drainage. *Chemosphere* **2001**, *42*, 9–18. [CrossRef] [PubMed]
32. Dauvalter, V.A.; Kashulin, N.A.; Denisov, D.B. Tendencies in the content change of heavy metals in lake sediments in Northern Fennoscandia over the last centuries. *Trans. Karelian Res. Cent. Russ. Acad. Sci.* **2015**, *9*, 62–75. (In Russian) [CrossRef]
33. Dauvalter, V. Influence of pollution and acidification on metal concentrations in Finnish Lapland lake sediments. *Water Air Soil Poll.* **1995**, *85*, 853–858. [CrossRef]
34. Dauvalter, V.F.; Dauvalter, M.V.; Kashulin, N.A.; Sandimirov, S.S. Chemical composition of bottom sedimentary deposits in lakes in the zone impacted by atmospheric emissions from the Severonikel plant. *Geochem. Int.* **2010**, *48*, 1148–1153. [CrossRef]
35. Ilyashuk, B.; Ilyashuk, E.; Dauvalter, V. Chironomid responses to long-term metal contamination: A paleolimnological study in two bays of Lake Imandra, Kola Peninsula, Northern Russia. *J. Paleolimnol.* **2003**, *30*, 217–230. [CrossRef]
36. Dauvalter, V.A.; Kashulin, N.A. Environmental and economic assessment of necessity of bottom sediment extraction from Lake Nudjavr, Monchegorskiy district, Murmansk region. *Vestn. MSTU* **2011**, *14*, 884–891. (In Russian)
37. Denisov, D.B. Changes in the Hydrochemical Composition and Diatomic Flora of Bottom Sediments in the Zone of Influence of Metal Mining Production (Kola Peninsula). *Water Resour.* **2007**, *34*, 682–692. [CrossRef]
38. Dauvalter, V.A.; Moiseenko, T.I.; Rodushkin, I.V. Geochemisty of rare earth elements in Lake Imandra, Murmansk region. *Geochemistry* **1999**, *4*, 376–383. (In Russian)
39. Dauvalter, V.; Slukovskii, Z.; Denisov, D.; Guzeva, A. A Paleolimnological Perspective on Arctic Mountain Lake Pollution. *Water* **2022**, *14*, 4044. [CrossRef]
40. Kashulin, N.A.; Vandysh, O.I. *Kola Peninsula on the Threshold of the Third Millenium: Environmental Problems*; Kola Science Center: Apatity, Russia, 2003. (In Russian)
41. Slukovskii, Z.I.; Guzeva, A.V.; Grigoriev, V.A.; Dauvalter, V.A.; Mitsukov, A.S. Paleolimnological reconstruction of the technogenic impact on the ecosystem of Lake Bolshoi Vudjavr (Kirovsk, Murmansk region): New geochemical data. *Ecol. Urban Territ.* **2020**, *4*, 96–107. (In Russian) [CrossRef]
42. Denisov, D.B.; Dauvalter, V.A.; Kashulin, N.A.; Kagan, L.Y. Long-term Changes of Diatom Assemblages in Subarctic Lakes under the Anthropogenic Influence (According to the Data of Diatom Analysis). *Inland Water Biol.* **2006**, *1*, 24–30.

43. Slukovskii, Z.I.; Mitsukov, A.S.; Dauvalter, V.A. Molybdenum in bottom sediments of Lake Bolshoi Vudjavr, Murmansk region: Vertical distribution and forms of this metal. *Proc. Fersman Sci. Sess. GI KSC RAS* **2019**, *16*, 534–538. (In Russian) [CrossRef]
44. Sulimenko, L.P.; Koshkina, L.B.; Mingaleva, T.A.; Svetlov, A.V.; Nekipelov, D.A.; Makarov, D.V.; Masloboev, V.A. *Molybdenum in Zone of Hypergenesis of Khibiny Mountain Massif*; MGTU Publisher: Murmansk, Russia, 2016. (In Russian)
45. Galakhov, A.V. *Petrology of the Khibiny Alkali Massif*; Nauka: Leningrad, Russia, 1975. (In Russian)
46. Dauvalter, V.A. Geochemistry of Lakes in a Zone Impacted by an Arctic Iron-Producing Enterprise. *Geochem. Int.* **2020**, *58*, 933–946. [CrossRef]
47. Dauvalter, V.A.; Kashulin, N.A.; Denisov, D.B. Dynamics of heavy metal content in bottom sediments of lakes from border area between Russia, Norway and Finland in the last centuries. *Proc. Fersman Sci. Sess. GI KSC RAS* **2015**, *12*, 366–368. (In Russian)
48. Quiñonez-Plaza, A.; Temores-Peña, J.; Garcia-Flores, E.; Rodriguez-Mendivil, D.D.; Pastrana-Corral, M.A.; Wakida, F.T. Assessment of heavy metal pollution of drain sediments in the urban area of Mexicali, Mexico. *Environ. Earth Sci.* **2020**, *79*, 447. [CrossRef]
49. Sun, G.; Li, Z.; Liu, T.; Chen, J.; Wu, T.; Feng, X. Rare Earth Elements in Street Dust and Associated Health Risk in a Municipal Industrial Base of Central China. *Environ. Geochem. Health* **2017**, *39*, 1469–1486. [CrossRef]
50. Wan, D.; Han, Z.; Yang, J.; Yang, G.; Liu, X. Heavy metal pollution in settled dust associated with different urban functional areas in a heavily air-polluted city in North China. *Int. J. Environ. Res. Publ. Health* **2016**, *13*, 1119. [CrossRef]
51. Christensen, A.M.; Nakajima, F.; Baun, A. Toxicity of water and sediment in a small urban river (Store Vejlea, Denmark). *Environ. Pollut.* **2006**, *144*, 621–625. [CrossRef] [PubMed]
52. Slukovskii, Z.; Dauvalter, V.; Guzeva, A.; Denisov, D.; Cherepanov, A.; Siroezhko, E. The Hydrochemistry and Recent Sediment Geochemistry of Small Lakes of Murmansk, Arctic Zone of Russia. *Water* **2020**, *12*, 1130. [CrossRef]
53. Filatov, N.N.; Kukharev, V.I. *Lakes of Karelia: Handbook*; Karelian Scientific Center of the Russian Academy of Sciences: Petrozavodsk, Russia, 2013. (In Russian)
54. Slukovskii, Z.I. Background concentrations of heavy metals and other chemical elements in the sediments of small lakes in the south of Karelia, Russia. *Vestn. MSTU* **2020**, *23*, 80–92. [CrossRef]
55. Slukovskii, Z.I.; Ilmast, N.V.; Sukhovskaya, I.V.; Borvinskaya, E.V.; Gogolev, M.A. The geochemical specifics of modern sedimentation processes on the bottom of a small Lake Lamba under technogenic impact. *Proc. Karelian Res. Cent. RAS* **2017**, *5*, 45. (In Russian) [CrossRef]
56. Slukovskii, Z.I. Accumulation level and fractions of heavy metals in sediments of small lakes of the urbanized area (Karelia). *Vestn. St. Petersburg Univ. Earth Sci.* **2020**, *65*, 171–192. (In Russian) [CrossRef]
57. Ganor, E.; Altshuller, S.; Foner, H.A.; Brenner, S.; Gabbay, J. Vanadium and nickel in dustfall as indicators of power plant pollution. *Water Air Soil Pollut.* **1988**, *42*, 241–252. [CrossRef]
58. Mejia, J.A.; Rodriguez, R.; Armienta, A. Aquifer Vulnerability Zoning, an Indicator of Atmospheric Pollutants Input? Vanadium in the Salamanca Aquifer, Mexico. *Water Air Soil Pollut.* **2007**, *185*, 95–100. [CrossRef]
59. Rybakov, D.S.; Krutskikh, N.V.; Shelekhova, T.S.; Lavrova, N.B.; Slukovskii, Z.I.; Krichevtsova, M.V.; Lazareva, O.V. *Climatic and Geochemical Aspects of Forming of Environmental Risks in the Republic of Karelia*; ElekSis: Saint Petersburg, Russia, 2013. (In Russian)
60. Komárek, M.; Ettler, V.; Chrastný, V.; Mihaljevi, M. Lead isotopes in environmental sciences: A review. *Environ. Int.* **2008**, *34*, 562–577. [CrossRef]
61. Thomas, V. The elimination of lead in gasoline. *Ann. Rev. Energy Environ.* **1995**, *20*, 301–324. [CrossRef]
62. Sayet, Y.E.; Revich, B.A.; Yanin, E.P.; Smirnova, R.S.; Basharkevich, I.L.; Onishchenko, T.L.; Pavlova, L.N.; Trefilova, N.Y.; Achkasov, A.I.; Sarkisyan, S.S. *Geochemistry of the Environment*; Nedra: Moscow, Russia, 1990; 335p. (In Russian)
63. Weiss, D.; Shotyk, W.; Appleby, P.G.; Cheburkin, A.K.; Kramers, J.D. Atmospheric Pb deposition since the Industrial Revolution recorded by five Swiss peat profiles: Enrichment factors, fluxes, isotopic composition, and sources. *Environ. Sci. Technol.* **1999**, *33*, 1340–1352. [CrossRef]
64. Krachler, M.; Zheng, J.; Koerner, R.; Zdanowicz, C.; Fisher, D.; Shotyk, W. Increasing atmospheric antimony contamination in the northern hemisphere: Snow and ice evidence from Devon Island, Arctic Canada. *J. Environ. Monit.* **2006**, *7*, 1169–1176. [CrossRef] [PubMed]
65. Medvedev, A.; Slukovskii, Z.; Novitcky, D. Heavy metals pollution of small urban lakes sediments within the Onego Lake catchment area. *Pol. J. Nat. Sci.* **2019**, *34*, 245–256.
66. Slukovskii, Z.I. Content of trace elements in bottom sediments of lakes as an indicator of appearance of environmental risks in condition of urbanization (the Republic of Karelia). *Water Manag. Russ.* **2018**, *6*, 70–82. (In Russian)
67. Slukovskii, Z.I.; Dauvalter, V.A. Features of Pb, Sb, Cd accumulation in sediments of small lakes in the south of the Republic of Karelia. *Trans. Karelian Res. Cent. Russ. Acad. Sci.* **2020**, *4*, 75–94. (In Russian) [CrossRef]
68. Potakhin, M.S.; Belkina, N.A.; Slukovskii, Z.I.; Novitsky, D.G.; Morozova, I.V. Change of bottom sediments of Vygozero as a result of multifactorial anthropogenic influence. *Soc. Environ. Dev.* **2018**, *3*, 107–117. (In Russian)
69. Slukovskii, Z.I.; Belkina, N.A.; Potakhin, M.S. Assessment of contamination of recent sediments of a large reservoir in the catchment area of Arctic Ocean, Northern Europe. *Pol. Polar Res.* **2021**, *42*, 25–43. [CrossRef]
70. Shevchenko, V.; Lisitzin, A.; Vinogradova, A.; Stein, R. Heavy metals in aerosols over the seas of the Russian arctic. *Sci. Total Environ.* **2003**, *306*, 11–25. [CrossRef]

71. Vinogradova, A.A.; Kotova, E.I.; Topchaya, V.Y. Atmospheric transport of heavy metals to the regions of the North of the European territory of Russia. *Geogr. Natl. Resour.* **2017**, *1*, 108–116. (In Russian) [CrossRef]
72. Guzeva, A.; Slukovskii, Z.; Dauvalter, V.; Denisov, D. Trace element fractions in sediments of urbanised lakes of the arctic zone of Russia. *Environ. Monit. Assess.* **2021**, *193*, 378. [CrossRef] [PubMed]
73. Postevaya, M.A.; Slukovskii, Z.I. Analysis of Atmospheric Emissions in Murmansk and Their Relationship with Pollution of Urban Lakes. *Vestn. MSTU* **2021**, *24*, 190–201. (In Russian) [CrossRef]
74. Postevaya, M.A.; Slukovskii, Z.I.; Dauvalter, V.A.; Bernadskaya, D.S. Estimation of Heavy Metal Concentrations in the Water of Urban Lakes in the Russian Arctic (Murmansk). *Water* **2021**, *13*, 3267. [CrossRef]
75. Slukovskii, Z.; Guzeva, A.; Dauvalter, V. Vanadium as an indicator of the impact of fuel oiled thermal power plants on the environment: Paleolimnological reconstructions. *Limnol. Freshw. Biol.* **2020**, *4*, 513–514. [CrossRef]
76. Minin, V.A. Heat supply of the cities of the Murmansk region. *Proc. Kola Sci. Cent. RAS* **2014**, *3*, 68–76. (In Russian)
77. Kuwae, M.; Tsugeki, N.K.; Agusa, T.; Toyoda, K.; Tani, Y.; Ueda, S.; Tanabe, S.; Urabe, J. Sedimentary records of metal deposition in Japanese alpine lakes for the last 250 years: Recent enrichment of airborne Sb and In in East Asia. *Sci. Total Environ.* **2013**, *442*, 189–197. [CrossRef]
78. Slukovskii, Z.I.; Guzeva, A.V.; Dauvalter, V.A. Rare earth elements in surface lake sediments of Russian arctic: Natural and potential anthropogenic impact to their accumulation. *Appl. Geochem.* **2022**, *142*, 105325. [CrossRef]
79. Faruque, A.M.; Hawa, B.H.I. Environmental assessment of Dhaka city (Bangladesh) based on trace metal contents in road dusts. *Environ. Geol.* **2007**, *51*, 975–985. [CrossRef]
80. Yuan, Y.; Cave, M.; Zhang, C. Using Local Moran's I to identify contamination hotspots of rare earth elements in urban soils of London. *Appl. Geochem.* **2018**, *88*, 167–178. [CrossRef]
81. Slukovskii, Z.I.; Dauvalter, V.A.; Denisov, D.B.; Siroezhko, E.V.; Cherepanov, A.A. Geochemistry features of sediments of small urban arctic Lake Komsomolskoye, Murmansk region. *IOP Confer. Series Earth Environ. Sci.* **2020**, *467*, 012004. [CrossRef]
82. Slukovskii, Z.; Medvedev, M.; Mitsukov, A.; Dauvalter, V.; Grigoriev, V.; Kudryavtzeva, L.; Elizarova, I. Recent geochemistry of Arctic small lakes (Russia): Geochemistry features and age. *Environ. Earth Sci.* **2021**, *80*, 302. [CrossRef]
83. Ivanov, D.V.; Osmelkin, E.V.; Ziganshin, I.I. A study of contemporary and historical sedimentation in waterbodies of the Volga upland and the low-lying trans-Volga region. *Proc. Karelian Res. Cent. RAS* **2018**, *9*, 31–43. (In Russian) [CrossRef]
84. Ivanov, D.V.; Ziganshin, I.I.; Osmelkin, E.V. Regional background concentrations of metals in bottom sediments of lakes of the Republic of Tatarstan. *Proc. Kazan State Univ.* **2010**, *152*, 185–191. (In Russian)
85. Valiev, V.S.; Ivanov, D.V.; Shagiullin, R.R.; Ziganshin, I.I.; Shamaev, D.E.; Khasanov, R.R. Types of distribution of pollutants in the water and bottom sediments of middle and lower Volga. *Water Manag. Russ.* **2017**, *2*, 94–107. (In Russian)
86. Ivanov, D.V. Bottom sediments of the Middle Kaban lake (Kazan, Russia). *Georesources* **2012**, *7*, 1–6. (In Russian)
87. McConnell, J.R.; Edwards, R. Coal burning leaves toxic heavy metal legacy in the Arctic. *Proc. Natl. Acad. Sci. USA* **2008**, *105*, 12140–12144. [CrossRef] [PubMed]
88. Udachin, V.N.; Bolshiyanov, D.Y.; Votyakov, S.L.; Kiseleva, D.V.; Khvorov, P.V.; Amincv, P.G.; Ivanov, Y.K. First data about geochemistry of bottom sediments of an artic lake Kenturku (Taimyr Peninsula). *Proc. IGG UB RAS* **2013**, *160*, 356–359. (In Russian)
89. Moiseenko, T.I.; Dinu, M.I.; Gashkina, N.A.; Kremleva, T.A.; Khoroshavin, V.Y. Geochemical features of elements distributions in the lake waters of the Arctic region. *Geochem. Int.* **2020**, *58*, 613–623. [CrossRef]
90. Tatsii, Y.G.; Moiseenko, T.I.; Razumovskii, L.V.; Borisov, A.P.; Khoroshavin, V.Y.; Baranov, D.Y. Bottom sediments of the West Siberian Arctic lakes as indicators of environmental changes. *Geochem. Int.* **2020**, *58*, 408–422. [CrossRef]
91. Denisov, D.B. The diatom-infer small subarctic waterbody ecosystem development reconstruction during the last 900 years (Akademicheskoye Lake, the Khibiny, the Kola peninsula). *Proc. Kola Sci. Cent. RAS* **2012**, *10*, 127–148. (In Russian)
92. Moiseenko, T.I.; Dauvalter, V.A.; Kagan, L.Y. Mountain lakes as an indicators of air pollution. *Water Resour.* **1997**, *24*, 600–608. (In Russian)
93. Dauvalter, V.; Kashulin, N. Chalcophile elements (Hg, Cd, Pb, As) in Lake Umbozero, Murmansk Province. *Water Resour.* **2010**, *37*, 497–512. [CrossRef]
94. Dauvalter, V.A.; Terentiev, P.M.; Denisov, D.B.; Udachin, V.N.; Filippova, K.A.; Borisov, A.P. Reconstruction of pollution of the territory of the Rybachy Peninsula of the Murmansk region with heavy metals. *Proc. Fersman Sci. Sess. Geolog. Inst. KSC RAS* **2018**, *15*, 441–444. (In Russian) [CrossRef]
95. Belkina, N.A.; Subetto, D.A.; Efremenko N.A.; Kulik, N.V. Peculiarities of Trace Elements Distribution in the Surface Layer of Sediments of Lake Onega. *Sci. Educ.* **2016**, *3*, 135–139. (In Russian)
96. Starodymova, D.P.; Shevchenko, V.P.; Kokryatskaya, N.M.; Aliev, R.A.; Bychkov, A.Y.; Zabelina, S.A.; Chupakov, A.V. Geochemistry of bottom sediments of a small lake (drainage area of Lake Onega, Arkhangelsk region). *Succes. Mod. Natl. Sci.* **2016**, *9*, 172–177. (In Russian)
97. Slukovskii, Z.; Medvedev, M.; Siroezhko, E. Long-range transport of heavy metals as a factor of the formation of the geochemistry of sediments in the southwest of the Republic of Karelia, Russia. *J. Elementol.* **2020**, *25*, 125–137. [CrossRef]
98. Slukovskii, Z.; Medvedev, M.; Siroezhko, E. The environmental geochemistry of recent sediments of small lakes in the southwest of Karelia, Russia. *Environ. Eng. Manag. J.* **2020**, *19*, 1043–1055. [CrossRef]
99. Virkutyte, J.; Vadakojyte, S.; Sinkevičius, S.; Sillanpää, M. Heavy metal distribution and chemical partitioning in Lake Saimaa (SE Finland) sediments and moss Pleurozium Schreberi. *Chem. Ecol.* **2008**, *24*, 119–132. [CrossRef]

100. Strakhovenko, V.D.; Belkina, N.A.; Efremenko, N.A.; Subetto, D.A.; Potakhin, M.S. Basic features of distribution of trace elements and phosphorus in the bottom sediments of Lake Onego in the Holocene (based on time graphs 137Cs, 210Pb). In Proceedings of the Russian Scientific Conference "Geological Evolution of Interaction Water with Rocks", Ulan-Ude, Russia, 17–20 August 2020. (In Russian). [CrossRef]
101. Strakhovenko, V.D.; Kabannik, V.G.; Malikova, I.N. Geochemical peculiars of ecosystem of Lake Kolyvanovskoe (Altai region) and influence of technogenesis on it. *Lithol. Miner.* **2014**, *3*, 220–234. (In Russian) [CrossRef]
102. Maltsev, A.E.; Leonova, G.A.; Bobrov, V.A.; Krivonogov, S.K. Geochemistry of Holocene core of sapropel of Lake Bolshie Toroki (Novosibirsk region). In Proceedings of the VII Russian Lithological Conference, Novosibirsk, Russia, 28–31 October 2013. (In Russian).
103. Shevchenko, V.P.; Starodymova, D.P.; Vorobyev, S.N.; Aliev, R.A.; Borilo, L.P.; Kolesnichenko, L.G.; Lim, A.G.; Osipov, A.I.; Trufanov, V.V.; Pokrovsky, O.S. Trace Elements in Sediments of Two Lakes in the Valley of the Middle Courses of the Ob River (Western Siberia). *Minerals* **2022**, *12*, 1497. [CrossRef]
104. Maltsev, A.E.; Leonova, G.A.; Bobrov, V.A.; Krivonogov, S.K. *Geochemistry of Holocene Sapropels from Small Lakes of the Southern Western Siberia and Eastern Baikal Regions*; Geo: Novosibirsk, Russia, 2019; p. 444. (In Russian)
105. Wedepohl, K.H. The composition of the continental crust. *Geochim. Cosmochim. Acta* **1995**, *59*, 1217–1232. [CrossRef]

**Disclaimer/Publisher's Note:** The statements, opinions and data contained in all publications are solely those of the individual author(s) and contributor(s) and not of MDPI and/or the editor(s). MDPI and/or the editor(s) disclaim responsibility for any injury to people or property resulting from any ideas, methods, instructions or products referred to in the content.

Article

# Different Adsorption Behaviors and Mechanisms of Anionic Azo Dyes on Polydopamine–Polyethyleneimine Modified Thermoplastic Polyurethane Nanofiber Membranes

Jiaoxia Sun *, Yao Zhou, Xueting Jiang and Jianxin Fan

School of River and Ocean Engineering, Chongqing Jiaotong University, Chongqing 400074, China
* Correspondence: sjx@cqu.edu.cn

**Abstract:** Considering the notable mechanical properties of thermoplastic polyurethane (TPU), polydopamine–polyethyleneimine (PEI) -modified TPU nanofiber membranes (PDA/PEI-TPU NFMs) have been developed successfully for removal of anionic azo dyes. The adsorption capacity of PDA/PEI-TPU NFMs was evaluated using three anionic dyes: congo red (CR), sunset yellow (SY), and methyl orange (MO). Interestingly, it exhibited different adsorption behaviors and mechanisms of CR on PDA/PEI-TPU NFMs compared with SY and MO. With the decrease in pH, leading to more positive charges on the PDA/PEI-TPU NFMs, the adsorption capacity of SY and MO increased, indicating electrostatic interaction as a main mechanism for SY and MO adsorption. However, wide pH range adaptability and superior adsorption have been observed during the CR adsorption process compared to SY and MO, suggesting a synergistic effect of hydrogen bonding and electrostatic interaction, likely as a critical factor. The adsorption kinetics revealed that chemical interactions predominate in the CR adsorption process, and multiple stages control the adsorption process at the same time. According to the Langmuir model, the maximum adsorption capacity of CR, SY and MO were reached 263, 17 and 23 mg/g, respectively. After six iterations of adsorption–desorption, the adsorption performance of the PDA/PEI-TPU NFMs did not decrease significantly, which indicated that the PDA/PEI-TPU NFMs have a potential application for the removal of CR molecules by adsorption from wastewater.

**Keywords:** polyethyleneimine; polydopamine; thermoplastic polyurethane; nanofiber membrane; anionic azo dyes; adsorption

**Citation:** Sun, J.; Zhou, Y.; Jiang, X.; Fan, J. Different Adsorption Behaviors and Mechanisms of Anionic Azo Dyes on Polydopamine–Polyethyleneimine Modified Thermoplastic Polyurethane Nanofiber Membranes. *Water* **2022**, *14*, 3865. https://doi.org/10.3390/w14233865

Academic Editor: Weiying Feng

Received: 17 October 2022
Accepted: 24 November 2022
Published: 27 November 2022

**Publisher's Note:** MDPI stays neutral with regard to jurisdictional claims in published maps and institutional affiliations.

**Copyright:** © 2022 by the authors. Licensee MDPI, Basel, Switzerland. This article is an open access article distributed under the terms and conditions of the Creative Commons Attribution (CC BY) license (https://creativecommons.org/licenses/by/4.0/).

## 1. Introduction

Dye contamination has become one of the most severe aspects of global water pollution [1,2]. Azo dyes account for more than two-thirds of all synthetic dyestuffs [3]. Presumably, at least 2000 different types and over 700,000 tons of azo dyes have been produced annually worldwide for the textile dyeing, pharmaceutical and paper industries [4]. However, about 10–15% of these azo dyes are unutilized and discharged into the environment through wastewater [5]. Congo red (CR), sunset yellow (SY), and methyl orange (MO) are three typical anionic dyes, which have been widely applied in the experimental analysis, food processing, plastics, rubber, printing and optoelectronics industries [6–8]. Under special circumstances, these azo dyes can be bio-transformed to release aromatic amines with toxicity, mutagenicity and persistence, which can alter DNA transcription, lead to lesions and bring on cancer once they are ingested by animals [9]. Hence, it is required to treat azo dye wastewater before its final discharge into water bodies.

Undoubtedly, adsorption technology has been widely used for the removal of dye stuffs, owing to its simplicity, high efficiency and low cost [10]. In recent decades, a multitude of adsorption nanoscale-materials, including carbon-based nanomaterials [11], nanocomposites [12] and nanofiber membranes [13,14], have been developed to remove organic dyestuffs. Among them, nanofiber membranes have attracted substantial interest for the removal of micropollutants due to their very large specific surface areas, high porosity and highly controlled surface characteristics [15]. Compared with other materials dispersed in the wastewater during the water treatment process, the nanofiber membrane, after adsorption, can be taken out from the wastewater as a whole or directly used as an adsorption filter membrane, thus avoiding the secondary removal of the adsorbed materials and improving the treatment efficiency. However, most of the polymer nanofiber membranes have poor mechanical properties that restrict their universal practicability [16,17].

Polyurethane (PU) is an engineered polymer material that is widely used in medical dressings, architectural coatings and fabric production. Some prominent features, such as durability, tensile fatigue resistance, shear resistance and elasticity are worth mentioning [18,19]. Therefore, it is possible that the mechanical properties of the nanofiber membranes prepared by PU can be improved compared to other membrane materials. In recent years, there has been an increasing amount of literature related to the application of PU foam-based adsorbents for oil–water separation [20–22], heavy metals [23,24] and dye [25,26] removal, which shows that this material has great potential for pollutant removal. As adsorbents, the external functional groups on PU play a vital role in adsorption, influencing the adsorption mechanisms. The molecule dopamine (DA) can securely connect to substrates by self-polymerization to produce a polydopamine (PDA) coating under mild alkaline conditions [27]. Therefore, DA has been used extensively in surface modification of the material to remove organic dyestuffs from polluted water [28,29]. Polyethyleneimine (PEI) is a typical hyperbranched polymer with excellent adsorption ability for organic dyes, as its polymer chains contain a large quantity of primary and secondary amine groups [30–32]. Functional PDA-PEI coating can be constructed by PEI-modified PDA coating, based on Michael addition or Schiff base reaction [33].

As our goal was to removal anionic azo dyes, TPU nanofiber membranes (TPU NFMs) were used, as they are soft and stretch-resistant, which can make up for the shortcomings of traditional film materials. PDA/PEI co-deposition can occur under mild conditions. This represented an easier, greener and less expensive material preparation process. Thus, we selected DA and PEI to construct a functional PEI-PDA coating in order to modify the TPU NFMs. We hypothesized that the modified TPU NFMs would exhibit good stability and adsorption capacity for anionic azo dyes. The TPU nanofiber membranes were, firstly, prepared by electrospinning. Then, TPU NFMs were modified by PDA/PEI co-deposition to obtain PDA/PEI-TPU NFMs. The membranes were characterized by X-ray photoelectron spectrometer (XPS), contact angle instruments, Fourier transform infrared spectrometer (FTIR) and scanning electron microscopy (SEM). Three typical anionic azo dyes, including CR, SY and MO, were selected as the main adsorption objects to explore the adsorption behavior of the prepared PDA/PEI-TPU NFMs.

## 2. Experimental Sections

### 2.1. Chemicals and Materials

All chemicals were used as obtained without additional purification: polyethyleneimine (PEI, $M_W$ = 10,000, 30% solution in water) and dopamine hydrochloride (DA, 98%) were purchased from J&K Scientific Co., Ltd. (Beijing, China). TPU powder was provided by BASF Co., Ltd. (Shanghai, China). N,N-Dimethylformamide (DMF, $C_3H_7NO$), acetone ($C_3H_6O$, AR), aqueous hydrochloric acid (HCl, AR), sodium hydroxide (NaOH, AR), congo red ($C_{32}H_{22}N_6Na_2O_6S_2$, AR) and methyl orange ($C_{14}H_{15}N_3NaO_2S$, AR) were obtained from Chuandong Chemicals Co., Ltd. (Chongqing, China). Sunset yellow ($C_{16}H_{10}N_2Na_2O_7S_2$, AR) and Tris(hydroxymethyl)aminomethane ($C_4H_{11}NO_3$, AR) were obtained from Chengdu Chron Chemicals Co., Ltd. (Chengdu, China).

### 2.2. Synthesis of TPU and PDA/PEI-TPU NFMs

Figure 1 depicts the adsorptive PDA/PEI-TPU NFMs preparation schematic. First, TPU NFMs were synthesized by the electrospinning method. A specific quantity of TPU powder was added to acetone (DMF ($v:v$ = 80:20)) and stirred for 12 h to prepare 15 wt%

TPU spinning solutions. The prepared electrospinning solution was injected into a 5 mL plastic syringe fitted with a steel spinneret (inner diameter: 0.25 mm), and electrospun for 5 h. Other optimized electrospun conditions were voltage (15 kV), injection speed rate (1 mL/h) and receiving distance (10 cm) [34]. The resulting TPU NFMs were cleaned with ethanol and vacuum-dried to use as the support membrane for subsequent modification.

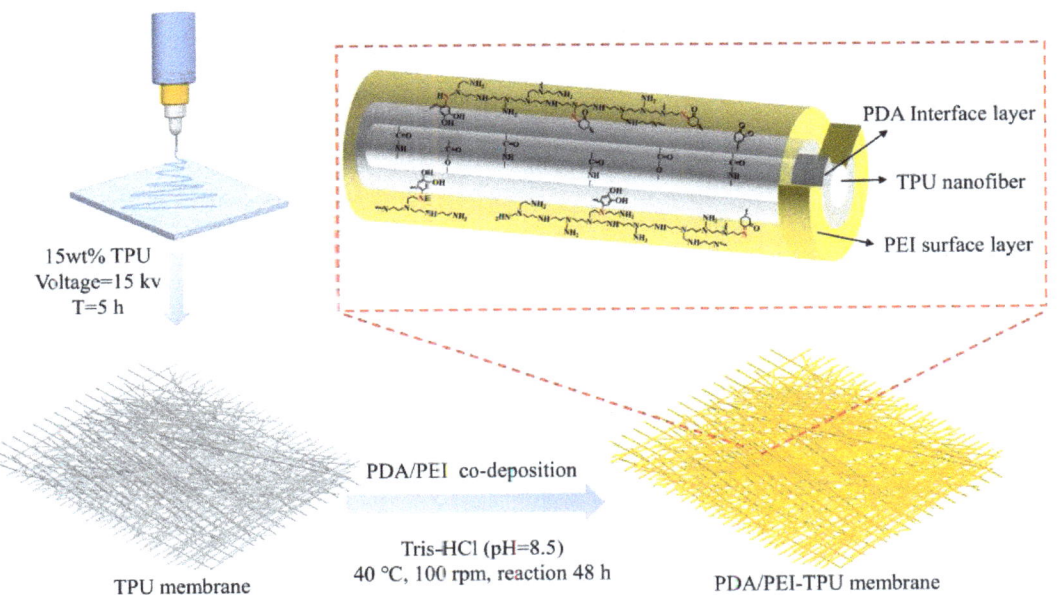

**Figure 1.** Schematic synthetic routes of the PDA/PEI-TPU NFMs.

Then, 0.2 g of DA and 0.2 g of PEI were dissolved in 100 mL of Tris buffer solution (pH = 8.5, 50 mM), and approximately 1.0 g of cleaned TPU NFMs were completely immersed in this DA/PEI mixture. The co-aggregation process was implemented at ambient temperature while being continuously shaken at 100 rpm for 12, 24, 36, 48, and 60 h. The time of co-aggregation affected the degree of PDA/PEI grafting on the surface of TPU NFMs. As the reaction proceeded, particles were gradually deposited on the surface, resulting in a darker color of the membranes. Pre-experiments also showed that the maximum adsorption capacity of dyes was found to be reached in the material co-deposited for 48 h, and as the deposition time increased, it did not significantly improve the adsorption of dyes. Therefore, the deposition time of this experiment was chosen to be 48 h. In addition, PDA/TPU NFMs were prepared as comparative adsorbents by adding 0.2 g of DA and 1.0 g of cleaned TPU NFMs to 100 mL of Tris buffer solution. Other conditions were consistent. After the co-aggregation, the obtained PDA-TPU NFMs and PDA/PEI-TPU NFMs were taken out, soaked and cleaned with deionized water for 24 h in order to remove unreacted substances on the surface. They were vacuum-dried at ambient temperature, then stored for subsequent characterization and measurement. Figure 1 also further shows the molecular structure of the co-aggregation reactions involving DA and PEI. Under alkaline conditions, DA oxidation auto-polymerization produces reactive sites, which react with the positively charged PEI molecules to form a stable cross-linked structure by Michael addition and Schiff base reaction [35].

## 2.3. Characterization of Adsorbents

The surface morphology of the membranes was examined by a scanning electron microscope (SEM, INSPECT F50, MA, USA). The surface chemical composition of the membranes was performed on X-ray photoelectron spectroscopy (XPS, Thermo Scientific K-Alpha, MA, USA). The functional groups of the membranes were obtained by FT-IR spectrometer (Thermo-Nicolet 670, MA, USA) equipped with an attenuated total reflection (ATR) unit. The static contact angles of the membranes were characterized by contact angle system (HARKE-SPCA-X3, Beijing, China). The concentration of dyes was quantitatively measured on an ultraviolet spectrophotometer (UV-vis-3150, Kyoto, Japan) from 600 to 200 nm, using deionized water for the background correction.

The net weight gain of the modified membranes was calculated by Equation (1):

$$W_g = \frac{m_1 - m_0}{m_0} \times 100\% \tag{1}$$

where $m_0$ (g) is the constant weight of pre-modified TPU NFMs, and $m_1$ (g) is the constant weight of the membranes after PDA and PEI modification.

## 2.4. Batch Adsorption Experiments

The initial and residual concentration of dyes were measured by a UV–vis spectrophotometer at the maximum wavelengths of ($\lambda_{max}$) 497 nm, 482 nm and 464 nm for CR, SY and MO, respectively. The adsorption capacity was calculated by Equation (2). Three parallel adsorption experiments were repeatedly carried out, and the average value was used to determine the final result.

$$Q = \frac{(C_0 - C_t)V}{m} \tag{2}$$

where m (g) is the weight of membrane, $C_0$ (mg/L) is the initial concentration, $C_t$ (mg/L) represents the residual concentrations at different time intervals, respectively, and V (L) is the volume of the dye solution.

### 2.4.1. The Impact of Initial Solution pH

The influences of original solution pH, which ranged from 4.0 to 10.0 on the capture of three azo dyes (CR, SY and MO) by PDA/PEI-TPU NFMs, were further investigated. The initial concentrations of CR, SY and MO were 100 mg/L each. Constant shaking was performed at room temperature, with 20 mg of adsorbents in 50 mL of CR solution, 20 mL of SY solution and 20 mL of MO solution, for 24 h. In the experiment, 1 M HCl or 1 M NaOH solution was used to adjust the pH.

### 2.4.2. Equilibrium Isotherms

The effect of initial concentrations of three dyes onto PDA/PEI-TPU NFMs were investigated under neutral conditions by isotherm adsorption experiment. The initial concentration of CR ranged from 50–400 mg/L, while SY and MO ranged from 5–100 mg/L. All the experiments were performed by adding 20 mg of PDA/PEI-TPU NFMs into 20 mL of dyes solution, and then shaking at 150 rpm for 24 h at room temperature to achieve adsorption equilibrium. Moreover, Langmuir and Freundlich adsorption isotherm models (Equations (3) and (4)) were implemented in order to simulate the experimental data and to define the maximum dye adsorption capacity of the PDA/PEI-TPU NFMs [36].

$$\text{Langmuir}: q_e = \frac{q_{max} K_L C_e}{1 + K_L C_e} \tag{3}$$

$$\text{Freundlich}: q_e = K_F C_e^{\frac{1}{n}} \tag{4}$$

In the present equation, $q_{max}$ (mg/g) is the maximum equilibrium adsorption capacity of the PDA/PEI-TPU NFMs; $K_L$ (L/mg) is Langmuir binding constant; $C_e$ (mg/L) and $q_e$

(mg/g) are the equilibrium concentration and adsorbent uptake capacity at equilibrium, respectively; and $K_f$ ($(mg/g)/(mg/L)^n$) and n are the empirical Freundlich constant and heterogeneity factor, respectively.

2.4.3. Kinetic Adsorption

The effect of different contact times of three dyes adsorbed onto PDA/PEI-TPU NFMs, from 0 to 48 h and under neutral conditions, were investigated in order to better understand the adsorption process. The initial dye concentration of CR was 300 mg/L; for both SY and MO, the concentration was 100 mg/L. Then, 200 mg of PDA/PEI-TPU NFMs were added to 200 mL of dye solution, and then shaken at 150 rpm. In order to further examine the adsorption process, the pseudo-first-order model, pseudo-second-order model and intra-particle diffusion model were used to analyze the kinetic data (Equations (5)–(7)) [37].

$$\text{Pseudo-first-order model}: q_t = q_e\left(1 - e^{-k_1 t}\right) \tag{5}$$

$$\text{Pseudo-second-order model}: q_t = \frac{k_2 q_e^2 t}{1 + k_2 q_e t} \tag{6}$$

$$\text{Intraparticle diffusion model}: q_t = k_i \times t^{\frac{1}{2}} + b \tag{7}$$

Herein, $q_e$ (mg/g) and $q_t$ (mg/g) are the adsorption capacity at the equilibrium time and the desired time, respectively. $k_1$ (1/h), $k_2$ (g/(mg h)) and $k_i$ (mg/(g h$^{1/2}$)) are the pseudo-first-order model rate constant, the pseudo-second-order model rate constant and the intraparticle diffusion rate constant, respectively, and b is the intercept of the linear portion of the equation.

2.4.4. Desorption Experiments

A desorption experiment was set up to evaluate the recycling performance of materials. In the experiment, 0.05 M NaOH solutions were employed as desorption solutions. The PDA/PEI-TPU NFMs, after reaching adsorption equilibrium, were taken out and placed in 20 mL of 0.05 M NaOH solution for ultrasonic inspection for 60 min, and then the CR concentration in the desorption solution was measured. The PDA/PEI-TPU NFMs, after desorption, were collected and rinsed sufficiently. Finally, they were dried in an oven before the next cycle. The adsorption capacity and desorption capacity were calculated using Equation (2).

# 3. Results and Discussion

## 3.1. The Grafted Yield and Stability of PDA/PEI-TPU NFMs

The grafted yields of PDA and PDA/PEI on TPU NFMs were calculated by the constant weight method. The weight loss of the membranes after ultrasonic treatment was also studied to roughly evaluate the stability of PDA-TPU NFMs and PDA/PEI-TPU NFMs. It can be seen in Figure 2a that when modified only by PDA, the grafted yield of the PDA-TPU NFMs was 5.7%. By adding PEI, the grafted yield of the PDA/PEI-TPU NFMs increased notably, to 10.2%. As shown in Figure 2b, no weight loss of PDA-TPU NFMs or PDA/PEI-TPU NFMs was observed even, after being ultrasonically treated at a power of 200 W for 60 min, demonstrating the high stability of PDA-TPU NFMs and PDA/PEI-TPU NFMs. Obviously, the PDA/PEI-TPU NFMs have good adsorption stability and a possibility of long-term application due to their high adhesion stability.

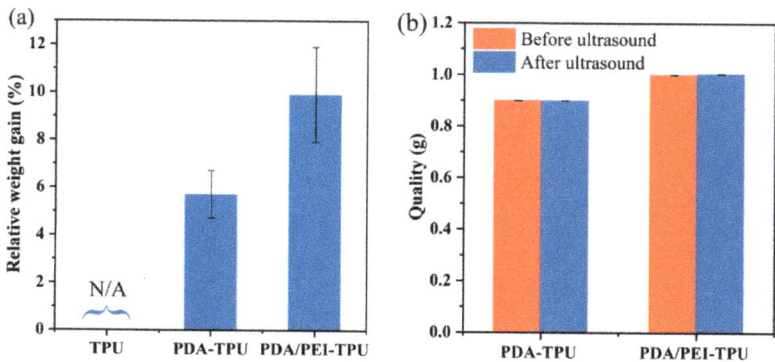

**Figure 2.** (**a**) The grafted yield of the membranes after modification. (**b**) Stability of modified membranes after ultrasound.

### 3.2. Morphology and Microstructure of the Nanofiber Membranes

The morphology and microstructure of the electrospun TPU NFMs, PDA-TPU NFMs and PDA/PEI-TPU NFMs were analyzed by SEM. As shown in Figure 3a,b, we can see that the pure TPU nanofibers exhibited quite smooth surfaces and relatively slim fiber contours, the average diameter of which was uniformly distributed at about 160 nm. After being coated by a PDA layer (Figure 3d,e), some nanoparticles were deposited on the surface, resulting in a rougher surface, and the average diameter of PDA-TPU nanofibers was slightly increased to 263 nm. Figure 3g,h show the morphology and microstructure of PDA/PEI-TPU NFMs. Obviously, PDA/PEI-TPU NFMs have a much different morphology compared with the TPU and PDA-TPU NFMs. A rougher fiber surface and larger average diameter (502 nm) were observed, which indicated significant adhesion of PDA/PEI cross-linked aggregates on the modified membrane surface. Some studies have reported that DA is able to trigger oxidative self-polymerization reactions in a weakly alkaline environment, leading to homogeneous coating on any polymer membranes or nanofibers [38–40]. PEI and DA are covalent cross-linked, based on both Michael addition and Schiff reaction, for forming robust aggregates on the TPU NFMs surface. In addition, PEI has a large number of cationic amino groups. While the intermediate product of PDA, catechol, has a strong conjugation system, the cation-π interaction between them will enhance the internal cohesion of co-aggregated coating, thus improving the PEI grafted amounts on the surface of TPU NFMs [39].

The ATR-FTIR was used to study the surface functional groups of the TPU NFMs, PDA-TPU NFMs and PDA/PEI-TPU NFMs; the results are shown in Figure 4. The typical characteristic absorption bands of TPU NFMs at 3355, 1727 and 1061 cm$^{-1}$ were assigned to the N-H stretching vibration, C = O stretching vibration and C-O-C stretching vibration, respectively [41]. After being deposited only with PDA, the intensity of a broad characterization absorption band at 3100–3650 cm$^{-1}$ increased compared with TPU NFMs, which can be attributed to the stretching vibration of the O-H and N-H groups. For the PDA/PEI-TPU NFMs, the new absorption band was observed at 1659 cm$^{-1}$, and can be ascribed to the formation of C = N bonds stretching vibrations between PEI and PDA [33]. The creation of a C = N bond suggested that PEI was successfully grafted onto TPU NFMs surface by reacting with the active site generated through DA oxidative self-polymerization.

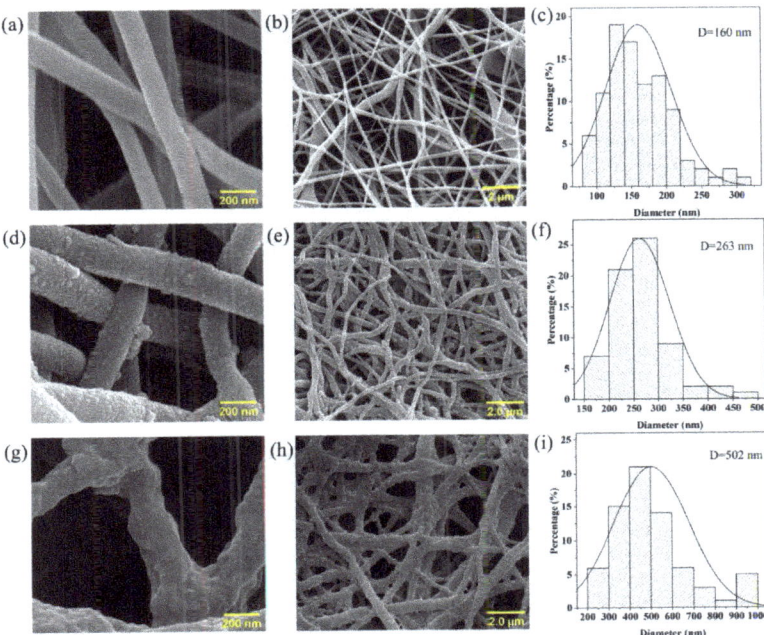

**Figure 3.** SEM images and nanofiber diameter distribution of (**a–c**) TPU NFMs, (**d–f**) PDA-TPU NFMs and (**g–i**) PDA/PEI-TPU NFMs.

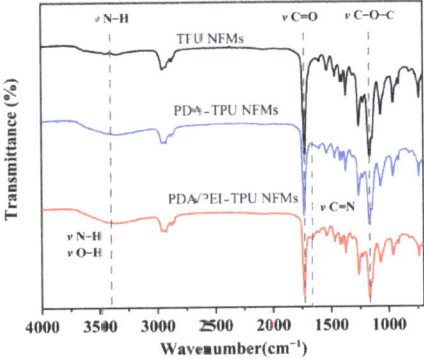

**Figure 4.** ATR-FTIR spectra of the TPU NFMs, PDA-TPU NFMs and PDA/PEI-TPU NFMs.

XPS investigation further showed the elemental compositions and chemical combined state of TPU NFMs, PDA-TPU NFMs and PDA/PEI-TPU NFMs. From the XPS full spectrum (Figure 5a) and the elemental percentages of C, O and N (Table 1), it can be clearly observed that the elemental N content in both PDA-TPU NFMs and PDA/PEI-TPU NFMs are increased compared with TPU NFMs, but the PEI modification has a considerably higher amount of the N element. This is mainly attributed to a small amount of terminal amino group in the PDA and the great quantity of amino groups in PEI. Meanwhile, the high-resolution spectrum of the element N in membranes is analyzed. The core spectral curves of N1s, fitted by XPS Peak software, are shown in Figure 5b,d. The original TPU NFMs surface contained few C-N structures due to the urethane and urea ester bonds in TPU NFMs [42,43]. After introducing PDA, a significant increase in the peak intensity of the C-N structure occurred. While PEI introduced the deposition process, the C-N structure intensity on the surface of PDA/PEI-TPU NFMs also dramatically increased (Figure 5d). A new peak also appeared, located in 402.1 eV, which was attributed to the C = N structures formed by PEI and DA [39]. These results further indicate that the co-aggregation process occurred successfully in the TPU NFMs.

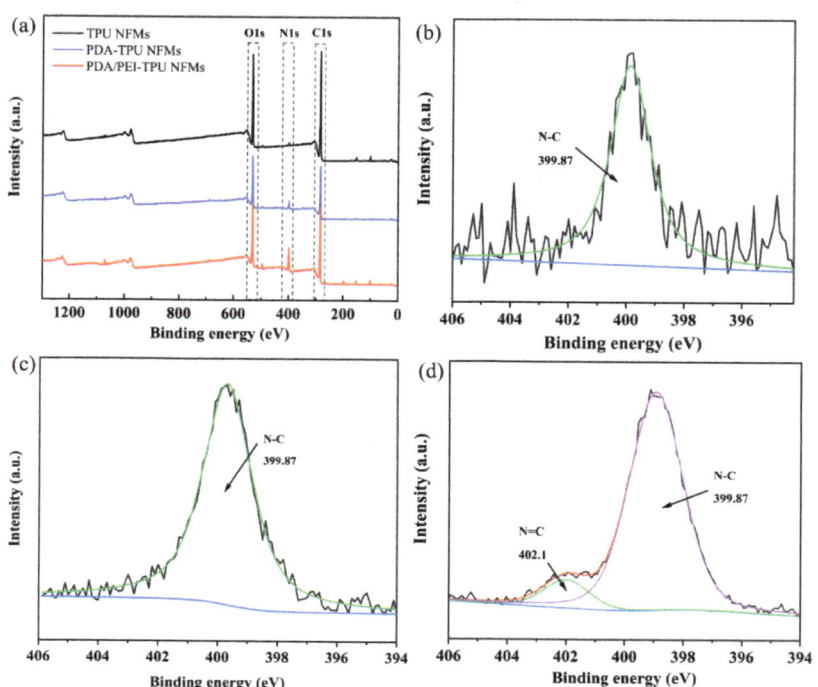

**Figure 5.** (a) XPS wide spectra and high-resolution XPS N1s spectra of (b)TPU NFMs, (c) PDA-TPU NFMs and (d) PDA/PEI-TPU NFMs.

**Table 1.** Percentage of surface element composition of the nanofiber membranes analyzed by XPS.

| Membranes | C1s | O1s | N1s | N/C Ratio | N/O Ratio |
|---|---|---|---|---|---|
| TPU NFMs | 71.77 | 26.81 | 1.42 | 0.02 | 0.052 |
| PDA-TPU NFMs | 68.78 | 25.71 | 5.51 | 0.08 | 0.21 |
| PDA/PEI-TPU NFMs | 66.46 | 22.35 | 11.19 | 0.168 | 0.500 |

PDA and PEI with hydrophilic properties, when introduced onto the PDA/PEI-TPU NFMs, can improve the hydrophilicity of the membrane surface. Therefore, contact angle measurement was used to confirm the change in hydrophilicity of the membrane surface. The images of water drop on the nanofiber membrane surfaces are shown in Figure 6. Within 3 s, the water contact angle of TPU NFMs changed insignificantly, while PDA-TPU NFMs and PDA/PEI-TPU NFMs showed better water permeability, with the water contact angle decreasing from 31.1° and 19.5° at the beginning to less than 10°. These phenomena indicate that the presence of PDA and PEI greatly improves the hydrophilicity of the PDA-TPU NFMs and PDA/PEI-TPU NFMs compared to TPU NFMs. This mainly contributed to the abundance of polar groups in PDA that can attract water molecules, as well as the many -$NH^+$ and -$NH_2$ groups induced by co-aggregation of PEI and PDA. The enhanced hydrophilicity of the PDA/PEI-TPU NFMs facilitates the removal of contaminants by this adsorbent in the aqueous environment.

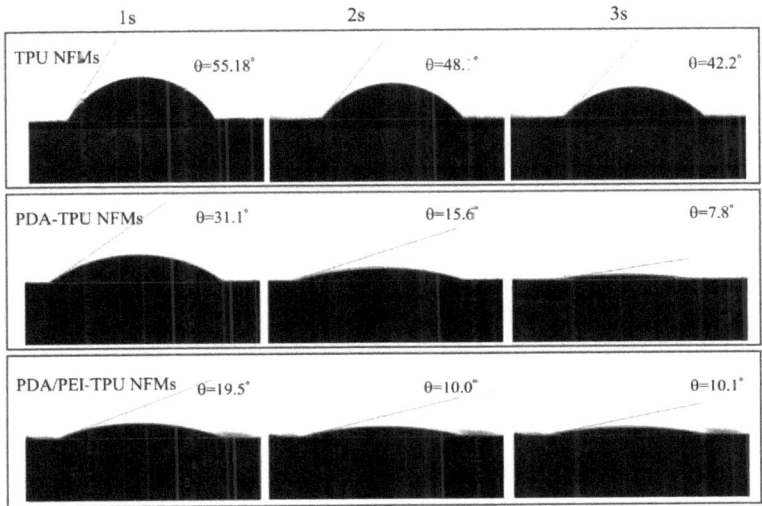

**Figure 6.** The contact angle of (**up**) TPU NFMs, (**middle**) PDA-TPU NFMs and (**bottom**) PDA/PEI-TPU NFMs.

### 3.3. Dye Adsorption Ability

In order to study the adsorption performance of the membranes (TPU NFMs, PDA-TPU NFMs and PDA/PEI-TPU NFMs) towards various anionic azo dyes, we chose three classical anionic dyes (CR, SY and MO) as model molecules (Figure 7). Compared with original TPU NFMs and PDA-TPU NFMs, the adsorption capacities of CR, SY and MO on to PDA/PEI-TPU NFMs were significantly enhanced, which indicated that cationic properties of PEI modification are an effective way to improve the adsorption ability for anionic dyes. Obvious differences were also found in the adsorption capacities of different azo dyes onto PDA/PEI-TPU NFMs. The adsorption ability of CR reached 232.55 mg/g, while for SY and MO, the values were, respectively, only 16.58 and 15.57 mg/g under the same neutral conditions, far less than CR. In addition, the pure TPU NFMs and PDA-TPU NFMs exhibited some adsorption for CR, but almost no adsorption for SY and MO. The different adsorption behavior may be attributed to the different adsorption mechanisms for CR and SY versus MO. The high adsorption capacity for CR also suggests that PDA/PEI-TPU NFMs have great potential in the treatment of CR-dyed wastewater.

Further, we investigated the effect of the solution pH on adsorption capacity of PDA/PEI-TPU NFMs for three dyes. The initial solution pH is very important for the adsorption process, as pH determines the degree of protonation of functional groups, as well as competitive adsorption between contaminants and hydrogen ions [44]. Since CR, MO and SY have good solubility in the pH range of 4–10, this pH range was chosen to study the adsorption performance of PDA/PEI-TPU NFMs. Figure 8 shows the significant differences in the adsorption of CR, SY and MO in the pH range of 4–10. CR has a wide pH adaptability, with more than 235 mg/g adsorption capacity and 95% removal efficiency in the pH 4–8. Even at pH 10, these values can reach 110 mg/g and 41%, respectively. In the determined pH range, the adsorption capacity of SY and MO by PDA/PEI-TPU NFMs was similar, but far lower than CR. With the pH increase, the adsorption capacity of SY and MO decreased gradually. The apparent difference in sorption capacity may be attributed to the difference in sorption mechanisms.

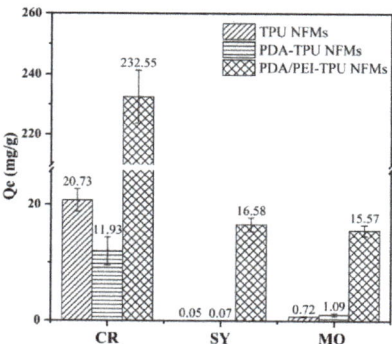

**Figure 7.** The adsorption capacity of original TPU, PDA-TPU and PDA/PEI-TPU NFMs for CR SY and MO dyes (dyes concentration: 100 mg/L, pH = 7).

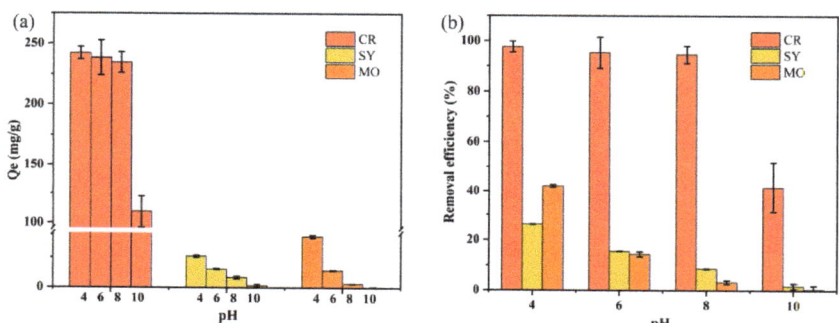

**Figure 8.** The effects of initial solution pH on adsorption capacity (**a**) and removal efficiency (**b**) of PDA/PEI-TPU NFMs toward CR, SY and MO (dyes concentration: 100 mg/L).

### 3.4. Possible Adsorption Mechanism

Theoretically, under the same conditions, the charge on the PDA/PEI-TPU NFMs surface is the same. Assuming that electrostatic adsorption is the main mechanism of adsorption, the PDA/PEI-TPU surface will adsorb the same number of anions, and will increase with the increase in positive charges on the PDA/PEI-TPU NFMs surface. In fact, with the increase in H$^+$, the amine and imine groups present in PEI of PDA/PEI-TPU NFMs gradually protonate into -NH$_3^+$ and -NH$_2^+$-, while the sulfonic acid groups of anionic dyes ionize hydrogen ions [45]. The adsorption of SY and MO on PDA/PEI-TPU NFMs is increased with positive charge on the PDA/PET-TPU NFMs, illustrating a positive correlation with electrostatic interaction. Nevertheless, CR is an exception. Unlike SY and MO, it has an excellent adsorption for CR, not only under acidic conditions, but also under neutral and alkaline conditions. By analyzing their chemical structure, we found that, in addition to the sulfonic acid groups (also present in SY and MO), CR has two another reactive amine groups. Considering the large amount of amino groups in PEI and the presence of -C = O(NH) and -COO- group in original TPU NFMs, it is speculated that intermolecular hydrogen bonding may occur between PDA/PEI-TPU NFMs and CR, which may possibly be the main reason for the highest CR adsorption capacity.

In order to confirm this speculation, the FTIR spectra of CR as well as PDA/PEI-TPU nanofibers before and after adsorbing CR molecules are compared in Figure 9, where some new peaks and characteristic bond shifts are observed. For example, the stretching vibrations of -N = N- and -S-O(SO$_3$-H), stretching vibrations belonging to the CR molecule, can be observed in PDA/PEI-TPU NFMs after adsorption of CR at 1592 cm$^{-1}$ and 1036 cm$^{-1}$ [46]. In addition, the stretching vibration peaks of C = N after adsorption of

CR shifted from 1659 cm$^{-1}$ to 1653 cm$^{-1}$, suggesting that N-H groups (amines) in CR have hydrogen bonding with C = N in PDA/PEI-TPU NFMs [30]. The C = O bond at 1727 cm$^{-1}$ is shifted to 1731 cm$^{-1}$, indicating that the carbonyl group in PDA/PEI-TPU NFMs is also a proton receptor [47]. In addition, the wide peak at 3355 cm$^{-1}$, corresponding to the stretching vibration of the N-H and O-H groups of PDA/PEI-TPU NFMs shifted to 3450 cm$^{-1}$, suggests that the hydroxyl and amino groups also take part in the adsorption as hydrogen bonding sites [47]. All of these changes confirm that there is hydrogen bonding between PDA/PEI-TPU NFMs and CR.

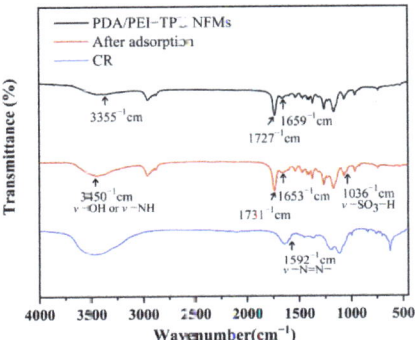

**Figure 9.** FTIR spectra of procured CR and PDA/PEI-TPU NFMs before and after adsorbed CR molecules.

In fact, according to the structures of adsorbent material and dye molecules, we also take into account π–π interaction and van der Waals force. However, the magnitude of these forces is much weaker than that of hydrogen bonding and electrostatic interaction during dye adsorption. Therefore, the mechanistic analysis focuses on hydrogen bonding and electrostatic adsorption. The main possible adsorption mechanism of PDA/PEI-TPU NFMs for three azo dyes is shown in Figure 10. Electrostatic interaction is considered a main mechanism for SY and MO adsorption. However, synergistic effect of hydrogen bonding and electrostatic interaction is likely a critical factor, leading to significant adsorption of PDA/PEI-TPU NFMs for CR, in a wide pH range.

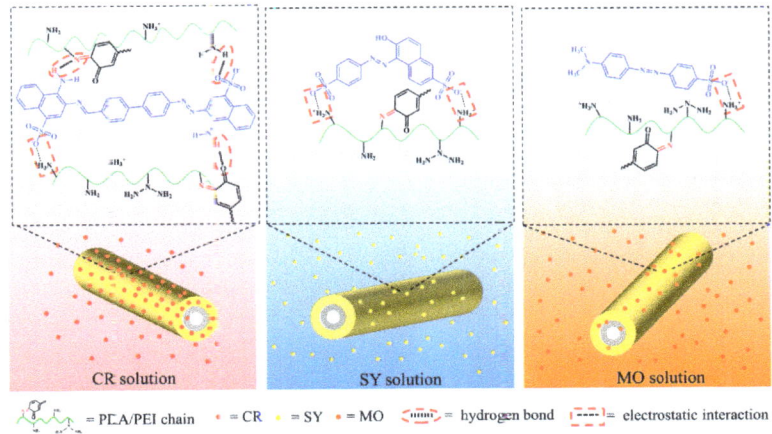

**Figure 10.** The possible main adsorption mechanism of PDA/PEI-TPU NFMs for three azo dyes.

## 3.5. Adsorption Kinetics and Adsorption Isotherm for Three Dyes

The adsorption kinetics and adsorption isotherm of PDA/PEI-TPU NFMs were conducted for three dyes in order to explore the possibility of PDA/PEI-TPU NFMs for adsorption in application. According to Figure 11a–c, the adsorption rates of CR, SY and MO were relatively fast in the first few hours. This can be explained by the fact that a great quantity of active adsorption sites on the surface of PDA/PEI-TPU NFMs are completely exposed to the aqueous solution in the initial stage, and the dye molecules in the solution occupy these active sites rapidly. Afterwards, as the adsorption time increases, the adsorption rate of the dye gradually decreases and finally reaches the adsorption equilibrium stage. To better describe the adsorption process, pseudo-first-order and pseudo-second-order kinetic models were used to fit the experimental data. The fit curves are shown in Figure 11a–c, and the fitted parameters are shown in Table 2. Based on the $R^2$ values, the pseudo-second-order model can better fit the experimental data, which indicates that chemisorption dominates the adsorption process of the three dyes. This result agrees with our previous assumptions that the hydrogen bonding and electrostatic effects of chemisorption were probably a critical factor leading to good adsorption of PDA/PEI-TPU NFMs for CR, SY and MO. The adsorption rate is controlled by the chemisorption mechanism, involving electron cooption and transfer. In addition, the theoretical $q_e$ values calculated from the pseudo-second-order model (217.99 mg/g, 13.24 mg/g and 8.25 mg/g for CR, SY and MO, respectively) are closer to the experimental $q_e$ values (220.34 mg/g, 14.33 mg/g and 8.42 mg/g, respectively), further indicating that the pseudo-second-order kinetics can have good corroboration with the experimental data.

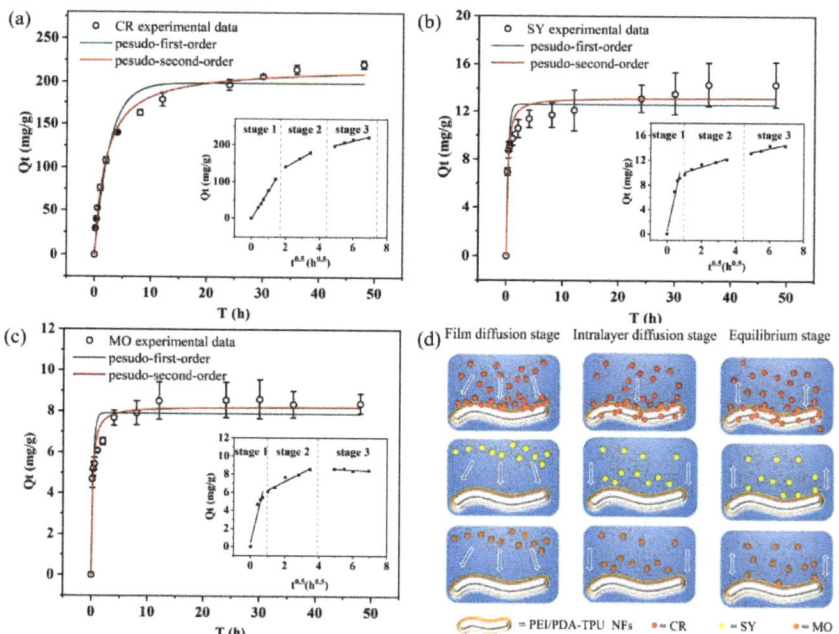

**Figure 11.** Adsorption kinetics of CR (**a**), SY (**b**) and MO (**c**) fitted with pseudo-first-order and pseudo-second-order models. Inset was intra-particle diffusion model. (**d**) Schematics of the diffusion process of CR, SY and MO onto PDA/PEI-TPU NFMs.

**Table 2.** Kinetic parameters for adsorption of three dyes onto the PDA/PEI-TPU NFMs.

| Kinetic Models | Parameters | CR | SY | MO |
|---|---|---|---|---|
| Pseudo-first order | $q_e$ (mg/g) | 198.84 | 12.63 | 7.90 |
| | $k_1$ (1/h) | 0.38 | 3.26 | 3.12 |
| | $R^2$ | 0.954 | 0.834 | 0.867 |
| Pseudo-second order | $q_e$ (mg/g) | 217.99 | 13.24 | 8.25 |
| | $k_2$ (g/(mg h)) | 0.0023 | 0.36 | 0.58 |
| | $R^2$ | 0.989 | 0.914 | 0.95 |
| Intraparticle diffusion | $k_{i\,1}$ (mg/(g h$^{1/2}$)) | 77.07 | 13.48 | 8.01 |
| | $b_1$ | −1.876 | 0.48 | 0.44 |
| | $R^2$ | 0.997 | 0.937 | 0.918 |
| | $k_{i\,2}$ (mg/(g h$^{1/2}$)) | 26.51 | 0.91 | 0.968 |
| | $b_2$ | 87.33 | 9.13 | 5.27 |
| | $R^2$ | 0.999 | 0.914 | 0.933 |
| | $k_{i\,3}$ (mg/(g h$^{1/2}$)) | 11.62 | 0.61 | −0.10 |
| | $b_3$ | 142.06 | 10.31 | 9.06 |
| | $R^2$ | 0.917 | 0.806 | 0.250 |

To explore the diffusion mechanism, an intraparticle diffusion model was applied (the insert of Figure 11a–c). The diffusive process is depicted as Figure 11d. Apparently, the adsorption process of CR, SY and MO can be roughly divided into three linear regions. The first sharp linear phase is related to surface adsorption, and indicates that the dye molecules diffuse from the liquid phase to the adsorbent surface, gradually occupying most of the adsorption sites on the PDA/PEI-TPU NFMs surface. The second linear phase is intraparticle diffusion, indicating that the molecules start to enter the inner surface from the outer, and are adsorbed internally. The last stage is the final adsorption–desorption equilibrium, during which the rate constant is greatly reduced due to the reduction in available adsorption sites and residual concentration of contaminants, which takes a long time to reach [48]. The fitted parameters of the intraparticle diffusion model are shown in Table 2, and the $K_{id}$ values of CR are much larger than those of SY and MO, which indicates that CR is subjected to intraparticle diffusion at a higher rate than SY and MO. In addition, none of the calculated C-values are past the zero point, indicating that the rate control of dye adsorption is determined by multiple stages, and that intraparticle diffusion is not the only determining stage.

Isothermal experiments were investigated for CR, SY and MO in varying concentrations, at room temperature. Two famous Langmuir and Freundlich isotherm models were able to characterize the equilibrium isotherm parameters. Figure 12 shows that the adsorption capacity of PDA/PEI-TPU NFMs gradually increased as the initial concentration of CR increased from 50 to 400 mg/L. SY increased from 5 to 100 mg/L, as did MO, and then reached adsorption equilibrium. This can be explained by the fact that adsorption sites on membranes are limited to hold only a specific quantity of contaminants, and no more active sites are available for the adsorption of additional contaminants. Although the overall trends for CR, SY and MO are similar, the adsorption capacities exhibited at low concentrations are significantly different. CR shows a vertical increase in adsorption capacity at low concentrations compared to SY and MO, indicating that PDA/PEI-TPU NFMs show high affinity for CR molecules [49]. Moreover, a small amount of CR was detected in the solution at a low concentration, indicating that CR was stably adsorbed on the membrane surface without desorption. It is noteworthy that the faster rate of increase in the adsorption capacity of SY compared to MO can be explained by the presence of a hydroxyl group in SY, which may have a weak hydrogen bonding interaction with the membranes. The fitting parameters of the isothermal models are listed in Table 3. The fitted $R^2$ values of the Langmuir model reached 0.999, which is higher than the Freundlich model, indicating that monolayer adsorption is more suitable for explaining the adsorption processes of CR, SY and MO onto PDA/PEI-TPU NFMs. The maximum adsorption capacities of CR (263 mg/g), SY (17 mg/g) and MO (23 mg/g) estimated by the Langmuir model are considerably different. This could be related to the hydrogen bond between amine group in CR or to the specific functional groups on the PDA/PEI-TPU NFMs surface.

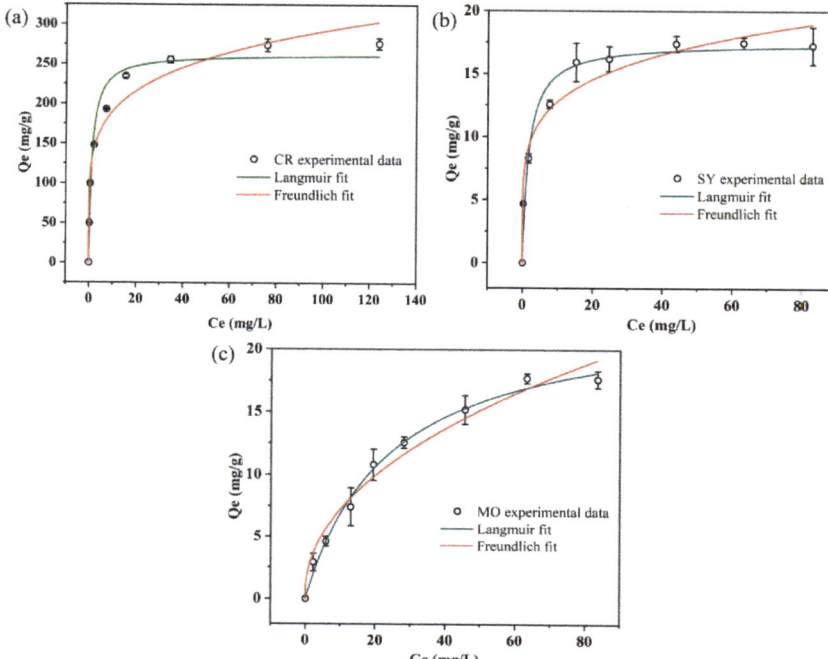

**Figure 12.** Different initial concentration of dyes (**a**) CR, (**b**) SY and (**c**) MO sorption onto PDA/PEI-TPU NFMs.

**Table 3.** Isotherm parameters for adsorption of three dyes onto the PDA/PEI-TPU NFMs.

| Dyes | Langmuir Isothermal Model | | | Freundlich Isothermal Model | | |
|---|---|---|---|---|---|---|
| | $K_L$ (L/mg) | $q_{max}$ (mg/g) | $R^2$ | $K_F$ (mg/g)/(mg/L)$^n$ | n | $R^2$ |
| CR | 0.772 | 262.95 | 0.969 | 123.39 | 5.36 | 0.955 |
| SY | 0.583 | 17.46 | 0.962 | 8.345 | 5.383 | 0.958 |
| MO | 0.042 | 23.34 | 0.992 | 2.509 | 2.174 | 0.974 |

In this work, the ultra-high adsorption capacity of CR was obtained by using PDA/PEI-TPU NFMs as adsorbents ($q_{max}$ = 263 mg/g). To compare the adsorption performance for CR removal among various adsorbents, the adsorption capacities, based on the adsorption isotherm experiment of separation membranes and other adsorbents, are listed in Table 4. It can be concluded that PDA/PEI-TPU NFMs have excellent adsorption and removal capacity for CR, and can be applied for the purpose of efficient removal of CR from aqueous solutions.

Table 4. Summary of adsorption capacities of CR on different nanofiber adsorbents or others.

| Materials | $q_m$ (mg/g) | Refs. |
|---|---|---|
| the coordination cluster $Zn_5(H_2Ln)_6(NO_3)_4$ | 166.91 | [50] |
| p-PEN/a-CNTs@TA/CC nanofibrous | 180 | [51] |
| (CEMNPs)-C/CoFe$_2$O$_4$ | 43.07 | [52] |
| (PCNFs) carbon nanofibers | 218 | [53] |
| PHMG-OCS-PVA nanofibers | 76.92 | [54] |
| Chitosan–alginate sponge | 121.95 | [55] |
| $Fe_3O_4$@bacteria | 320.1 | [56] |
| PDA/PEI-TPU nanofibers | 262.95 | This work |

*3.6. Regeneration of the PDA/PEI-TPU NFMs*

The desorption and regeneration ability of the adsorption membrane is one of the most important factors used to determine its recyclability [57]. After 24 h adsorption, the membranes were placed in 0.05 M NaOH solution for ultrasound for 60 min in order to desorb CR completely, and then dried in the oven and reused for the next adsorption. After six iterations of adsorption–desorption, the adsorption performance of the PDA/PEI-TPU NFMs did not decrease significantly (Figure 13), which indicates that the PDA/PEI-TPU NFMs has potential application value for the removal of CR molecules by adsorption from water.

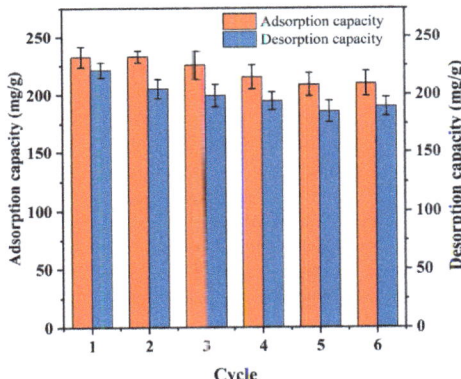

Figure 13. Repeated adsorption–desorption performance of PDA/PEI-TPU NFMs on CR dye.

## 4. Conclusions

In this work, TPU NFMs were prepared by electrostatic spinning and then modified by PDA/PEI co-aggregation to obtain an environmentally friendly and widely available adsorption method: PDA/PEI-TPU NFMs. The PEI coating provided the membranes with high adsorption sites and covalent binding stability, while the PDA offered a strong adhesion, allowing the PEI modified layer to be grafted onto the surface of TPU support membranes. Electrostatic interaction is deemed mainly as a mechanism for SY and MO adsorption on PDA/PEI-TPU NFMs, because the adsorption was changed with the positive charge of the PDA/PEI-TPU NFMs. However, the synergistic effect of hydrogen bonding and electrostatic interaction are likely critical factors, resulting in good adsorption of PDA/PEI-TPU NFMs for CR within a wide pH range (4–10). The maximum adsorption capacity of CR, SY and MO at pH = 7 were 263, 17 and 23 mg/g, respectively. Moreover, the PDA/PEI-TPU NFMs exhibit excellent desorption regeneration performance in NaOH solution. After the 6 iterations of repeated adsorption and desorption, the PDA/PEI-TPU NFMs still maintained high adsorption capacity levels. Eventually, the PDA/PEI-TPU NFMs could be suggested as potential adsorbents for CR wastewater treatment.

**Author Contributions:** J.S.: conceptualization, methodology, supervision. Y.Z.: investigation, data curation, writing—original draft. X.J.: investigation, methodology. J.F.: supervision, writing—reviewing and editing. All authors have read and agreed to the published version of the manuscript.

**Funding:** This research was funded by Natural Science Foundation of Chongqing, China (Project No. cstc2020jcyj-msxm X0928) and Natural Science Foundation of China (Project No. 41977337).

**Data Availability Statement:** Not applicable.

**Conflicts of Interest:** This research received no external funding.

## References

1. Usman, M.; Ahmed, A.; Yu, B.; Wang, S.; Shen, Y.; Cong, H. Simultaneous Adsorption of Heavy Metals and Organic Dyes by β-Cyclodextrin-Chitosan Based Cross-Linked Adsorbent. *Carbohydr. Polym.* **2021**, *255*, 117486. [CrossRef] [PubMed]
2. Huang, Z.; Wu, P.; Yin, Y.; Zhou, X.; Fu, L.; Wang, L.; Chen, S.; Tang, X. Preparation of Pyridine-Modified Cotton Fibers for Anionic Dye Treatment. *React. Funct. Polym.* **2022**, *172*, 105155. [CrossRef]
3. Chung, K.-T. Azo Dyes and Human Health: A Review. *J. Environ. Sci. Health Part C* **2016**, *34*, 233–261. [CrossRef] [PubMed]
4. Puvaneswari, N.; Muthukrishnan, J.; Gunasekaran, P. Toxicity Assessment and Microbial Degradation of Azo Dyes. *Indian J. Exp. Biol.* **2006**, *44*, 618–626. [PubMed]
5. Cui, M.-H.; Liu, W.-Z.; Tang, Z.-E.; Cui, D. Recent Advancements in Azo Dye Decolorization in Bio-Electrochemical Systems (BESs): Insights into Decolorization Mechanism and Practical Application. *Water Res.* **2021**, *203*, 117512. [CrossRef]
6. Zheng, Y.; Cheng, B.; Fan, J.; Yu, J.; Ho, W. Review on Nickel-Based Adsorption Materials for Congo Red. *J. Hazard. Mater.* **2021**, *403*, 123559. [CrossRef]
7. Oladoye, P.O.; Bamigboye, M.O.; Ogunbiyi, O.D.; Akano, M.T. Toxicity and Decontamination Strategies of Congo Red Dye. *Groundw. Sustain. Dev.* **2022**, *19*, 100844. [CrossRef]
8. Moradi, O.; Pudineh, A.; Sedaghat, S. Synthesis and Characterization Agar/GO/ZnO NPs Nanocomposite for Removal of Methylene Blue and Methyl Orange as Azo Dyes from Food Industrial Effluents. *Food Chem. Toxicol.* **2022**, *169*, 113412. [CrossRef]
9. Brüschweiler, B.J.; Merlot, C. Azo Dyes in Clothing Textiles Can Be Cleaved into a Series of Mutagenic Aromatic Amines Which Are Not Regulated Yet. *Regul. Toxicol. Pharmacol.* **2017**, *88*, 214–226. [CrossRef]
10. Wang, X.; Jiang, C.; Hou, B.; Wang, Y.; Hao, C.; Wu, J. Carbon Composite Lignin-Based Adsorbents for the Adsorption of Dyes. *Chemosphere* **2018**, *206*, 587–596. [CrossRef]
11. Bichave, M.S.; Kature, A.Y.; Koranne, S.V.; Shinde, R.S.; Gongle, A.S.; Choudhari, V.P.; Topare, N.S.; Raut-Jadhav, S.; Bokil, S.A. Nano-Metal Oxides-Activated Carbons for Dyes Removal: A Review. *Mater. Today Proc.* **2022**, in press. [CrossRef]
12. Easwaran, G.; Packialakshmi, J.S.; Syed, A.; Elgorban, A.M.; Vijayan, R.; Sivakumar, K.; Bhuvaneswari, K.; Palanisamy, G.; Lee, J. Silica Nanoparticles Derived from Arundo Donax L. Ash Composite with Titanium Dioxide Nanoparticles as an Efficient Nanocomposite for Photocatalytic Degradation Dye. *Chemosphere* **2022**, *307*, 135951. [CrossRef] [PubMed]
13. Sun, Z.; Feng, T.; Zhou, Z.; Wu, H. Removal of Methylene Blue in Water by Electrospun PAN/β-CD Nanofibre Membrane. *e-Polymers* **2021**, *21*, 398–410. [CrossRef]
14. Wang, Y.; Ma, F.F.; Zhang, N.; Wei, X.; Yang, J.; Zhou, Z.W. Blend-Electrospun Poly(Vinylidene Fluoride)/Polydopamine Membranes: Self-Polymerization of Dopamine and the Excellent Adsorption/Separation Abilities. *J. Mater. Chem. A* **2017**, *5*, 14430–14443. [CrossRef]
15. Abd Halim, N.S.; Wirzal, M.D.H.; Hizam, S.M.; Bilad, M.R.; Nordin, N.A.H.M.; Sambudi, N.S.; Putra, Z.A.; Yusoff, A.R.M. Recent Development on Electrospun Nanofiber Membrane for Produced Water Treatment: A Review. *J. Environ. Chem. Eng.* **2021**, *9*, 104613. [CrossRef]
16. Chen, H.; Huang, M.; Liu, Y.; Meng, L.; Ma, M. Functionalized Electrospun Nanofiber Membranes for Water Treatment: A Review. *Sci. Total Environ.* **2020**, *739*, 139944. [CrossRef] [PubMed]
17. Ren, L.-F.; Al Yousif, E.; Xia, F.; Wang, Y.; Guo, L.; Tu, Y.; Zhang, X.; Shao, J.; He, Y. Novel Electrospun TPU/PDMS/PMMA Membrane for Phenol Separation from Saline Wastewater via Membrane Aromatic Recovery System. *Sep. Purif. Technol.* **2019**, *212*, 21–29. [CrossRef]
18. Selvasembian, R.; Gwenzi, W.; Chaukura, N.; Mthembu, S. Recent Advances in the Polyurethane-Based Adsorbents for the Decontamination of Hazardous Wastewater Pollutants. *J. Hazard. Mater.* **2021**, *417*, 125960. [CrossRef] [PubMed]
19. Wang, B.; Sun, Z.; Liu, T.; Wang, Q.; Li, C.; Li, X. $NH_2$-Grafting on Micro/Nano Architecture Designed PS/TPU@$SiO_2$ Electrospun Microfiber Membrane for Adsorption of Cr(VI). *Desalin. Water Treat.* **2019**, *154*, 82–91. [CrossRef]
20. Anju, M.; Renuka, N.K. Magnetically Actuated Graphene Coated Polyurethane Foam as Potential Sorbent for Oils and Organics. *Arab. J. Chem.* **2020**, *13*, 1752–1762. [CrossRef]
21. Santos, O.S.H.; Coelho da Silva, M.; Silva, V.R.; Mussel, W.N.; Yoshida, M.I. Polyurethane Foam Impregnated with Lignin as a Filler for the Removal of Crude Oil from Contaminated Water. *J. Hazard. Mater.* **2017**, *324*, 406–413. [CrossRef] [PubMed]
22. Vieira Amorim, F.; José Ribeiro Padilha, R.; Maria Vinhas, G.; Ramos Luiz, M.; Costa de Souza, N.; Medeiros Bastos de Almeida, Y. Development of Hydrophobic Polyurethane/Castor Oil Biocomposites with Agroindustrial Residues for Sorption of Oils and Organic Solvents. *J. Colloid Interface Sci.* **2021**, *581*, 442–454. [CrossRef] [PubMed]

23. Hong, H.-J.; Lim, J.S.; Hwang, J.Y.; Kim, M.; Jeong, H.S.; Park, M.S. Carboxymethlyated Cellulose Nanofibrils(CMCNFs) Embedded in Polyurethane Foam as a Modular Adsorbent of Heavy Metal Ions. *Carbohydr. Polym.* **2018**, *195*, 136–142. [CrossRef]
24. Xue, D.; Li, T.; Liu, Y.; Yang, Y.; Zhang, Y.; Cui, J.; Guo, D. Selective Adsorption and Recovery of Precious Metal Ions from Water and Metallurgical Slag by Polymer Brush Graphene–Polyurethane Composite. *React. Funct. Polym.* **2019**, *136*, 138–152. [CrossRef]
25. Jin, L.; Gao, Y.; Yin, J.; Zhang, X.; He, C.; Wei, Q.; Lu, X.; Liang, F.; Zhao, W.; Zhao, C. Functionalized Polyurethane Sponge Based on Dopamine Derivative for Facile and Instantaneous Clean-up of Cationic Dyes in a Large Scale. *J. Hazard. Mater.* **2020**, *400*, 123203. [CrossRef]
26. Li, Z.; Chen, K.; Chen, Z.; Li, W.; Biney, B.W.; Guo, A.; Liu, D. Removal of Malachite Green Dye from Aqueous Solution by Adsorbents Derived from Polyurethane Plastic Waste. *J. Environ. Chem. Eng.* **2021**, *9*, 104704. [CrossRef]
27. Cheng, W.; Zeng, X.; Chen, H.; Li, Z.; Zeng, W.; Mei, L.; Zhao, Y. Versatile Polydopamine Platforms: Synthesis and Promising Applications for Surface Modification and Advanced Nanomedicine. *ACS Nano* **2019**, *13*, 8537–8565. [CrossRef]
28. Feng, X.; Yu, Z.; Long, R.; Sun, Y.; Wang, M.; Li, X.; Zeng, G. Polydopamine Intimate Contacted Two-Dimensional/Two-Dimensional Ultrathin Nylon Basement Membrane Supported RGO/PDA/MXene Composite Material for Oil-Water Separation and Dye Removal. *Sep. Purif. Technol.* **2020**, *247*, 116945. [CrossRef]
29. Lefebvre, L.; Agusti, G.; Bouzeggane, A.; Edouard, D. Adsorption of Dye with Carbon Media Supported on Polyurethane Open Cell Foam. *Catal. Today* **2018**, *301*, 98–103. [CrossRef]
30. Li, Z.; Hanafy, H.; Zhang, L.; Sellaoui, L.; Schadeck Netto, M.; Oliveira, M.L.S.; Seliem, M.K.; Luiz Dotto, G.; Bonilla-Petriciolet, A.; Li, Q. Adsorption of Congo Red and Methylene Blue Dyes on an Ashitaba Waste and a Walnut Shell-Based Activated Carbon from Aqueous Solutions: Experiments, Characterization and Physical Interpretations. *Chem. Eng. J.* **2020**, *388*, 124263. [CrossRef]
31. Guo, R.; Guo, W.; Pei, H.; Wang, B.; Guo, X.; Liu, N.; Mo, Z. Polypyrrole Deposited Electrospun PAN/PEI Nanofiber Membrane Designed for High Efficient Adsorption of Chromium Ions (VI) in Aqueous Solution. *Colloids Surf. A Physicochem. Eng. Asp.* **2021**, *627*, 127183. [CrossRef]
32. Pashaei-Fakhri, S.; Peighambardoust, S.J.; Foroutan, R.; Arsalani, N.; Ramavandi, B. Crystal Violet Dye Sorption over Acrylamide/Graphene Oxide Bonded Sodium Alginate Nanocomposite Hydrogel. *Chemosphere* **2021**, *270*, 129419. [CrossRef] [PubMed]
33. Yang, H.-C.; Liao, K.-J.; Huang, H.; Wu, Q.-Y.; Wan, L.-S.; Xu, Z.-K. Mussel-Inspired Modification of a Polymer Membrane for Ultra-High Water Permeability and Oil-in-Water Emulsion Separation. *J. Mater. Chem. A* **2014**, *2*, 10225–10230. [CrossRef]
34. Sundaran, S.P.; Reshmi, C.R.; Sagitha, P.; Manaf, O.; Sujith, A. Multifunctional Graphene Oxide Loaded Nanofibrous Membrane for Removal of Dyes and Coliform from Water. *J. Environ. Manag.* **2019**, *240*, 494–503. [CrossRef] [PubMed]
35. Zhang, W.; Song, H.; Zhu, L.; Wang, G. Zeng, Z.; Li, X. High Flux and High Selectivity Thin-Film Composite Membranes Based on Ultrathin Polyethylene Porous Substrates for Continuous Removal of Anionic Dyes. *J. Environ. Chem. Eng.* **2022**, *10*, 107202. [CrossRef]
36. Chen, X.; Hossain, M.F.; Duan, C.; Lu, J.; Tsang, Y.F.; Islam, M.S.; Zhou, Y. Isotherm Models for Adsorption of Heavy Metals from Water—A Review. *Chemosphere* **2022**, *307*, 135545. [CrossRef]
37. Zhang, M.; Zhang, Z.; Peng, Y.; Feng, L.; Li, X.; Zhao, C.; Sarfaraz, K. Novel Cationic Polymer Modified Magnetic Chitosan Beads for Efficient Adsorption of Heavy Metals and Dyes over a Wide PH Range. *Int. J. Biol. Macromol.* **2020**, *156*, 289–301. [CrossRef]
38. Almasian, A.; Jalali, M.L.; Fard, G.C.; Maleknia, L. Surfactant Grafted PDA-PAN Nanofiber: Optimization of Synthesis, Characterization and Oil Absorption Property. *Chem. Eng. J.* **2017**, *326*, 1232–1241. [CrossRef]
39. Fang, J.; Chen, Y.; Fang, C.; Zhu, L. Regenerable Adsorptive Membranes Prepared by Mussel-Inspired Co-Deposition for Aqueous Dye Removal. *Sep. Purif. Technol.* **2022**, *281*, 119876. [CrossRef]
40. Mavukkandy, M.O.; Ibrahim, Y.; Almarzooqi, F.; Naddeo, V.; Karanikolos, G.N.; Alhseinat, E.; Banat, F.; Hasan, S.W. Synthesis of Polydopamine Coated Tungsten Oxide@ Poly(Vinylidene Fluoride-Co-Hexafluoropropylene) Electrospun Nanofibers as Multifunctional Membranes for Water Applications. *Chem. Eng. J.* **2022**, *427*, 131021. [CrossRef]
41. Tang, C.Y.; Kwon, Y.-N.; Leckie, J.O. Effect of Membrane Chemistry and Coating Layer on Physiochemical Properties of Thin Film Composite Polyamide RO and NF Membranes: I. FTIR and XPS Characterization of Polyamide and Coating Layer Chemistry. *Desalination* **2009**, *242*, 149–167. [CrossRef]
42. Liu, L.; Jiang, W.; Song, X.; Duan, Q.; Zhu, E. A Novel Strategy of Lock-in Effect between Conjugated Polymer and $TiO_2$ towards Dramatic Enhancement of Photocatalytic Activity under Visible Light. *Sci. Rep.* **2020**, *10*, 6513. [CrossRef] [PubMed]
43. Liu, C.; Xu, W.Z.; Charpentier, P.A. Synthesis and Photocatalytic Antibacterial Properties of Poly [2,11′-Thiopheneethylenethiophene-Alt-2,5-(3-Carboxyl)Thiophene]. *ACS Appl. Polym. Mater.* **2020**, *2*, 1886–1896. [CrossRef]
44. Zúñiga-Zamora, A.; García-Mena, J.; Cervantes-González, E. Removal of Congo Red from the Aqueous Phase by Chitin and Chitosan from Waste Shrimp. *Desalin. Water Treat.* **2016**, *57*, 14674–14685. [CrossRef]
45. Kuang, Y.; Zhang, Z.; Wu, D. Synthesis of Graphene Oxide/Polyethyleneimine Sponge and Its Performance in the Sustainable Removal of Cu(II) from Water. *Sci. Total Environ.* **2022**, *806*, 151258. [CrossRef]
46. Kim, U.-J.; Kimura, S.; Wada, M. Highly Enhanced Adsorption of Congo Red onto Dialdehyde Cellulose-Crosslinked Cellulose-Chitosan Foam. *Carbohydr. Polym.* **2019**, *214*, 294–302. [CrossRef] [PubMed]
47. Zhou, X.; Wei, J.; Liu, K.; Liu, N.; Zhou, B. Adsorption of Bisphenol A Based on Synergy between Hydrogen Bonding and Hydrophobic Interaction. *Langmuir* **2014**, *30*, 13861–13868. [CrossRef]

48. Ahmad, Z.U.; Yao, L.; Wang, J.; Gang, D.D.; Islam, F.; Lian, Q.; Zappi, M.E. Neodymium Embedded Ordered Mesoporous Carbon (OMC) for Enhanced Adsorption of Sunset Yellow: Characterizations, Adsorption Study and Adsorption Mechanism. *Chem. Eng. J.* **2019**, *359*, 814–826. [CrossRef]
49. Wang, S.; Sun, H.; Ang, H.M.; Tadé, M.O. Adsorptive Remediation of Environmental Pollutants Using Novel Graphene-Based Nanomaterials. *Chem. Eng. J.* **2013**, *226*, 336–347. [CrossRef]
50. Wu, X.; Wang, X.; Hu, Y.; Chen, H.; Liu, X.; Dang, X. Adsorption Mechanism Study of Multinuclear Metal Coordination Cluster $Zn_5$ for Anionic Dyes Congo Red and Methyl Orange: Experiment and Molecular Simulation. *Appl. Surf. Sci.* **2022**, *586*, 152745. [CrossRef]
51. Zhao, S.; Zhan, Y.; Feng, Q.; Yang, W.; Dong, H.; Sun, A.; Wen, X.; Chiao, Y.-H.; Zhang, S. Easy-Handling Carbon Nanotubes Decorated Poly(Arylene Ether Nitrile)@tannic Acid/Carboxylated Chitosan Nanofibrous Composite Absorbent for Efficient Removal of Methylene Blue and Congo Red. *Colloids Surf. A Physicochem. Eng. Asp.* **2021**, *626*, 127069. [CrossRef]
52. Lu, M.; Wu, Q.; Guan, X.-H.; Zheng, Q.-Y.; Wang, G.-S. Preparation of $C/CoFe_2O_4$ Nanocomposites Based on Membrane Dispersion-Hydrothermal Carbonization and Their Application for Dyeing Removal. *Desalin. Water Treat.* **2019**, *148*, 285–295. [CrossRef]
53. Wang, J.; Cai, C.; Zhang, Z.; Li, C.; Liu, R. Electrospun Metal-Organic Frameworks with Polyacrylonitrile as Precursors to Hierarchical Porous Carbon and Composite Nanofibers for Adsorption and Catalysis. *Chemosphere* **2020**, *239*, 124833. [CrossRef] [PubMed]
54. Chen, S.; Li, C.; Hou, T.; Cai, Y.; Liang, L.; Chen, L.; Li, M. Polyhexamethylene Guanidine Functionalized Chitosan Nanofiber Membrane with Superior Adsorption and Antibacterial Performances. *React. Funct. Polym.* **2019**, *145*, 104379. [CrossRef]
55. Zhang, Q.; Xie, M.; Guo, X.; Zeng, L.; Luo, J. Fabrication and Adsorption Behavior for Congo Red of Chitosan and Alginate Sponge. *Integr. Ferroelectr.* **2014**, *151*, 61–75. [CrossRef]
56. Pi, Y.; Duan, C.; Zhou, Y.; Sun, S.; Yin, Z.; Zhang, H.; Liu, C.; Zhao, Y. The Effective Removal of Congo Red Using a Bio-Nanocluster: $Fe_3O_4$ Nanoclusters Modified Bacteria. *J. Hazard. Mater.* **2022**, *424*, 127577. [CrossRef]
57. Kamran, U.; Bhatti, H.N.; Noreen, S.; Tahir, M.A.; Park, S.-J. Chemically Modified Sugarcane Bagasse-Based Biocomposites for Efficient Removal of Acid Red 1 Dye: Kinetics, Isotherms, Thermodynamics, and Desorption Studies. *Chemosphere* **2022**, *291*, 132796. [CrossRef]

Article

# Distribution, Sources, and Risk of Polychlorinated Biphenyls in the Largest Irrigation Area in the Yellow River Basin

Qi Zhang [1,†], Yafang Li [1,†], Qingfeng Miao [1], Guoxia Pei [1], Yanxia Nan [1], Shuyu Yu [1], Xiaole Mei [1] and Weiying Feng [2,*]

[1] Water Conservancy and Civil Engineering College, Inner Mongolia Agricultural University, Huhhot 010018, China
[2] School of Space and Environment, Beihang University, Beijing 100191, China
* Correspondence: fengweiying@buaa.edu.cn
† These authors contributed equally to this work.

**Abstract:** To investigate the contamination of PCBs in agricultural soils irrigated chronically with polluted water and the distribution and migration of PCBs under long-term irrigation, 100 farmland soil profile samples were collected in the Yellow River irrigation area in Inner Mongolia, China, to determine PCB content. Cluster analysis was used to identify possible sources of PCBs products, and the USEPA Health Risk Evaluation Model assessed the health risks posed by PCBs to humans. The results showed that the detection rates of eight monomers in the different soil layers of each sample site ranged from 5% to 90%, and the concentration ranged from not detected to 87.71 $ng \cdot g^{-1}$. The PCBs content showed a vertical distribution rule of accumulation in the shallow layer, sudden decrease in the middle layer. Low-chlorinated PCBs were dominant in each soil profile. Source identification indicated that PCB pollution in the study area originated mainly from the Aroclor1242, Aroclor1248, Aroclor1016, Aroclor1252, and Aroclor1221 industrial products and domestic transformer oil. Finally, a health risk assessment demonstrated that child and adult groups in study area were exposed to negligible carcinogenic and noncarcinogenic risks.

**Keywords:** Yellow River irrigation area; polychlorinated biphenyl; distribution characteristics; source apportionment; health risks

## 1. Introduction

It has been reported that more than 80 countries are currently facing freshwater shortages, and global water quality is deteriorating [1]. Global freshwater use has increased by a factor of six over the past 100 years, Agriculture currently accounts for 69% of global water withdrawals, which are mainly used for irrigation [2]. Pollutants in irrigation water, particularly persistent organic pollutants (POPs), would be inevitably deposited in the farmland, absorbed by crops, and finally transferred to the human body, pose potential threats to the soil-crop system and human health [3]. The control of non-point source pollution of POPs in the farmland is facing great challenges. The Yellow River is the fifth-largest river in the world, and the largest water source in northern China, and it is responsible for the irrigation of farmland in the Yellow River irrigation area. The Yellow River irrigation area in Inner Mongolia is an important commodity grain and oil production base for the Inner Mongolia Autonomous Region and for China in general. Ensuring food security and sustainable agriculture in this area is critical to achieving sustainable development goals. In recent years, there have been numerous reports on water pollution in the Yellow River. A study on the distribution and migration of POPs in the Inner Mongolia section of the Yellow River Polychlorinated biphenyls (PCBs) were detected [4].

PCBs are noted for their persistence, ubiquity, long-range atmospheric transport, bioaccumulation [5], and adverse effects on the environment [6] and human health [7]. PCBs have been extensively used in various industrial and commercial applications, such as transformers, capacitors, and plasticizers [8]. After entering the farmland with water, PCBs will gradually migrate to the deeper layers of the soil, eventually causing groundwater pollution, and even the entire ecosystem pollution. Although PCBs were banned in 2004 by the Stockholm Convention, they are still frequently detected in farmland soils in many countries because of their persistence and non-degradable nature [9]. China is a large agricultural country with a long history and large farmlands. Due to the significant historical production and usage of PCBs, residues of these contaminants have been found at high levels in farmland soil in some places in China [10–12].

To date, many scholars have studied the PCB pollution in the irrigation farming area such as farmland sewage irrigation, reclaimed water irrigation, and clear irrigation areas [13–16]. In the study on the effect of irrigation years on the concentration of PCBs in agrarian soils, Abrahao et al. found that during the first years of irrigation in the Lerma basin, irrigation has not significantly influenced the concentrations of PCBs in the soils [17]. Teng et al. collected soil samples from paddy fields in Liaohe River Plain and monitored the PCBs contents, and verified that there is a significant contribution of long-term irrigation of polluted river water to PCBs contaminations in paddy soils [18]. Ngweme at al investigated the levels of PCBs in irrigation water and soils from the vegetable growing sites in Congo, and evaluated the potential human health risks; the results showed that PCBs did not pose a carcinogenic risk to the local population [19]. These studies have focused on PCBs' pollution of the soil surface. However, there have been few reports on the PCBs' pollution vertical distribution characteristics in soils with polluted water. To investigate the contamination of PCBs in agricultural soils irrigated chronically with polluted water and the distribution and migration of PCBs under long-term irrigation with the Yellow River irrigation area of Inner Mongolia in China as the study area, this study investigated PCB contamination in agricultural soils chronically irrigated with contaminated Yellow River water. The vertical and spatial distribution characteristics of PCBs in soils of irrigated areas under the effect of long-term irrigation were studied. The possible product sources of PCBs in the study area were identified through cluster analysis, and the health risk was evaluated. To contribute a theoretical basis for the distribution and migration of PCBs in soil and to provide reference for the study of persistent organic pollutants in irrigated soils by contaminated water, it is of great significance to seek safe irrigation modes, study food security, and promote the agricultural sustainable development.

## 2. Materials and Methods

### 2.1. Study Area

The Yellow River irrigation area of Inner Mongolia is located at the northern end of the Yellow River Basin, with a total land area of 19.27 thousand km$^2$. The existing arable land is 12.52 thousand km$^2$, and the effective irrigation area is approximately 7.3 thousand km$^2$. The soil type is mainly salinized light color meadow soil and saline soil, and the soil texture is mainly light sandy loam. The Yellow River irrigation area of Inner Mongolia is composed of the Hetao irrigation district, Tumochuan irrigation district, and the South Bank of the Yellow River irrigation district. The Hetao-Tumochuan Plain is the main agricultural production area in the Yellow River irrigation area of Inner Mongolia. Therefore, the Hetao (mainly in the Jiefangzha and Wulanbuhe irrigation area) and Tumochuan (Including the Madihao, Dengkou and National Unity Pumping irrigation area) irrigation areas were selected as the study area; the study area location is shown in Figure 1. The main crops grown are wheat, corn, and sunflower. The irrigated area has flat terrain and sufficient heat, with annual sunshine of 2900–3200 h and frost-free period of 140–180 d. Rainfall in this area is scarce, with an annual average precipitation of only 130–400 mm and annual average evaporation of 1200–1600 mm. Agricultural production in this area is therefore entirely dependent on irrigation, with the Yellow River as the main water source.

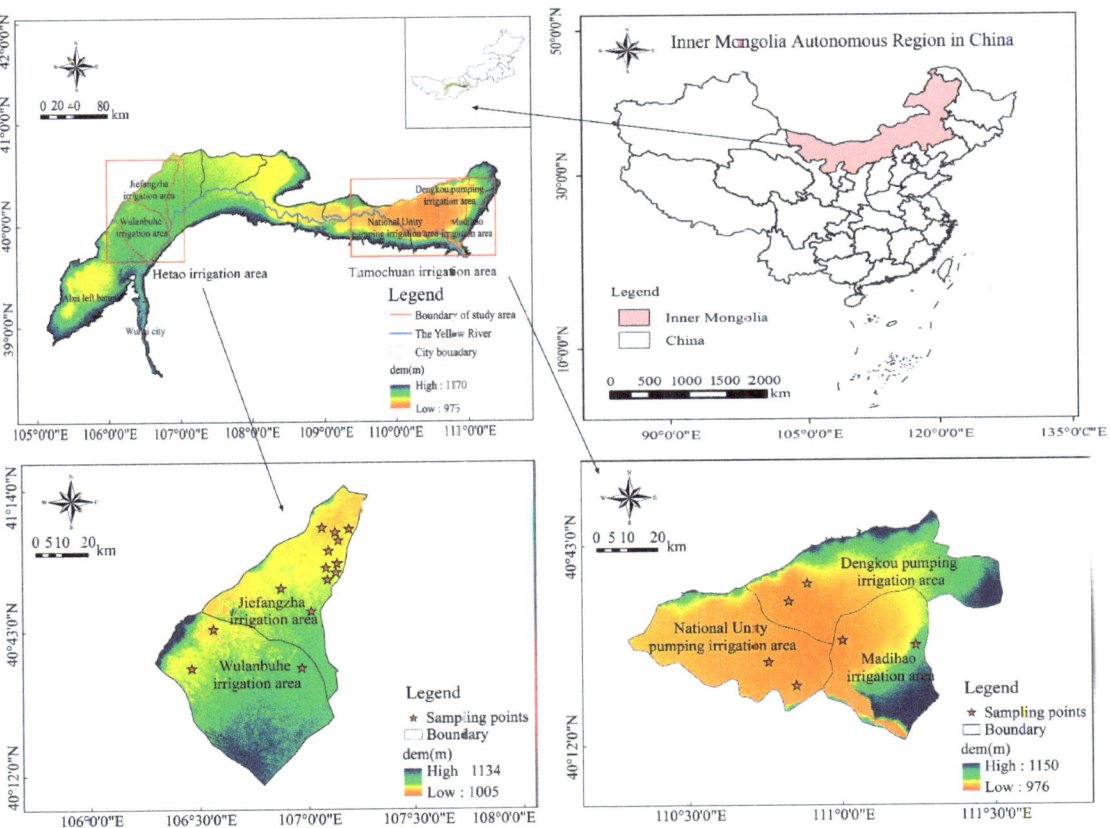

**Figure 1.** The Yellow River irrigation area in Inner Mongolia and distribution of sampling points

### 2.2. Sample Collection

In this study, a total of 20 sampling sites were laid in differing irrigation districts along the Yellow River in Inner Mongolia (Figure 1), and Global Positioning System was used to record the specific locations. At each site, a profile (0–100 cm) soil was taken using a hand soil auger. One sample was collected every 20 cm starting from the soil surface, i.e., 0–20, 20–40, 40–60, 60–80, 80–100 cm, meaning 5 samples were collected for each site and a total of 100 samples at 20 sites. The samples were labelled, stored in clean special aluminum boxes, sealed, and frozen at 4 °C until analysis. All the equipment was carefully washed with acetone and n-hexane and dried before use.

### 2.3. Material and Reagents

All solvents used pesticide grade for the chromatographic analysis. External calibration standards of PCBs #2, 5, 29, 47, 98, 154, 171, and 201 were purchased from Accu Standard (New Haven, CT, USA) and all have a highly pure (>98%). The Florisil column was obtained from Supelco (Bellefonte, PA, USA).

### 2.4. Sample Extraction and Analysis

After removing pebbles and plant roots, the fresh soils were placed in a freeze-dryer that removed all water over the course of 24 h, then ground and sieved through sieve. Next, 4.00 g of soil and diatomite (1.00 g) were accurately weighed and mixed before being adding to the extract pond for accelerated solvent extraction with n-hexane and acetone (V = 1:1).

The optimized operational variables of the accelerated solvent extraction were as follows: static extraction time, 6 min; extracting pressure, 1700 psi; extraction temperature, 100 °C; flush volume, 90%; cycle number, 3. The Florisil columns were first preconditioned with 30 mL of acetone and then rinsed with 20 mL of n-hexane. This mixture was subsequently concentrated in a preconditioned Florisil column with n-hexane and hexane-acetone as the eluent, as outlined in the USEPA Method 8082A.

For PCB analysis, a Varian 450 GC system with a CP-Sil 5CB (30 m × 0.25 mm; 0.39 μm thick) and an electron capture detector was used, and the external standard calibration of the peak area versus the concentration was also determined. The resulting mixture was further concentrated to approximately 2 mL under a gentle stream of dry nitrogen gas. The injection port was operated at 280 °C in split-less mode, nitrogen (99.999% pure) was used as the carrier gas, and the column flow was 1.0 mL·min$^{-1}$. The linear temperature program of the GC oven was as follows: the initial temperature was maintained at 80 °C for 2 min, then heated to 180 °C at 10 °C·min$^{-1}$ and then 250 °C for 10 min at 3 °C·min$^{-1}$ before being set to a temperature program from 250 to 280 °C at 20 °C·min$^{-1}$ and maintained for 8 min.

Quality assurance and quality control were performed using a five-level multipoint calibration standard based on the peak area. Laboratory blank, laboratory blank spike, and matrix spike samples were run for the eight PCBs in parallel with the samples to check for any contaminants during concentration and detection. The concentration of the lowest calibration standard was taken as the limit of quantitation (LOQ). The LOQs of PCBs ranged from 0.1 to 0.3 ng·g$^{-1}$, and recoveries between 84.25% and 109.25%. In all the blank samples, all values were lower than the detection limit. These results indicate that the method was accurate and replicable.

*2.5. Statistical Analysis*

Analysis of data using SPSS 25.0. The Kolmogorov–Smirnov test was used to test the data and determined that all data conformed to the normal distribution. The results were explained via cluster analysis, which was used to categorize $\Sigma_8$PCBs values in similar sets and clarify their sources.

*2.6. Health Risk Assessment Methods*

2.6.1. Exposure Assessment

The USEPA health risk assessment model was used to assess the exposure of PCBs in farmland soil in the Yellow River irrigation area from three exposure routes: ingestion, dermal, and inhalation, according to the chronic daily intake (CDI). The CDI in soil was computed by following Equations (1)–(3) [20]:

$$CDI_{ing} = \frac{C \times IngR \times EF \times ED \times CF}{BW \times AT} \quad (1)$$

$$CDI_{der} = \frac{C \times SA \times AF \times EF \times ED \times ABS \times CF}{BW \times AT} \quad (2)$$

$$CDI_{inh} = \frac{C \times InhR \times EF \times ED}{PEF \times BW \times AT} \quad (3)$$

where $CDI_{ing}$, $CDI_{der}$, and $CDI_{inh}$ refer to the exposure amount (mg·(kg·d)$^{-1}$) of soil pollutants under ingestion, dermal, and inhalation, respectively. C is the concentration of PCBs in the soil (mg·kg$^{-1}$). Other parameters and their meanings are listed in Table 1 [21–23].

Table 1. Exposure evaluation parameter values.

| Symbol | Meaning of Parameters | Selected Values | |
|---|---|---|---|
| | | Adults | Children |
| CF | Conversion factor (kg·mg$^{-1}$) | $10^{-6}$ | $10^{-6}$ |
| EF | Exposure frequency (d·a$^{-1}$) | 350 | 350 |
| ED | Exposure frequency (a) | 24 | 6 |
| EW | Body weight (kg) | 70 | 15 |
| AT | Averaging time, non-carcinogenic/carcinogenic (d) | 8760/25,550 | 2190/25,550 |
| IngR | Ingestion rate of soil particle (mg·d$^{-1}$) | 100 | 200 |
| SA | Contact surface area of skin with soil (cm$^2$) | 5700 | 2800 |
| AF | Soil-to-skin adherence factor (mg·cm$^{-2}$) | 0.07 | 0.2 |
| ABS | Dermal absorption factor (%) | 0.13 | 0.13 |
| InhR | Inhalation rate (m$^3$·d$^{-1}$) | 16 | 8 |
| PEF | Particle emission factor (m$^3$·kg$^{-1}$) | $1.36 \times 10^9$ | $1.36 \times 10^9$ |

2.6.2. The Carcinogenic Risk

The carcinogenic risk ($R_T$) is calculated as follows:

$$R_i = CDI_i \times SF_i \quad (4)$$

$$R_T = \sum R_i \quad (5)$$

where $R_i$ is the carcinogenic risk generated by different routes, dimensionless; $R_T$ is the total risk of cancer under multiple routes, without dimensionality; and $SF_I$ is the slope factor of carcinogenesis via different routes (kg·d·mg$^{-1}$). The SF value of PCBs is 2 kg·d·mg$^{-1}$ for ingestion and the dermal route, and $2.18 \times 10^{-3}$ kg·d·mg$^{-1}$ for inhalation [24]. According to the US Environmental Protection Agency [20], a range of $10^{-6}$ to $10^{-4}$ is considered an acceptable level of cancer risk.

2.6.3. The Non-Carcinogenic Risk

The non-carcinogenic risk ($HI$) is calculated as follows:

$$HQ_i = \frac{CDI_i}{RfD_i} \quad (6)$$

$$HI = \sum HQ_i \quad (7)$$

where $HQ$ is the non-carcinogenic risk generated by different approaches, dimensionless; $HI$ is the total noncarcinogenic risk across multiple pathways, nondimensional; and $RfD_i$ is the reference dose of non-carcinogenic pollutants in different channels, mg·(kg·d)$^{-1}$. The $RfD$ value of PCBs is $2.3 \times 10^{-5}$ mg (kg·d)$^{-1}$ [20]. An HI value of 1 is the acceptable hazard quotient [20].

3. Results

*3.1. Distribution Characteristics of Polychlorinated Biphenyls (PCBs) in Soil Profiles*

3.1.1. Vertical Distribution of PCBs

The statistical analysis results for PCBs in the 0–100 cm soil profile in the Yellow River irrigation area of Inner Mongolia are shown in Table 2. The detection rates of eight monomers in different soil layers from each sample point ranged from 5% to 90%, and their concentrations ranged from Nd-87.71 ng·g$^{-1}$. The highest concentration was found in the 20–40 cm soil layer, and the lowest concentration was concentrated in the 40–60 cm soil layer. Overall, the content of PCBs in the deeper soil profile in the study area increased and the coefficient of variation was high.

Table 2. Statistical characteristics of PCBs in soil.

| Soil Profiles | Statistical Characteristics | Low-Chlorinated PCBs | | | | High-Chlorinated PCBs | | | |
|---|---|---|---|---|---|---|---|---|---|
| | | PCB1 | PCB5 | PCB29 | PCB47 | PCB98 | PCB154 | PCB171 | PCB201 |
| 0–20 cm | Concentration range/ng·g$^{-1}$ | 0–71.13 | 0–9.40 | 0–10.21 | 0–18.19 | 0–15.23 | 0–3.32 | 0–0.88 | 0–1.05 |
| | Mean of monomer/ng·g$^{-1}$ | 6.55 | 1.71 | 2.37 | 5.77 | 2.29 | 0.37 | 0.11 | 0.11 |
| | Coefficient of variation/% | 229.01 | 135.35 | 107.24 | 83.13 | 183.07 | 214.81 | 264.7 | 272.45 |
| | Detection rate/% | 90 | 60 | 85 | 80 | 50 | 35 | 10 | 10 |
| | Mean/ng·g$^{-1}$ | | | 16.4 | | | | 2.75 | |
| 20–40 cm | Concentration range/ng·g$^{-1}$ | 0–56.14 | 0–8.22 | 0–20.14 | 0–22.42 | 0–18.15 | 0–1.12 | 0–1.84 | 0–3.25 |
| | Mean of monomer/ng·g$^{-1}$ | 10.23 | 2.24 | 2.53 | 5.22 | 2.82 | 0.23 | 0.21 | 0.34 |
| | Coefficient of variation/% | 185.63 | 127.43 | 166.93 | 114.87 | 170.96 | 146.85 | 248.48 | 255.18 |
| | Detection rate/% | 85 | 75 | 85 | 85 | 60 | 40 | 20 | 20 |
| | Mean/ng·g$^{-1}$ | | | 20.23 | | | | 3.61 | |
| 40–60 cm | Concentration range/ng·g$^{-1}$ | 0–22.73 | 0–5.20 | 0–11.13 | 0–12.82 | 0–6.51 | 0–2.74 | 0–1.22 | 0–2.74 |
| | Mean of monomer/ng·g$^{-1}$ | 3.9 | 1.33 | 2.93 | 3.59 | 0.72 | 0.25 | 0.17 | 0.06 |
| | Coefficient of variation/% | 125.84 | 115.01 | 99.78 | 100.68 | 201.29 | 244.33 | 191.84 | 435.89 |
| | Detection rate/% | 90 | 60 | 90 | 70 | 40 | 30 | 25 | 5 |
| | Mean/ng·g$^{-1}$ | | | 11.76 | | | | 1.21 | |
| 60–80 cm | Concentration range/ng·g$^{-1}$ | 0–87.71 | 0–19.49 | 0–6.08 | 0–21.92 | 0–22.49 | 0–6.69 | 0–0.63 | 0–1.69 |
| | Mean of monomer/ng·g$^{-1}$ | 8.73 | 2.76 | 1.44 | 4.53 | 2.24 | 0.59 | 0.05 | 0.17 |
| | Coefficient of variation/% | 216.41 | 164.73 | 110.48 | 111.34 | 223.26 | 258.25 | 272.44 | 237.49 |
| | Detection rate/% | 80 | 80 | 80 | 75 | 50 | 25 | 20 | 25 |
| | Mean/ng·g$^{-1}$ | | | 17.46 | | | | 3.05 | |
| 80–100 cm | Concentration range/ng·g$^{-1}$ | 0–36.11 | 0–11.24 | 0–77.09 | 0–15.56 | 0–9.53 | 0–1.17 | 0–1.25 | 0–1.21 |
| | Mean of monomer/ng·g$^{-1}$ | 4.82 | 2.3 | 5.4 | 4.72 | 1.73 | 0.2 | 0.14 | 0.12 |
| | Coefficient of variation/% | 166.79 | 140.04 | 306.98 | 98.44 | 155.22 | 155.13 | 224.77 | 255.94 |
| | Detection rate/% | 80 | 65 | 75 | 70 | 50 | 40 | 25 | 15 |
| | Mean/ng·g$^{-1}$ | | | 17.24 | | | | 2.2 | |

The detection rate of $\sum_8$PCBs in the study area was 100% and decreased gradually along the vertical section. The contents of $\sum_8$PCBs ranged from not detected to 120.65 ng·g$^{-1}$. The $\sum_8$PCBs in each soil layer was in the order 20–40 > 60–80 > 80–100 > 0–20 > 40–60 cm. Apart from the average concentration of 40–60 cm at 12.97 ng·g$^{-1}$, the average concentrations of the other soil layers were all within 5 ng·g$^{-1}$, and the vertical distribution was relatively uniform in all profiles (Figure 2). The detection rate and content of low-chlorinated PCBs were higher than those of high-chlorinated PCBs in all profiles. The detection rate of low-chlorinated PCBs ranged from 65% to 90%, and the content of low-chlorinated PCBs in each layer accounted for 85.13–90.67% of the total PCB content. It is noteworthy that the concentration order of low-chlorinated PCBs in each layer was the same as $\sum_8$PCBs, and the vertical distribution trend was similar to that of $\sum_8$PCBs (Figure 2). The distribution of PCBs in the study area was dominated by low-chlorinated PCBs. The detection rate of high-chlorinated PCBs ranged from 5% to 60%. In contrast with low-chlorinated PCBs, the content of high-chlorinated PCBs in the 20–40 cm soil layer was higher than that in the 80–100 cm soil layer, indicating that it was more difficult for high-chlorinated PCBs to migrate to deeper soil than low-chlorinated PCBs.

The composition of PCBs in each soil layer of the study area consisted mainly of PCB1 and PCB47. The proportion of high-chlorinated PCBs in each layer was lower than 15%, the monomer of high-chlorinated PCBs was mainly PCB98, and the content of 6–8 PCBs in each layer was lower than 1 ng·g$^{-1}$. Pentachlorobiphenyl is often used as domestic paint additives, it is speculated that PCBs in the study area were related to domestic paint additives. Our results indicate that the use of and residue from PCB-containing products are potential sources of PCBs in farmland soil in the Yellow River irrigation area of Inner Mongolia. The composition characteristics of PCBs in the topsoil were as follows: PCB1 (34.21%) > PCB47 (30.10%) > PCB29 (12.37%) > PCB98 (11.94%) > PCB5 (8.94%) > PCB154 (1.55%) > PCB171 (0.45%) > PCB201 (0.44%) [4]. This is essentially consistent with the composition structure of PCBs in the Inner Mongolia section of the Yellow River (PCB47 > PCB29 > PCB5 > PCB154 >

PCB201 > PCB98 > PCB171). The component distribution of PCBs in the soil profile (Figure 2) suggested that the types of PCBs in the deep soil were similar to those in the shallow soil.

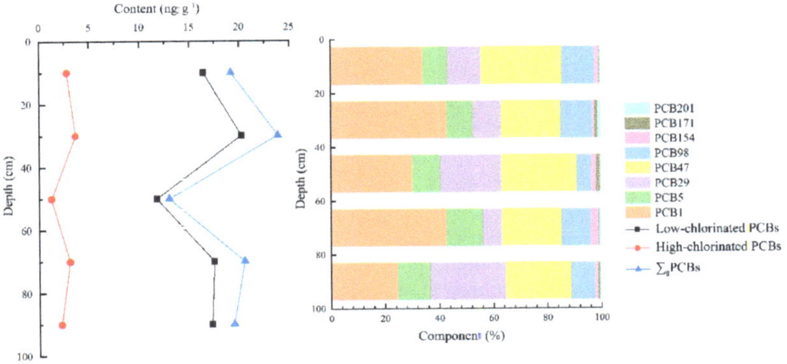

**Figure 2.** Concentration and component distribution of PCBs in soil profiles.

### 3.1.2. Spatial Distribution of PCBs

The distribution trend of PCBs in the Yellow River irrigation area of Inner Mongolia is shown in Table 3, the order of mean $\Sigma_8$PCB concentrations in different soil profiles in different irrigation areas was Wulanbuhe > Dengkou and National Unity > Madihao > Jiefangzha. The degree of pollution was the most serious in the Wulanbuhe irrigation area, and the mean concentration of $\Sigma_8$PCBs in each soil core in this irrigation area ranged from 24.24 to 68.50 ng·g$^{-1}$, which was approximately 3–6 orders of magnitude higher than those in other irrigation areas. The residual PCBs in the other irrigation areas were low.

**Table 3.** Contents of PCBs in different irrigation areas (ng·g$^{-1}$).

| Homologue | Soil Profiles | Jiefangzha Irrigation | | Wulanbuhe Irrigation | | Madihao Irrigation | | Dengkou and National Unity Irrigation | |
|---|---|---|---|---|---|---|---|---|---|
| | | Range | Mean | Range | Mean | Range | Mean | Range | Mean |
| Low-chlorinated PCBs | 0–20 cm | 4.21–29.98 | 12.37 | 4.28–74.01 | 28.03 | 9.77–14.48 | 12.13 | 10.23–26.10 | 20.88 |
| | 20–40 cm | 3.21–49.16 | 10.82 | 54.82–57.39 | 55.74 | 7.82–14.58 | 11.2 | 17.11–37.99 | 23.96 |
| | 40–60 cm | ND–14.01 | 8.24 | 12.00–29.56 | 22.07 | 10.76–12.39 | 11.57 | 4.24–23.23 | 13.79 |
| | 60–80 cm | ND–20.93 | 7.99 | 3.56–96.51 | 43.5 | 12.79–17.94 | 15.37 | 9.12–40.58 | 25.02 |
| | 80–100 cm | ND–23.46 | 8.8 | 0.28–120.65 | 42.95 | 5.73–16.44 | 11.09 | ND–35.91 | 24.24 |
| High-chlorinated PCBs | 0–20 cm | ND–2.40 | 0.83 | 9.68–15.23 | 12.92 | 2.33–2.57 | 2.45 | ND–2.29 | 0.57 |
| | 20–40 cm | ND–8.75 | 1.93 | 6.39–18.15 | 12.77 | 2.36–6.53 | 4.45 | ND–37.00 | 0.92 |
| | 40–60 cm | ND–3.71 | 1.2 | ND–6.51 | 2.17 | 2.06–2.38 | 2.22 | ND | ND |
| | 60–80 cm | ND–12.58 | 1.97 | ND–22.49 | 10.17 | 3.88–4.87 | 4.37 | ND | ND |
| | 80–100 cm | ND–6.38 | 1.28 | ND–9.53 | 5.67 | 5.36–5.20 | 5.78 | ND–1.40 | 0.35 |
| $\Sigma_8$PCBs | 0–20 cm | 4.72–29.98 | 13.2 | 15.46–89.24 | 40.95 | 12.10–17.05 | 14.57 | 10.23–26.10 | 21.45 |
| | 20–40 cm | ND–57.91 | 12.75 | 61.39–75.54 | 68.5 | 10.19–21.11 | 15.65 | 17.11–37.99 | 24.88 |
| | 40–60 cm | ND–17.72 | 9.44 | 18.51–29.56 | 24.24 | 13.13–14.45 | 13.79 | 4.24–23.23 | 13.79 |
| | 60–80 cm | ND–30.44 | 9.97 | 11.58–119.00 | 53.67 | 16.67–22.81 | 19.74 | 9.12–40.58 | 25.02 |
| | 80–100 cm | ND–29.82 | 10.07 | 7.77–120.65 | 48.62 | 11.09–22.64 | 16.86 | ND–35.91 | 24.59 |

The content of PCBs in different irrigation plots was dominated by low-chlorinated PCBs, and the highest concentrations of low- and high-chlorinated PCBs were observed in the Wulanbuhe irrigation area soil profile. The distribution characteristics of low-chlorinated biphenyls in different irrigation areas were basically consistent with $\Sigma_8$PCEs; high-chlorinated PCBs were significantly different, and the mean concentrations of high-chlorinated PCBs from high to low were Wulanbuhe > Madihao > Jiefangzha > Dengkou and National Unity. The high-chlorinated PCBs in the Dengkou and National Unity pumping irrigation areas were lower than 1 ng·g$^{-1}$, and were not detected in the 40–80 cm soil layers.

Figure 3 shows the concentration and component distribution of PCBs in the soil pro-files of the different irrigation fields in the study area. The trend of low-chlorinated PCBs and $\sum_8 PCBs$ was consistent across all irrigation fields. The concentrations of low-chlorinated PCBs and $\sum_8 PCBs$ in the surface layer of the Jiefangzha irrigation area showed accumulation characteristics, gradually decreasing from 0–60 cm, and slightly increasing from 60–100 cm, with a small variation range. Diametrically opposite, in Madihao irrigation area, the PCBs distribution of 0–60 cm was stable, and the concentration of 60–80 cm increased to a certain extent, which accumulated in deep soil. Variation characteristics in the Wulanbuhe irrigation area were similar to Dengkou and National unity pumping irrigation areas. Moving down the soil profile, the concentration first increased, then decreased, and then increased, with relatively significant changes; the concentration range was wider in the Wulanbuhe irrigation area. The content of high-chlorinated PCBs in different sections in various irrigation fields showed little difference, and the vertical distribution was uniform without obvious change characteristic patterns.

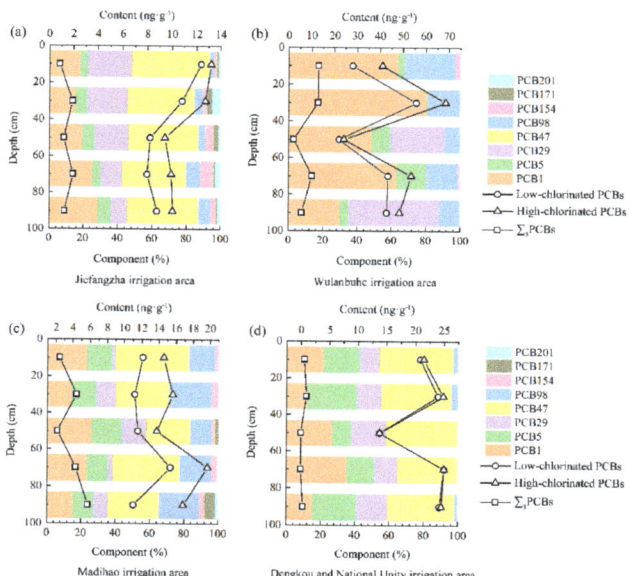

**Figure 3.** Concentration and component distribution of PCBs in soil profiles of different irrigation areas. (**a**) Jiefangzha irrigation areas; (**b**) Wulanbuhe irrigation areas; (**c**) Madihao irrigation areas; (**d**) Dengkou and National Unity pumping irrigation areas.

The vertical distribution of each component of PCBs in the soil was analyzed. All monomers were detected in different soil layers in the Jiefangzha irrigation area, with a small degree of dispersion. Among them, PCB1 and PCB29 showed obvious trends. With increasing depth, the proportion of PCB1 gradually increased, whereas that of PCB29 gradually decreased. The soil layers in the Wulanbuhe irrigation area were dominated by PCB1, PCB29, and PCB98, indicating that the pollution of PCBs in this area was related to the domestic PCBs products, which was consistent with the above speculation. PCB1 was widely distributed from 0–80 cm, PCB29 was concentrated in deep soil, and PCB98 was concentrated in shallow soil. The vertical distribution of each monomer in the profiles of Madihao, Dengkou and National Unity pumping irrigation area were uniform.

### 3.2. Source Apportionment of PCBs in Shallow Soil

PCBs are synthetic compounds, and the main source of environmental PCB pollution is the production and manufacture of products containing PCBs. After entering soil, air, water, and other media, PCBs can undergo metabolic transformation through physical, chemical, and biological processes. Their migration and transformation behaviors include adsorption and desorption [25], abiotic degradation (photolysis [26], hydrolysis [27], volatilization [28], biodegradation [29], and bioaccumulation [30]. The soil in the present study received PCBs mainly through atmospheric deposition, water irrigation, and biological decay.

The Aroclor series in the United States is the most common PCB product in the global PCBs production history; typical Aroclor products include [31] Aroclor1016, Aroclor 1221, Aroclor1232, Aroclor1242, Aroclor1248, Aroclor1254, Aroclor1260, Aroclor1262, and Aroclor1268. Of PCBs produced in China, 90% are trichlorinated biphenyls, which are mostly used in the production of capacitors and transformers, and 10% are pentachlorinated biphenyls, which are commonly used as paint additives [23]. The main components of these PCB products and domestic transformer oils are summarized in Table 4 [32,33].

Table 4. Main components of some PCB products (%).

| Homologue | Aroclor | | | | | | | | | Domestic Transformers |
| --- | --- | --- | --- | --- | --- | --- | --- | --- | --- | --- |
| | 1016 | 1221 | 1232 | 1242 | 1248 | 1254 | 1260 | 1262 | 1268 | |
| Monochlorobiphenyl | 1.1 | 63 | 35 | 1.4 | 2 | | | | | |
| Dichlorobiphenyl | 17 | 31 | 19 | 18 | 18 | 0.3 | 0.24 | 0.33 | | 9 |
| Trichlorinated biphenyl | 48 | 4.3 | 19 | 37 | 4 | 0.7 | 0.41 | 1.1 | | 63 |
| Tetrachlorobiphenyl | 33 | 1.4 | 21 | 37 | 36 | 14 | 0.95 | 1.3 | 0.1 | 24 |
| Pentachlorinated biphenyl | 0.77 | 0.18 | 4.2 | 6.5 | 4 | 54 | 8.2 | 2.9 | 0.15 | 4 |
| Hexachlorinated biphenyl | | | 0.56 | | | 26 | 39 | 22 | 0.49 | |
| Heptachlorobiphenyl | | | | | | 4.1 | 43 | 53 | 5.6 | |
| Octachlorobiphenyl | | | | | | 0.43 | 7.3 | 17 | 41 | |

The sources of PCB pollution in agricultural soil in the Yellow River irrigation area of Inner Mongolia are complicated owing to the influence of various factors. According to the above analysis, the PCBs content in the study area was higher in the shallow soil, and they did not readily migrate; the PCBs in the deeper soil were related to the migration of PCBs in the shallow zone. Therefore, the main product sources of PCBs in the shallow soil were identified. Cluster analysis was used to analyze the relationship of PCBs components in soil samples from the Hetao and Tumochuan irrigation areas, domestic transformer oil, and Aroclor series PCB products to determine the possible sources of PCBs in the study area.

The clustering results of soil samples in the Hetao irrigation area, domestic transformer oil, and Aroclor series PCB products are shown in Figure 4, which can be divided into five categories. The first category included 11 shallow soil samples from the Hetao irrigation area, four types of Aroclor industrial products (Aroclor1242, Aroclor1248, Aroclor1016, and Aroclor1232), and domestic transformer oil. This grouping implied that the PCBs in these 11 samples may come from the four types of Aroclor industrial products and domestic transformer oil. The second, third, and fourth categories did not include any soil samples: the second category included Aroclor1260 and Aroclor1262; the third category was only Aroclor1268; and the fourth category was Aroclor1254. The fifth category included three soil samples and Aroclor1221, which, therefore, may be the source of PCB contamination in these three samples. In conclusion, PCBs in 78.57% of samples in shallow soil in the Hetao irrigation area originated from Aroclor1242, Aroclor1248, Aroclor1016, and Aroclor1232 products and domestic transformer oil. On the other hand, PCBs in 21.43% of the samples were obtained from Aroclor1221 products.

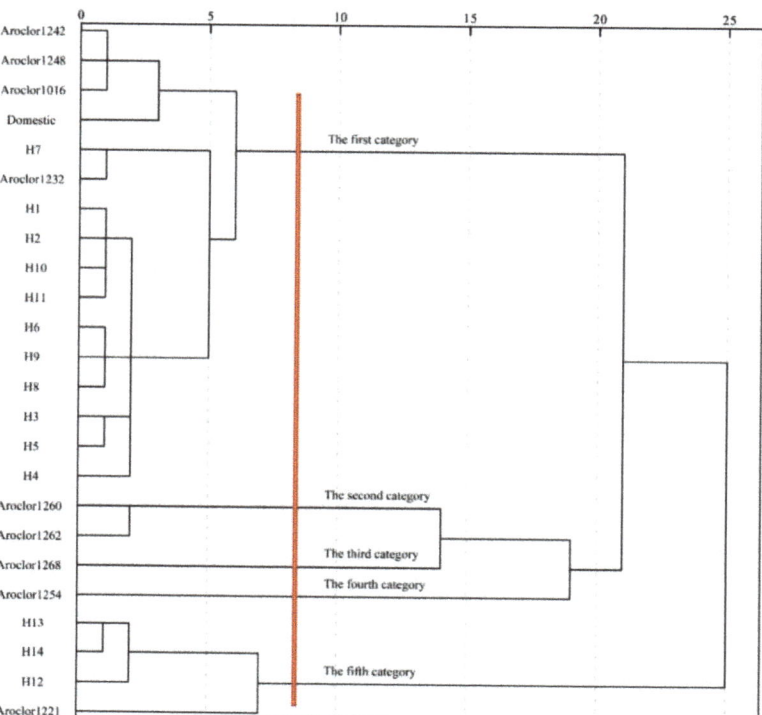

**Figure 4.** Hierarchial dendogram for soil samples in Hetao irrigation area, Aroclor PCBs and Chinese transformer oil.

The clustering results of soil samples in the Tumochuan irrigation area, domestic transformer oil, and Aroclor series PCB products are shown in Figure 5, which can also be divided into five categories. The first category included six samples of shallow soil in the Hetao irrigation area, four types of Aroclor industrial products (Aroclor1242, Aroclor1248, Aroclor1016, and Aroclor1232), and domestic transformer oil. Therefore, the PCBs of the above six sample points may come from these four types of Aroclor industrial products and domestic transformer oils. The second category was Aroclor1221, the third included Aroclor1260 and a Aroclor1262, the fourth was Aroclor1268, and the fifth was Aroclor1254. In conclusion, PCBs in shallow soil of Tumochuan irrigation area originated from Aroclor1242, Aroclor1248, Aroclor1016, and Aroclor1232 industrial products and domestic transformer oil.

Combined with the above analysis results, the PCB pollution in shallow soil in the Yellow Irrigation area of Inner Mongolia mainly came from Aroclor1242, Aroclor1248, Aroclor1016, and Aroclor1232 products and domestic transformer oil, and a small part came from Aroclor1221 products.

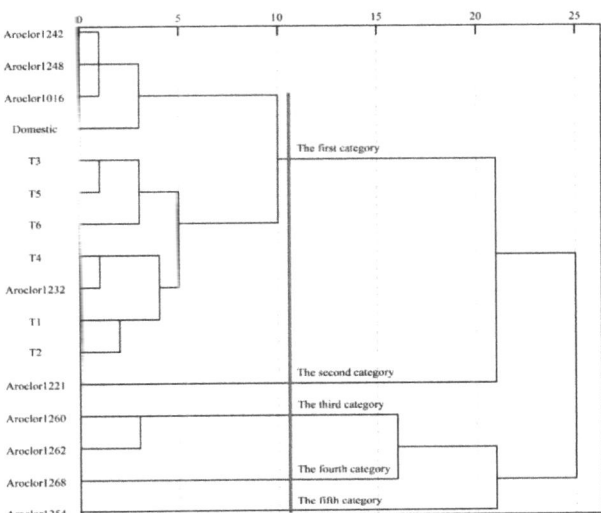

**Figure 5.** Hierarchial dendogram for soil samples in Tumochuan irrigation area, Aroclor PCBs and Chinese transformer oil.

### 3.3. Health Risk Assessment of PCBs in Shallow Soil

The carcinogenic and non-carcinogenic risk of PCBs in 0–20 and 20–40 cm soil through ingestion, inhalation, and dermal contact was assessed and is presented in Tables 5 and 6. PCBs in the study area pose a certain risk of cancer in children. The cumulative carcinogenic risks to adults and children in the 0–20 cm soil were $5.46 \times 10^{-7}$ and $1.15 \times 10^{-6}$, respectively, and $6.80 \times 10^{-7}$ and $1.42 \times 10^{-6}$, respectively, in 20–40 cm soil. The carcinogenic risk values of PCBs to children in both soil layers were higher than the minimum acceptable risk values stipulated by the USEPA, and the potential threat of PCBs to children could not be ignored. From the perspective of different exposure pathways, the carcinogenic risk to adults and children is in the order of oral ingestion > skin contact > respiratory inhalation. The carcinogenic risk to children caused by oral ingestion in soil of 20–40 cm is greater than $10^{-6}$, and is thus the main pathway of carcinogenic harm to children. The contribution rate of respiratory inhalation to carcinogenic risk in the two groups was very small.

The cumulative hazard quotients of adults and children in 0–20 cm soil were 0.035 and 0.290, respectively, and 0.043 and 0.361, respectively, in 20–40 cm soil, which did not cause non-carcinogenic harm to adults and children. Under three exposure modes, the non-carcinogenic risk of oral exposure was the highest, followed by skin contact, and the respiratory route hardly had any risk. It is worth paying attention to that, regardless of the exposure route, the carcinogenic and non-carcinogenic risks in children (who are less physically strong than adults) are higher than those in adults; thus, the potential threat posed by PCBs to children should be carefully considered.

The carcinogenic and non-carcinogenic risks of different monomers in the 0–20 cm soil were as follows: PCB1 > PCB47 > PCB29 > PCB98 > PCB5 > PCB154 > PCB171 > PCB201; and in 20–40 cm soil were as follows: PCB1 > PCB47 > PCB98 > PCB29 > PCB5 > PCB201 > PCB154 > PCB171, which corresponds to the occurrence level of each monomer in the soil, and the health risk caused by 20–40 cm soil was greater than that caused by 0–20 cm soil.

Table 5. Carcinogenic and non-carcinogenic risks of PCBs in 0–20 cm soil.

| Receptors | Monomer | Carcinogenic | | | | Non-Carcinogenic | | | |
|---|---|---|---|---|---|---|---|---|---|
| | | $R_{ing}$ | $R_{der}$ | $R_{inh}$ | $R_T$ | $HQ_{ing}$ | $HQ_{der}$ | $HQ_{inh}$ | HI |
| Adults | PCB1 | $1.23 \times 10^{-7}$ | $6.38 \times 10^{-8}$ | $1.58 \times 10^{-14}$ | $1.87 \times 10^{-7}$ | $7.80 \times 10^{-3}$ | $4.05 \times 10^{-3}$ | $9.18 \times 10^{-7}$ | $1.19 \times 10^{-2}$ |
| | PCB5 | $3.22 \times 10^{-8}$ | $1.67 \times 10^{-8}$ | $4.13 \times 10^{-15}$ | $4.89 \times 10^{-8}$ | $2.04 \times 10^{-3}$ | $1.06 \times 10^{-3}$ | $2.40 \times 10^{-7}$ | $3.10 \times 10^{-3}$ |
| | PCB29 | $4.45 \times 10^{-8}$ | $2.31 \times 10^{-8}$ | $5.71 \times 10^{-15}$ | $6.76 \times 10^{-8}$ | $2.82 \times 10^{-3}$ | $1.46 \times 10^{-3}$ | $3.32 \times 10^{-7}$ | $4.29 \times 10^{-3}$ |
| | PCB47 | $1.08 \times 10^{-7}$ | $5.62 \times 10^{-8}$ | $1.39 \times 10^{-14}$ | $1.64 \times 10^{-7}$ | $6.87 \times 10^{-3}$ | $3.56 \times 10^{-3}$ | $8.08 \times 10^{-7}$ | $1.04 \times 10^{-2}$ |
| | PCB98 | $4.30 \times 10^{-8}$ | $2.23 \times 10^{-8}$ | $5.51 \times 10^{-15}$ | $6.53 \times 10^{-8}$ | $2.72 \times 10^{-3}$ | $1.41 \times 10^{-3}$ | $3.20 \times 10^{-7}$ | $4.14 \times 10^{-3}$ |
| | PCB154 | $5.56 \times 10^{-9}$ | $2.89 \times 10^{-9}$ | $7.14 \times 10^{-16}$ | $8.45 \times 10^{-9}$ | $3.53 \times 10^{-4}$ | $1.83 \times 10^{-4}$ | $4.15 \times 10^{-8}$ | $5.36 \times 10^{-4}$ |
| | PCB171 | $1.61 \times 10^{-9}$ | $8.36 \times 10^{-10}$ | $2.07 \times 10^{-16}$ | $2.45 \times 10^{-9}$ | $1.02 \times 10^{-4}$ | $5.30 \times 10^{-5}$ | $1.20 \times 10^{-8}$ | $1.55 \times 10^{-4}$ |
| | PCB201 | $1.60 \times 10^{-9}$ | $8.28 \times 10^{-10}$ | $2.05 \times 10^{-16}$ | $2.43 \times 10^{-9}$ | $1.01 \times 10^{-4}$ | $5.25 \times 10^{-5}$ | $1.19 \times 10^{-8}$ | $1.54 \times 10^{-4}$ |
| | $\Sigma_8$PCBs | $3.60 \times 10^{-7}$ | $1.87 \times 10^{-7}$ | $4.61 \times 10^{-14}$ | $5.46 \times 10^{-7}$ | $2.28 \times 10^{-2}$ | $1.18 \times 10^{-2}$ | $2.68 \times 10^{-6}$ | $3.47 \times 10^{-2}$ |
| Children | PCB1 | $2.87 \times 10^{-7}$ | $1.05 \times 10^{-7}$ | $9.21 \times 10^{-15}$ | $3.92 \times 10^{-7}$ | $7.28 \times 10^{-2}$ | $2.65 \times 10^{-2}$ | $2.14 \times 10^{-6}$ | $9.93 \times 10^{-2}$ |
| | PCB5 | $7.51 \times 10^{-8}$ | $2.73 \times 10^{-8}$ | $2.41 \times 10^{-15}$ | $1.02 \times 10^{-7}$ | $1.90 \times 10^{-2}$ | $6.93 \times 10^{-3}$ | $5.60 \times 10^{-7}$ | $2.60 \times 10^{-2}$ |
| | PCB29 | $1.04 \times 10^{-7}$ | $3.78 \times 10^{-8}$ | $3.33 \times 10^{-15}$ | $1.42 \times 10^{-7}$ | $2.63 \times 10^{-2}$ | $9.59 \times 10^{-3}$ | $7.75 \times 10^{-7}$ | $3.59 \times 10^{-2}$ |
| | PCB47 | $2.53 \times 10^{-7}$ | $9.20 \times 10^{-8}$ | $8.10 \times 10^{-15}$ | $3.45 \times 10^{-7}$ | $6.41 \times 10^{-2}$ | $2.33 \times 10^{-2}$ | $1.89 \times 10^{-6}$ | $8.74 \times 10^{-2}$ |
| | PCB98 | $1.00 \times 10^{-7}$ | $3.65 \times 10^{-8}$ | $3.21 \times 10^{-15}$ | $1.37 \times 10^{-7}$ | $2.54 \times 10^{-2}$ | $9.26 \times 10^{-3}$ | $7.48 \times 10^{-7}$ | $3.47 \times 10^{-2}$ |
| | PCB154 | $1.30 \times 10^{-8}$ | $4.73 \times 10^{-9}$ | $4.16 \times 10^{-16}$ | $1.77 \times 10^{-8}$ | $3.29 \times 10^{-3}$ | $1.20 \times 10^{-3}$ | $9.68 \times 10^{-8}$ | $4.49 \times 10^{-3}$ |
| | PCB171 | $3.76 \times 10^{-9}$ | $1.37 \times 10^{-9}$ | $1.21 \times 10^{-16}$ | $5.13 \times 10^{-9}$ | $9.54 \times 10^{-4}$ | $3.47 \times 10^{-4}$ | $2.80 \times 10^{-8}$ | $1.30 \times 10^{-3}$ |
| | PCB201 | $3.73 \times 10^{-9}$ | $1.36 \times 10^{-9}$ | $1.19 \times 10^{-16}$ | $5.08 \times 10^{-9}$ | $9.45 \times 10^{-4}$ | $3.44 \times 10^{-4}$ | $2.78 \times 10^{-8}$ | $1.29 \times 10^{-3}$ |
| | $\Sigma_8$PCBs | $8.40 \times 10^{-7}$ | $3.06 \times 10^{-7}$ | $2.69 \times 10^{-14}$ | $1.15 \times 10^{-6}$ | $2.13 \times 10^{-1}$ | $7.75 \times 10^{-2}$ | $6.26 \times 10^{-6}$ | $2.90 \times 10^{-1}$ |

Table 6. Carcinogenic and non-carcinogenic risks of PCBs in 20–40 cm soil.

| Receptors | Monomer | Carcinogenic | | | | Non-Carcinogenic | | | |
|---|---|---|---|---|---|---|---|---|---|
| | | $R_{ing}$ | $R_{der}$ | $R_{inh}$ | $R_T$ | $HQ_{ing}$ | $HQ_{der}$ | $HQ_{inh}$ | HI |
| Adults | PCB1 | $1.92 \times 10^{-7}$ | $9.97 \times 10^{-8}$ | $2.47 \times 10^{-14}$ | $2.92 \times 10^{-7}$ | $1.22 \times 10^{-2}$ | $6.32 \times 10^{-3}$ | $1.43 \times 10^{-6}$ | $1.85 \times 10^{-2}$ |
| | PCB5 | $4.20 \times 10^{-8}$ | $2.18 \times 10^{-8}$ | $5.38 \times 10^{-15}$ | $6.38 \times 10^{-8}$ | $2.66 \times 10^{-3}$ | $1.38 \times 10^{-3}$ | $3.13 \times 10^{-7}$ | $4.04 \times 10^{-3}$ |
| | PCB29 | $4.76 \times 10^{-8}$ | $2.47 \times 10^{-8}$ | $6.10 \times 10^{-15}$ | $7.23 \times 10^{-8}$ | $3.02 \times 10^{-3}$ | $1.57 \times 10^{-3}$ | $3.55 \times 10^{-7}$ | $4.58 \times 10^{-3}$ |
| | PCB47 | $9.81 \times 10^{-8}$ | $5.09 \times 10^{-8}$ | $1.26 \times 10^{-14}$ | $1.49 \times 10^{-7}$ | $6.22 \times 10^{-3}$ | $3.23 \times 10^{-3}$ | $7.32 \times 10^{-7}$ | $9.45 \times 10^{-3}$ |
| | PCB98 | $5.31 \times 10^{-8}$ | $2.75 \times 10^{-8}$ | $6.80 \times 10^{-15}$ | $8.06 \times 10^{-8}$ | $3.36 \times 10^{-3}$ | $1.74 \times 10^{-3}$ | $3.96 \times 10^{-7}$ | $5.11 \times 10^{-3}$ |
| | PCB154 | $4.27 \times 10^{-9}$ | $2.21 \times 10^{-9}$ | $5.47 \times 10^{-16}$ | $6.48 \times 10^{-9}$ | $2.70 \times 10^{-4}$ | $1.40 \times 10^{-4}$ | $3.18 \times 10^{-8}$ | $4.11 \times 10^{-4}$ |
| | PCB171 | $4.04 \times 10^{-9}$ | $2.09 \times 10^{-9}$ | $5.18 \times 10^{-16}$ | $6.13 \times 10^{-9}$ | $2.56 \times 10^{-4}$ | $1.33 \times 10^{-4}$ | $3.01 \times 10^{-8}$ | $3.89 \times 10^{-4}$ |
| | PCB201 | $6.39 \times 10^{-9}$ | $3.32 \times 10^{-9}$ | $8.20 \times 10^{-16}$ | $9.71 \times 10^{-9}$ | $4.05 \times 10^{-4}$ | $2.10 \times 10^{-4}$ | $4.77 \times 10^{-8}$ | $6.15 \times 10^{-4}$ |
| | $\Sigma_8$PCBs | $4.48 \times 10^{-7}$ | $2.32 \times 10^{-7}$ | $5.74 \times 10^{-14}$ | $6.80 \times 10^{-7}$ | $2.84 \times 10^{-2}$ | $1.47 \times 10^{-2}$ | $3.34 \times 10^{-6}$ | $4.31 \times 10^{-2}$ |
| Children | PCB1 | $4.49 \times 10^{-7}$ | $1.63 \times 10^{-7}$ | $1.44 \times 10^{-14}$ | $6.12 \times 10^{-7}$ | $1.14 \times 10^{-1}$ | $4.14 \times 10^{-2}$ | $3.35 \times 10^{-6}$ | $1.55 \times 10^{-1}$ |
| | PCB5 | $9.80 \times 10^{-8}$ | $3.57 \times 10^{-8}$ | $3.14 \times 10^{-15}$ | $1.34 \times 10^{-7}$ | $2.49 \times 10^{-2}$ | $9.05 \times 10^{-3}$ | $7.31 \times 10^{-7}$ | $3.39 \times 10^{-2}$ |
| | PCB29 | $1.11 \times 10^{-7}$ | $4.04 \times 10^{-8}$ | $3.56 \times 10^{-15}$ | $1.52 \times 10^{-7}$ | $2.82 \times 10^{-2}$ | $1.03 \times 10^{-2}$ | $8.29 \times 10^{-7}$ | $3.84 \times 10^{-2}$ |
| | PCB47 | $2.29 \times 10^{-7}$ | $8.33 \times 10^{-8}$ | $7.34 \times 10^{-15}$ | $3.12 \times 10^{-7}$ | $5.81 \times 10^{-2}$ | $2.11 \times 10^{-2}$ | $1.71 \times 10^{-6}$ | $7.92 \times 10^{-2}$ |
| | PCB98 | $1.24 \times 10^{-7}$ | $4.51 \times 10^{-8}$ | $3.97 \times 10^{-15}$ | $1.69 \times 10^{-7}$ | $3.14 \times 10^{-2}$ | $1.14 \times 10^{-2}$ | $9.23 \times 10^{-7}$ | $4.28 \times 10^{-2}$ |
| | PCB154 | $9.95 \times 10^{-9}$ | $3.62 \times 10^{-9}$ | $3.19 \times 10^{-16}$ | $1.36 \times 10^{-8}$ | $2.52 \times 10^{-3}$ | $9.19 \times 10^{-4}$ | $7.42 \times 10^{-8}$ | $3.44 \times 10^{-3}$ |
| | PCB171 | $9.42 \times 10^{-9}$ | $3.43 \times 10^{-9}$ | $3.02 \times 10^{-16}$ | $1.28 \times 10^{-8}$ | $2.39 \times 10^{-3}$ | $8.70 \times 10^{-4}$ | $7.03 \times 10^{-8}$ | $3.26 \times 10^{-3}$ |
| | PCB201 | $1.49 \times 10^{-8}$ | $5.43 \times 10^{-9}$ | $4.78 \times 10^{-16}$ | $2.03 \times 10^{-8}$ | $3.78 \times 10^{-3}$ | $1.38 \times 10^{-3}$ | $1.11 \times 10^{-7}$ | $5.16 \times 10^{-3}$ |
| | $\Sigma_8$PCBs | $1.04 \times 10^{-6}$ | $3.80 \times 10^{-7}$ | $3.35 \times 10^{-14}$ | $1.42 \times 10^{-6}$ | $2.65 \times 10^{-1}$ | $9.64 \times 10^{-2}$ | $7.79 \times 10^{-6}$ | $3.61 \times 10^{-1}$ |

## 4. Discussion

In this study, the vertical distribution of PCBs is aggregation in the shallow layer and a sudden decrease in the middle layer. The soil PCBs in the process of vertical migration are more likely to be enriched in shallow soil rich in organic matter: in this regard, the 20–40 cm range of long-term soil is disturbed by farming tools, relatively tight, with low porosity, large soil bulk density, and poor permeability. All of this elevates the concentration of PCBs in the soil. The crops developed strong roots at 40–60 cm. On the other hand, crop roots have the ability to absorb PCBs in the soil rhizosphere; additionally, roots secrete biosurfactants that can promote mass transfer and the degradation of PCBs in soil. In addition, their metabolic activities provide a suitable micro-ecological environment for microorganisms to survive, which can enhance indigenous microorganisms with the ability to degrade PCBs in soil [34,35]. Therefore, the PCB content in the 40–60 cm soil layer was low. The low concentration of PCBs in the surface soil could be attributed to the continuous loss of surface soil water via evaporation and erosion, and some of the

PCBs were transferred to the atmosphere via water evaporation. However, some PCBs migrated to the lower soil because of gravity, irrigation, and precipitation, among other factors. Increased concentrations of PCBs in deep soils may pose a threat to groundwater. The combination of the long history of irrigation with Yellow River water in this area; the influence of irrigation leaching is far-reaching, and the distribution and migration of PCBs will be affected to a certain extent by soil tillage. The detection rate and content of low-chlorinated PCBs were higher than those of high-chlorinated PCBs in all profiles.

Compared with high-chlorinated PCBs, low-chlorinated PCBs had lower octanol water distribution coefficients (logKow), weaker hydrophobicity, and were more likely to migrate with soil moisture. The soil profiles of the study area were dominated by low-chlorinated PCBs, this characteristic is consistent with the composition profile of PCBs in farmland soil in China [36] and with the low-chlorinated PCBs content in PCB-containing products (24.7% of tetrachlorobiphenyls). PCB98 is the dominant high-chlorinated PCBs in each soil layer, Pentachlorobiphenyl is often used as domestic paint additives, it is speculated that PCBs in the study area were related to domestic paint additives. The composition characteristics of PCBs in the topsoil is essentially consistent with the composition structure of PCBs in the Inner Mongolia section of the Yellow River, suggesting that irrigation using Yellow River water is one of the reasons for the accumulation of soil PCBs in the study area. Meanwhile, compared with a study on PCBs in the urban atmosphere of Inner Mongolia [37], it was found that the composition of PCBs in the soil of the Yellow River irrigation area was consistent with that in the atmosphere. Therefore, the channels for PCBs entering the soil in the Yellow River irrigation area of Inner Mongolia include irrigation with Yellow River water, dechlorination degradation, and atmospheric deposition over the Inner Mongolia autonomous region. The PCBs in the deep soil may be related to the migration of PCBs in the shallow zone.

The distribution trend of PCBs in the Yellow River irrigation area of Inner Mongolia was higher in the west and lower in the east. The degree of pollution was the most serious in the Wulanbuhe irrigation area. The concentration of mining companies, energy companies, and substations in the Wulanbuhe irrigation area once produced a large amount of tar, plasticizer, and other waste raw materials and waste power equipment, the volatilization and leakage of waste gas and waste liquid led to an increase in PCB content in the soil. The Wulanbuhe irrigation area is close proximity to the entrance of the Inner Mongolia section of the Yellow River. The pollutants discharged into the Yellow River from industrial parks in Wuhai City and the Alxa League, along with irrigation with Yellow River water, enter the farmland soil in the Wulanbuhe irrigation area in large quantities, resulting in high soil pollution. The residual PCBs in the other irrigation areas were low, indicating that there was no centralized point source pollution. The Yellow River has a large sand content and coarse sand, which readily absorb PCBs present in water [38]. Due to adsorption in the higher reaches of the river, the contents of PCBs in the middle and lower reaches of the river were reduced to a certain extent, and the content of PCBs in the irrigation areas was also greatly reduced. In addition, PCBs content is also related to soil organic matter content, physical and chemical properties, and tillage methods; climate change can also cause changes in the soil environment [39], which in turn could impact PCBs contents. The high-chlorinated PCBs in the Dengkou and National Unity pumping irrigation areas were low. It was inferred that there was no historical residue of PCB products, and the degree of transformation from high- to low-chlorinated PCBs by degradation was higher than that of other irrigation fields.

The vertical distribution of components in each irrigation tract in the study area differed; the overall characteristics were as follows:(1) PCB1 is a non-negligible component in all irrigation tracts, with contribution rates ranging from 12–81%, indicating that PCBs in the study area are highly biodegradable. (2) Tetrachlorobiphenyl (26–45%) accounted for the highest proportion in all soil layers of the irrigation areas, except for the Wulanbuhe irrigation area; dichlorobiphenyl (5–29%) and trichlorinated biphenyl (3–25%) also accounted for significant proportions, which agrees with distribution results of PCBs obtained in a study of rural soil in China [36]. (3) Low-chlorinated PCBs migrated to the deep soil more easily, whereas high-chlorinated PCBs tended to accumulate in the upper soil. This finding is related to the differences in the physical and chemical properties, as well as the migration abilities of different monomers.

Analysis of the results of the health risk evaluation found that both carcinogenic and non-carcinogenic risks caused by PCBs in the 0–20 and 20–40 cm soils corresponded to the concentration of each monomer, indicating that the concentration of pollutants in the soil had a great impact on the health risk assessment. Wu analyzed the sensitivity of all risk indices for health risk assessment of nitrogen pollution in groundwater in the Songnen Plain [40]. Their results indicated that pollutant concentration was the most sensitive of all indexes, contributing more than 90% to the risk value, and playing a decisive role in determining the risk value.

## 5. Conclusions

This study investigated the distribution, source, and risk of PCBs in agricultural soils of the Yellow River irrigation area in China. The results indicate the widespread occurrence of PCBs, even though these compounds have been banned for over 30 years, due to their extensive historical usage and persistent nature. Low-chlorinated PCBs were dominant in each section. Source identification indicated that PCB pollution in the study mainly originated from Aroclor1242, Aroclor1248, Aroclor1016, Aroclor1232, and Aroclor1221 industrial products and domestic transformer oil. The lifetime carcinogenic and non-carcinogenic risks of PCBs through ingestion, inhalation, and dermal contact indicate that PCB residues in agricultural soils were at a low-risk level.

**Author Contributions:** Q.Z.: Methodology, Investigation, Data Curation, Formal Analysis, Visualization, Writing—Original Draft; Y.L.: Conceptualization, Resources, Project Administration, Data Curation, Methodology and Writing—Review and Editing; Q.M.: Software, Investigation, Data Curation, Formal Analysis, Visualization, Writing—Original Draft; G.P.: Conceptualization, Resources, Funding Acquisition, Project Administration, Validation, Supervision, Writing—Review and Editing; Y.N.: Visualization and Supervision; S.Y.: Validation and Writing—Review and Editing; X.M.: Investigation; W.F.: Modified, editing and funding. All authors have read and agreed to the published version of the manuscript.

**Funding:** This work was supported by the National Natural Science Foundation of China (52009056, 51469023), Inner Mongolia Agricultural University High-level Talents Introduction Scientific Research Startup Project (NDYB2016-22), the Inner Mongolia Autonomous Region Science and Technology Department (2021MS04012), and the Science and Technology Plan Project of Inner Mongolia Autonomous Region (2022YFHH0044).

**Data Availability Statement:** The data that support the findings of this study are available on request from the corresponding author. The data are not publicly available due to privacy restrictions.

**Conflicts of Interest:** The authors declare no conflict of interest.

**Consent to Participate:** Not applicable.

**Consent for publish:** Not applicable.

## References

1. United Nations. World Water Development Report. 2019. Available online: https://www.unwater.org/publications/world-water-development-report2019/ (accessed on 17 October 2022).
2. United Nations. World Water Development Report. 2021. Available online: https://www.unwater.org/publications/world-water-development-report2021/ (accessed on 17 October 2022).
3. Wu, F.; Li, F.; Zhao, X.; Bolan, N.S.; Fu, P.; Lam, S.S.; Mašek, O.; Ong, H.C.; Pan, B.; Qiu, X.; et al. Meet the challenges in the "Carbon Age". *Carbon Res.* **2022**, *1*, 1. [CrossRef]
4. Pei, G.X.; Zhang, Y.; Ma, T.L.; Tian, C.Y.; Ren, Z.H. Distribution of HCHs and PCBs in water body of Inner Mongolia section of Yellow River. *J. Water Resour. Water Eng.* **2010**, *21*, 25–27, 33.

5. Net, S.; Henry, F.; Rabodonirina, S.; Diop, M.; Merhaby, D.; Mahfouz, C.; Amara, R.; Ouddane, B. Accumulation of PAHs, Me-PAHs, PCBs and total Mercury in sediments and Marine Species in Coastal Areas of Dakar, Senegal: Contamination level and impact. *Int. J. Environ. Res.* **2015**, *9*, 419–432.
6. Ranjbaran, S.; Sobhanardakani, S.; Cheraghi, M.; Lorestani, B.; Sadr, M.K. Ecological and human health risks assessment of some polychlorinated biphenyls (PCBs) in surface soils of central and southern parts of city of Tehran, Iran. *J. Environ. Health Sci. Eng.* **2021**, *19*, 1491–1503. [CrossRef]
7. Donato, F.; Moneda, M.; Portolani, N.; Rossini, A.; Molfino, S.; Ministrini, S.; Contessi, G.B.; Pesenti, S.; De Palma, G.; Gaia, A.; et al. Polychlorinated biphenyls and risk of hepatocellular carcinoma in the population living in a highly polluted area in Italy. *Sci. Rep.* **2021**, *11*, 3064. [CrossRef]
8. Khalid, F.; Hashmi, M.Z.; Jamil, N.; Qadir, A.; Ali, M.I. Microbial and enzymatic degradation of PCBs from e-waste-contaminated sites A review. *Environ. Sci. Pollut. Res.* **2021**, *28*, 10474–10487. [CrossRef]
9. Agbo, I.A.; Abaye, D. Levels of Polychlorinated Biphenyls in Plastic Resin Pellets from Six Beaches on the Accra-Tema Coastline, Ghana. *J. Health Pollut.* **2016**, *6*, 9–17. [CrossRef]
10. Liu, C.; Wei, B.K.; Bao, J.S.; Wang, Y.; Hu, J.C.; Tang, Y.E.; Chen, T.; Jin, J. Polychlorinated biphenyls in the soil-crop-atmosphere system in e-waste dismantling areas in Taizhou: Concentrations, congener profiles, uptake, and translocation. *Environ. Pollut.* **2020**, *257*, 113622. [CrossRef]
11. Sun, L.X.; Mao, J.; Liu, T.F.; Yang, D.F. Analysis of polychlorinated biphenyls pollution status in farmland soils of south Jiangsu under different land-use types. *J. Food Saf. Qual.* **2019**, *10*, 5615–5620. [CrossRef]
12. Lu, Y.T.; Liu, M.L.; Wang, J.; Zhang, S.C.; Yao, H.; Sun, S.B. Distribution characteristics and ecological risk assessment of polychlorinated biphenyls in farmland soil of Tongliao City. *J. Beijing Jiaotong Univ.* **2017**, *41*, 61–69. [CrossRef]
13. Cetin, B. Investigation of PAHs, PCBs and PCNs in soils around a Heavily Industrialized Area in Kocaeli, Turkey: Concentrations, distributions, sources and toxicological effects. *Sci. Total Environ.* **2016**, *560*, 160–169. [CrossRef] [PubMed]
14. Haddaoui, I.; Mahjoub, O.; Mahjoub, B.; Boujelben, A.; Di Bella, G. Occurrence and distribution of PAHs, PCBs, and chlorinated pesticides in Tunisian soil irrigated with treated wastewater. *Chemosphere* **2016**, *146*, 195–205. [CrossRef] [PubMed]
15. Han, S.L.; Wang, B.S.; Ruan, T.; Wang, Y.W.; Fu, J.J.; Hu, J.T.; Jiang, G.B. Within-field spatial distribution of polychlorinated biphenyls and polybrominated diphenyl ethers in farm soils with different irrigation sources. *Environ. Chem.* **2012**, *31*, 958–965.
16. Kumar, B.; Mishra, M.; Verma, V.K.; Rai, P.; Kumar, S. Organochlorines in urban soils from Central India: Probabilistic health hazard and risk implications to human population. *Environ. Geochem. Health* **2018**, *40*, 2465–2480. [CrossRef]
17. Abrahao, R.; Sarasa, J.; Causape, J.; Garcia-Garizabal, I.; Ovelleiro, J.L. Influence of irrigation on the occurrence of organic and inorganic pollutants in soil, water and sediments of a Spanish agrarian basin (Lerma). *Span. J. Agric. Res.* **2011**, *9*, 124–134. [CrossRef]
18. Teng, M.; Zhang, H.; Fu, Q.; Lu, X.; Chen, J.; Wei, F. Irrigation-induced pollution of organochlorine pesticides and polychlorinated biphenyls in paddy field ecosystem of Liaohe River Plain, China. *Chin. Sci. Bull.* **2013**, *58*, 1751–1759. [CrossRef]
19. Ngweme, G.N.; Al Salah, D.M.o.h.a.m.m.e.d.M.; Laffite, A.; Sivalingam, P.; Grandjean, D.; Konde, J.N.; Mulaji, C.K.; Breider, F.; Poté, J. Occurrence of organic micropollutants and human health risk assessment based on consumption of Amaranthus viridis, Kinshasa in the Democratic Republic of the Congo. *Sci. Total Environ.* **2021**, *754*, 142175. [CrossRef] [PubMed]
20. United States Environmental Protection Agency (USEPA). *Risk Assessment Guidance for Superfund (Volume 1) Human Health Evaluation Manual*; EPA/540/189/002; Office of Emergency and Remedial Response: Washington, DC, USA, 1989.
21. United States Environmental Protection Agency (USEPA). Regional Screening Levels (RSL) for Chemical Contaminants at Superfund Sites. 2021. Available online: http://www.epa.gov/region9/superfund/prg/ (accessed on 25 May 2022).
22. Li, Y.; Huang, G.H.; Gu, H.; Huang, Q.Z.; Li, L.; Liu, H.L. Assessment of Contamination Risk of PCBs in Soils and Agricultural Products in Typical Irrigation District in Beijing. *Trans. Chin. Soc. Agric. Mach.* **2018**, *49*, 313–322. [CrossRef]
23. Lu, Y.T.; Liu, M.L.; Liu, Y.Z.; Zhang, S.C.; Xiang, X.X.; Yao, H. Characteristics and health risk assessment of polychlorinated biphenyls in surface soil of the Yangtze River. *China Environ. Sci.* **2018**, *38*, 4617–4624. [CrossRef]
24. Chen, X.R.; Wang, Y.; Liu, Q.; Zhang, J.J.; Rui, Y.U.; Cui, Z.W.; Liu, J.S. Residual Characteristics and Health Risk Assessment of Polychlorinated Biphenyls in Suburban Vegetable Soils in Different Industrial Cities. *Soils Crops* **2016**, *5*, 14–23. [CrossRef]
25. Adeyinka, G.C.; Moodley, B. Kinetic and thermodynamic studies on partitioning of polychlorinated biphenyls (PCBs) between aqueous solution and modeled individual soil particle grain sizes. *J. Environ. Sci.* **2019**, *76*, 100–110. [CrossRef] [PubMed]
26. Wu, N.N.; Cao, W.M.; Qu, R.J.; Zhou, D.M.; Sun, C.; Wang, Z.Y. Photochemical transformation of decachlorobiphenyl (PCB-209) on the surface of microplastics in aqueous solution *Chem. Eng. J.* **2021**, *420*, 129813. [CrossRef]
27. Sako, T.; Sugeta, T.; Otake, K.; Kamizawa, C.; Okaro, M.; Negishi, A.; Tsurumi, C. Dechlorination of PCBs with Supercritical Water Hydrolysis. *J. Chem. Eng. Jpn.* **1999**, *32*, 830–832. [CrossRef]
28. Lohmann, R.; Klanova, J.; Kukucka, P.; Yonis, S.; Bollinger, K. PCBs and OCPs on a East-to-West Transect: The Importance of Major Currents and Net Volatilization for PCBs in the Atlantic Ocean. *Environ. Sci. Technol.* **2012**, *46*, 10471–10479. [CrossRef] [PubMed]
29. Nair, S.; Abraham, J. Biodegradation of Polychlorinated Biphenyls. *Microb. Metab. Xenobiotic Compd.* **2019**, *10*, 263–284. [CrossRef]
30. Palladini, J.; Bagnati, R.; Passoni, A.; Davoli, E.; Lanno, A.; Terzaghi, E.; Falakdin, P.; Di Guardo, A. Bioaccumulation of PCBs and their hydroxy and sulfonated metabolites in earthworms: Comparing lab and field results. *Environmental Pollution.* **2021**, *293*, 11507. [CrossRef]

31. Uhler, A.D.; Hardenstine, J.H.; Edwards, D.A.; Lotufo, G.R. Leaching Rate of Polychlorinated Biphenyls (PCBs) from Marine Paint Chips. *Arch. Environ. Contam. Toxicol.* **2021**, *81*, 324–334. [CrossRef]
32. Frame, G.M.; Cochran, J.W.; Bowadt, S.S. Complete PCB congener distributions for 17 aroclor mixtures determined by 3 HRGC systems optimized for comprehensive, quantitative, congener-specific analysis. *J. High Resolut. Chromatogr.* **1996**, *19*, 657–668. [CrossRef]
33. Jiang, Q.L.; Zhou, H.Y.; Xu, D.D.; Chai, Z.F.; Li, Y.F. Characteristics of PCB congenersand homologues in Chinese transformer oil. *China Environ. Sci.* **2007**, *27*, 608–612. [CrossRef]
34. Gomathy, M.; Sabarinathan, K.G.; Subramanian, K.S.; Ananthi, K.; Kalaiyarasi, V.; Jeyshri, M.; Dutta, P. Rhizosphere: Niche for microbial rejuvenation and biodegradation of pollutants. In *Microbial Rejuvenation of Polluted Environment*; Springer: Singapore, 2021; Volume 25, pp. 1–22. [CrossRef]
35. Cai, Z.; Yan, X.; Gu, B. Applying C:N ratio to assess the rationality of estimates of carbon sequestration in terrestrial ecosystems and nitrogen budgets. *Carbon Res.* **2022**, *1*, 2. [CrossRef]
36. Zhang, Z. *Polychlorinated Biphenyls in Chinese Air and Surface Soil: Spatial Distribution Characteristics and Thrie Inherent Causes*; Harbin Institute of Technology: Shenyang, China, 2010; pp. 41–43.
37. Zhang, X.H. *Distribution Characteristics of Polychlorinated Biphenyls (PCBs) in Main Urban Atmospheric Particles of Inner Mongolia*; Inner Mongolia normal University: Hohhot, China, 2015.
38. Zhang, Q.; Pei, G.X.; Liu, G.Y.; Zhang, Y. Temporal Distribution of PCBs in River Water at Toudaoguai Section of the Yellow River. *Arid. Zone Res.* **2014**, *31*, 937–942. [CrossRef]
39. Zhang, Z.; Li, M.; Song, X.L.; Xue, Z.S.; Lv, X.G.; Jiang, M.; Wu, H.T.; Wang, X.H. Effects of Climate Change on Molecular Structure and Stability of Soil Carbon Pool: A General Review. *Acta Pedol. Sin.* **2018**, *55*, 273–282. [CrossRef]
40. Wu, J.; Bian, J.; Wan, H.; Ma, Y.; Sun, X. Health risk assessment of groundwater nitrogen pollution in Songnen Plain. *Ecotoxicol. Environ. Saf.* **2021**, *207*, 111245. [CrossRef] [PubMed]

*Review*

# A Review of Groundwater Contamination in West Bank, Palestine: Quality, Sources, Risks, and Management

Ashraf Zohud and Lubna Alam *

The Institute for Environment and Development (LESTARI), Universiti Kebangsaan Malaysia (The National University of Malaysia), Bangi 43600, Selangor, Malaysia
* Correspondence: lubna@ukm.edu.my

**Abstract:** The contamination and shortages of drinking water in the West Bank are among the most important challenges facing the Palestinian National Authority (PA) and the population residing in all sectors. In general, the contamination of water sources makes it difficult to obtain a sufficient quantity of drinking water of suitable quality, since contaminated water has a harmful effect on health, which profoundly impairs the quality of life. Despite knowledge of the adverse health effects of chemical and biological groundwater contamination, few studies have been conducted to suggest measures that can be taken to overcome the contamination and shortages of water. In our review, four levels of domains are used to evaluate the groundwater situation/condition in the West Bank, including (i) assessing the groundwater quality in the West Bank, (ii) identifying the sources of groundwater pollution, (iii) determining the degree of health risks associated with groundwater pollution, and (iv) determining the role of groundwater management in maintaining the quality and sustainability of these sources. To this end, the previous literature on groundwater status was reviewed for the past 27 years. In order to analyze the existing literature, a review matrix based on these four core domains was developed. Our findings revealed only 5 studies corresponding to the first nine years and 9 and 16 studies in the second and third periods, respectively. Furthermore, we found that only a few studies have examined the degree of health risk of groundwater in the West Bank. Although the government of Palestine has made access to safe drinking water a priority for its population, the PA struggles to provide sufficient and clean water to its residents, with a number suffering from water shortages, especially in dry seasons.

**Keywords:** groundwater; health risks; management; West Bank; Palestinian National Authority

**Citation:** Zohud, A.; Alam, L. A Review of Groundwater Contamination in West Bank, Palestine: Quality, Sources, Risks, and Management. *Water* 2022, 14, 3417. https://doi.org/10.3390/w14213417

Academic Editors: Weiying Feng, Fang Yang and Jing Liu

Received: 15 September 2022
Accepted: 24 October 2022
Published: 27 October 2022

**Publisher's Note:** MDPI stays neutral with regard to jurisdictional claims in published maps and institutional affiliations.

**Copyright:** © 2022 by the authors. Licensee MDPI, Basel, Switzerland. This article is an open access article distributed under the terms and conditions of the Creative Commons Attribution (CC BY) license (https://creativecommons.org/licenses/by/4.0/).

## 1. Introduction

Most of the world's population suffers from a lack of safe water supplies [1]. As the population of the world grows and the environment becomes further affected by human activity, access to fresh drinking water dwindles, and Palestinians in the West Bank, similarly to the rest of the world, suffer from a lack of potable water [2]. PA in the West Bank use groundwater as the main source of water, representing more than approximately 90% of the total water supply [3]. Groundwater, in the form of wells and springs, comprises the main sources of water in the West Bank, the land of which is limestone with karstic characteristics [4]. Spring water is naturally found where ground water emerges from the Earth's surface in a defined flow [5]. While water from natural spring represents an important source of drinking water, its quality is currently being seriously threatened by microbiological and chemical contamination. A spring's water can be described as any natural occurrence where water flows onto the surface of the Earth from below. Springs are key elements of the natural environment that respond sensitively to any changes occurring in natural ecosystems and can therefore be classified as important hydrogeological indicators [6]. However, this form of groundwater is incredibly vulnerable to pollution given the karstic nature of the aquifer and due to various human activities resulting in untreated wastewater, pesticides, chemical fertilizers, livestock farm waste, and unsanitary landfills [2]. Importantly, an increase in any one of the physico-chemical and biological parameters in groundwater beyond the permissible limits indicated by the World Health Organization (WHO) guidelines and the Palestinian national standards (PSI) may result in damage to human health [5,7,8] (Table 1).

Table 1. The Palestinian national standards (PSI) and World Health Organization (WHO) permissible limits for water [7,8].

| WHO | PSI | Unit | Water Tests (Physicochemical Parameters) |
|---|---|---|---|
| - | - | µS/cm | Conductivity of water (EC) |
| 1.5 | 1.5 | mg/L | Fluoride in water (F) |
| 50 | 50 | mg/L | Nitrate in water ($NO_3$) |
| NA | NA | mg/L | $PO_4$ (as P) |
| 6.5–8.5 | 6.5–8.8 | | pH of water |
| 250 | 200 | mg/L | Sulfate in water ($SO_4$) |
| 1000 | 1000 | mg/L | Total dissolved solids (TDS) in water |
| 500 | 500 | mg/L | Total hardness of water (TH) |
| 1.5 | - | mg/L | Ammonia in water ($NH_3$) |
| 250 | 250 | mg/L | Chloride (Cl) |
| - | 100 | mg/L | Calcium (Ca) |
| - | 100 | mg/L | Magnesium (Mg) |
| - | 200 | mg/L | Sodium (Na) |
| - | 10 | mg/L | Potassium (K) |
| - | 0.2 | mg/L | Aluminum (Al) |
| - | 0.3 | mg/L | Iron (Fe) |
| - | 0.1 | mg/L | Manganese (Mn) |
| 2 | 1 | mg/L | Cupper (Cu) |
| - | 5 | mg/L | Zinc (Zn) |
| 0.05 | 0.05 | mg/L | Total chromium (Cr) |
| 0.003 | 0.005 | mg/L | Cadmium (Cd) |
| 0.07 | 0.05 | mg/L | Nickel (Ni) |
| 0.7 | - | mg/L | Barium (Ba) |

Several studies have shown that there are rising concentrations of some chemicals, such as nitrate, in groundwater, which affects the sustainability of this limited water resource [9–11]. Monitoring the sources of pollution in the water recharge areas of the aquifers is extremely important and would help in managing water resources in a highly effective way, thereby enhancing governance of the entire sector [12]. The Oslo I and II Accords provide broad guidelines for how water is used in the West Bank, including pollution management, waste treatment, the extraction and use of natural resources, and the prevention of harm to water infrastructure [13]. It has become necessary to protect water sources from pollution in order to maintain the quality and sustainability of these sources.

The assessment of water quality provides baseline information on water safety, and continuous monitoring of water is essential because water quality in any source of water and at the point of use can change with time and other factors [14]. However, the list of parameters to be tested in any water assessment and monitoring program may vary according to the local conditions of the area. Parameters that are basic and generally considered priorities in any water quality assessment may include physico-chemical, harmful chemicals, and microbiological parameters [15].

Groundwater contamination is often the result of human activity. In areas where population density is high and there is intensive human land use, groundwater is especially vulnerable. Generally, most activities whereby chemicals or wastes may be released to the environment, either intentionally or accidentally, has the potential to pollute groundwater. When groundwater becomes contaminated, cleanup or remediation become difficult and expensive [16]. Biological and chemical pollutants, when they reach groundwater, may cause harm to humans and the environment. An increase in the incidence of waterborne human diseases, such as diarrhea and emesis, occurs due to drinking of polluted water [17]. These waterborne diseases can lead to death if correct treatment is not provided [18]. Assuring the safety of drinking water has been a crucial challenge for public health. Water contamination with pathogenic microorganisms represents a seriously increased threat to human health [6]. Agricultural activities, such as the addition of pesticides and fertilizers, soil washing, and evaporation processes [16], could lead to the emergence of many pollutants, such as nitrate, in groundwater. Moreover, groundwater quality is widely affected by various factors, including human—i.e., agricultural and industrial—activities and, importantly, insufficiently treated sewage [19]. In West Bank, Palestine, cesspits are considered an important source of groundwater pollution, mainly in rural areas, where connection to the mains sewerage network system is inaccessible, impractical, and costly. Cesspit effluents contain a wide variety of chemical and biological pollutants [2,20].

Population growth and urban expansion affect the quantity, quality, and sustainability of water resources. The pressure on water resources in the coming years due to the expected population increase will affect the quantity and quality of water in the sources, and this will undoubtedly affect the sustainability of these sources [21]. Furthermore, the hydrological status in Palestine is unique due to both political and natural conditions. The main natural conditions include (i) scarcity and uneven distribution of rainfall due to extreme topographic variations within the region and (ii) the hydrogeological location of the West Bank, which extends from the upstream portion of the Shared Carbonate Aquifer System to downstream of the Jordan River Basin.

According to the Palestinian Central Bureau of Statistics (PCBS) and Palestinian water authority PWA, the per capita use of water in the West Bank is 73 L/capita/day, which is the share of water for Palestinians and is considerably lower than the 100 L/capita/day minimum recommended by WHO. In some communities in area C, Palestinians survive on as little as 20 L/capita/day [22,23]. All water resources in the Occupied Palestinian Territory were placed under military control by Israel when it occupied the Palestinian territories in 1967 (Military Order No. 92, 1967); these orders are still in effect, but they only apply to Palestinians and not to Israeli settlers who are subject to Israeli law [24]. Some water management responsibilities were issued to the Palestinian Authority in accordance with the Oslo Accords, in accordance with Article 40 of the environmental provisions in the Oslo II Accord, where approximately 80% of the waters pumped from the aquifers were allocated for Israeli use and the remaining 20% for Palestinian use [25]. Despite the lack of water, meeting the basic needs of the Palestinian population is a national priority, and the government is struggling to expand access to safe drinking water and sanitation. The proportion of the population using safely managed drinking water services in Palestine was 59.1% in 2017 [26].

Monitoring water quality is an essential step to enhance its public use. Therefore, this study aims to shed light on groundwater pollution in the West Bank in terms of quality, sources, risks, management, and the relationships among them in addition to providing a description of the groundwater situation. An analysis of the existing literature was conducted on the past 27 years of literature, which was reviewed using a review matrix based on the four core categories developed for this purpose. Importantly, this review matrix will provide an idea of the number of studies that have been completed during the specified period, which is expected to offer a clear idea about groundwater contamination in the West Bank, the degree of health risks, and the role of sound management in maintaining the quality and sustainability in groundwater. Studies were classified according to the four abovementioned categories, which can offer an idea of the sort of studies required in the future (Table 2). It is impossible to guarantee adequate groundwater quality if there is no appropriate environmental management of pollution sources. To ensure the sustainability and sustainable availability of fresh water sources in the future, more water quality studies need to be conducted over a period of time. The availability of water quality monitoring data for periods of at least a few years is very useful for comparisons of different ion concentrations [27].

Table 2. Review matrix of groundwater research in West Bank, Palestine, for the years 1994–2022.

| Author(s) | Year | Groundwater Quality | Sources of Pollution | Health Risks | Groundwater Management |
|---|---|---|---|---|---|
| J. Isaac, V. Qumsieh, and M. Owewi [28] | 1995 | x | | | x |
| J. Isaac and W. Sabbah [29] | 1997 | | | | x |
| PWA [30] | 2000 | x | | x | x |
| L. J. Froukh [31] | 2003 | | | | x |
| M. Ghanem [32] | 2005 | | | | x |
| R. M. Stephan [33] | 2007 | | | | x |
| A. Mohammad S. Juaidi [34] | 2008 | | | | x |
| F. M. Anayah and M. N. Almasri [10] | 2009 | x | x | | x |
| G. A. Daghrah [35] | 2010 | x | | | |
| M. Ghanem, S. Samhan, E. Carlier, and W. Ali [36] | 2011 | x | x | x | |
| Z. A. Mimi, N. Mahmoud, and M. A. Madi [37] | 2012 | | x | | |
| D. A. Shreim [38] | 2012 | | x | | |
| B. Borst, N. J. Mahmoud, N. P. van der Steen, and P. N. L. Lens [39] | 2013 | | x | | |
| H. Malassa, M. Hadidoun, M. Al-Khatib, F. Al-Rimawi, and M. Al-Qutob [40] | 2014 | x | | | |
| A. Aliewi and I. A. Al-Khatib [11] | 2015 | x | x | x | |
| T. Judeh, M. Haddad, and G. Özerol [12] | 2017 | | | | x |
| A. H. D. M. G. Atta [41] | 2017 | x | x | | |
| World Bank [42] | 2018 | | | | x |
| H. Jebreen, A. Banning, S. Wohnlich, A. Niedermayr, M. Ghanem, and F. Wisotzky [9] | 2018 | x | | | |
| A. Daghara, I. A. Al-Khatib, and M. Al-Jabari [43] | 2019 | x | | | |
| M. Rudolph [44] | 2020 | | | | x |
| M. N. Almasri, T. G. Judeh, and S. M. Shadeed [45] | 2020 | x | x | | x |
| B. Hejaz, I. A. Al-Khatib, and N. Mahmoud [2] | 2020 | x | | | |
| D. H. M. A. Daajna [46] | 2020 | x | | | x |
| M. N. Almasri, T. G. Judeh, and S. M. Shadeed [45] | 2020 | x | x | | x |
| W. Ahmad and M. Ghanem [47] | 2021 | x | x | | |
| T. Judeh, H. Bian, and I. Shahrour [48] | 2021 | x | | | x |
| M. Ghanem, W. Ahmad, Y. Keilani, F. Sawaftah, L. Schelter, and H. Schuettrumpf, [49] | 2021 | x | | | |
| N. Mahmoud, O. Zayed, and B. Petrusevski [27] | 2022 | x | | | |
| R. A. Thaher, N. Mahmoud, I. A. Al-Khatib, and Y. T. Hung [50] | 2022 | | x | | |

## 2. Groundwater Quality in West Bank, Palestine

Groundwater contamination is the addition of undesirable substances to groundwater caused by human activities, and this contamination can render groundwater unsuitable for use [51]. The major contaminants in the West Bank are $NO_3$, Cl, Na, $NH_4$, $PO_4$, TDS, salinity, and FC [27]. There are many classes of contaminants detected in groundwater, but chemical and biological contaminants are the most important. These contaminants can come from natural and anthropogenic sources [52]. Anthropogenic activities are increasingly threatening groundwater quality due to the large amounts of nitrogen, phosphorus, and heavy metals that infiltrate the soil when precipitation and irrigation and reach the groundwater [53]. Unsanitary landfill and agricultural areas scattered in the northern West Bank may lead to the deterioration of groundwater [54]. In recent years, nitrate contamination in groundwater has been reported in many countries (e.g., Spain [55], Italy, Morocco, Tunisia [56], Syria [57], Iran [58], Pakistan [59], Thailand [60], China [61], Mexico, and Brazil [62]), and Palestine also suffers from the same problem [2]. Importantly, nitrate ($NO_3$) is one of the most important chemical parameters by which water quality is measured. A recent study in the northern West Bank has shown that 18% of the samples examined from groundwater were above the permissible limit for the WHO guidelines and PS Table 1. Likewise, fecal coliform (FC) and total coliform (TC) results showed that 1.3% of the samples were low-risk [2]. Concentrations of nitrate in groundwater can generally be affected by wastewater, cesspits, farming activities (fertilizers), septic tanks, and animal manure [10]. In the West Bank, the high to very high level of phosphate additionally confirmed that the occurrence of groundwater pollution with untreated or insufficiently treated wastewater is extremely likely [27]. In different regions of the world, such as the rural areas of northern China, groundwater is polluted by large amounts of nitrogen fertilizers, which are used by humans in agricultural activities [63]. On the other hand, the large-scale use of fertilizers in the West Bank led to an increase in the concentration of nitrates in groundwater.

The presence of a certain concentration of nitrogen in groundwater is an indicator of the pollution of this water. Nitrogen leaks into the groundwater from various sources, such as cesspits, urban sewage collection lines, dry and wet sedimentation, the flow of treated and raw wastewater into valleys and waterbodies, fertilizers (chemical and manure), and irrigation with polluted water [45]. According to the study of Qana valley springs in Salfit governorate, Palestine biological tests indicated that all the springs are not suitable for drinking purposes because of their contamination with E. coli and fecal bacteria, and the reason for this contamination is untreated wastewater [64]. Previous studies in the West Bank have shown that 10% of the samples had a hardness value above the permissible limit of the PSI, 15% had a sodium content exceeding the permissible limit of the PSI, and only a very small fraction of the samples (3%) were contaminated with fecal coliforms [43]. Mahmoud pointed out in his study that the concentration of ions and treatments (such as $Cl^-$, $Na^+$, $NH_4^+$, TDS, and $NO_3$) affect the quality of water related to aesthetics and health, and those of heavy metals (such as Cr, Cu, Fe, Mn, Pb, Cd, and As), are within the recommended limits for drinking water (Table 3). However, signs of contamination, namely elevated levels of nitrate and ammonium, have been observed even in some deep wells [27]. About 41% of the selected wells in the Tulkarm and Qalqiliya were demonstrated to have nitrate concentrations higher than the limits in the WHO and Palestinian standards, while the chloride concentration was within the acceptable limit [4] (Figure 1).

Table 3. Physicochemical parameters of the groundwater from 29 wells in West Bank, Palestine [27].

| Parameter | Average (STD) | Range | PSI | WHO |
|---|---|---|---|---|
| pH | 7.4 (0.2) | 6.8–7.9 | 6.5–8.5 | NA |
| TDS (mg/L) | 340 (56) | 265–449 | 1000 | NA |
| F (mg/L) | 0.3 (0.2) | 0.1–1.2 | 1.5 | 1.5 |

Table 3. Cont.

| Parameter | Average (STD) | Range | PSI | WHO |
|---|---|---|---|---|
| Cl (mg/L) | 59.8 (27.3) | 33–132 | 250 | NA |
| $SO_4$ (mg/L) | 17.1 (8.7) | 8–48 | 200 | NA |
| $HCO_3$ (mg/L) | 246 (8.8) | 226–259 | NA | NA |
| $NO_3$ (mg/L) | 21.5 (10.9) | 0–46.2 | 50 | 50 |
| $PO_4$ (mg/L as P) | 0.8 (0.7) | 0.0–3.0 | NA | NA |
| Ca (mg/L) | 50.7 (3.3) | 46–59 | 100 | NA |
| Mg (mg/L) | 20 (1.7) | 17–25 | 100 | NA |
| Na (mg/L) | 39.8 (18.8) | 21–91 | 200 | NA |
| K (mg/L) | 2.7 (4.5) | 0–19 | 10 | NA |
| TH (mg/L) | 208.5 (13.4) | 187.2–250 | 500 | NA |
| $NH_4$ (mg/L) as N | 1.6 (2.4) | 0–8.5 | NA | |

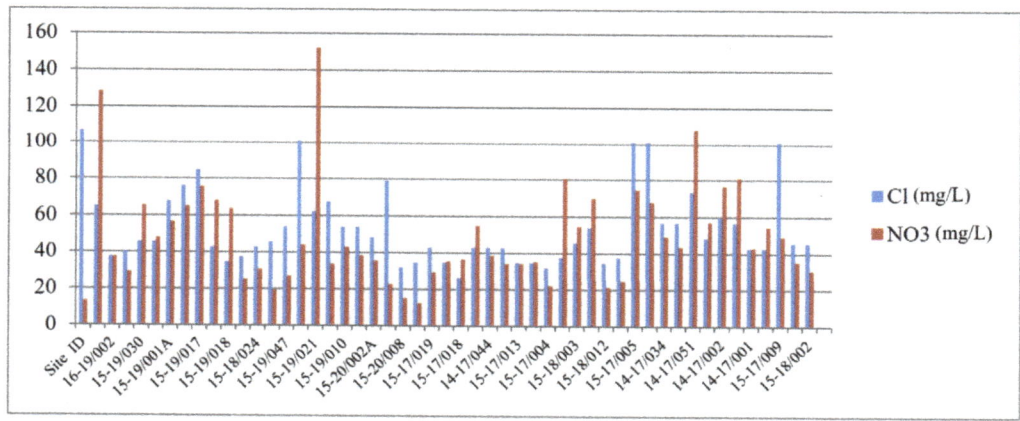

Figure 1. Annual average chloride and nitrate content in selected wells in Qalqilia and Tulkarm in the West Bank 2013 [4].

## 3. Sources of Groundwater Pollution in West Bank, Palestine

Raw wastewater is considered one of the potential sources of groundwater pollution in the West Bank, where sewage flows into the nearby valleys and waterbodies, leaving behind large amounts of pollution [11]. Cesspits, agricultural activities, and the random dumping of solid waste are also considered sources of groundwater pollution [11,39,45,50] The high concentrations of chemical and biological parameters of groundwater above the permissible limits of WHO and PSI will have direct effects on public health [65,66].

More than 200 Israeli settlements and outposts discharge a large quantity of wastewater into the valleys of the West Bank every year [29,46,47,64,67] (Figure 2). In a recent study that assessed the impact of untreated wastewater discharged to Sarida Valley in the West Bank, a strong relation was found between the wastewater flow in Sarida valley and the spring water quality system in the drainage catchment [47]. Another study in the West Bank found that the reason behind the high nitrate pollution in spring water could be attributed to agricultural activities in addition to the high groundwater recharge. However, leaking septic and sewer systems are considerably causing the nitrate contamination of groundwater in populated areas [68]. In another study of the quality of spring water in the central West Bank, the cause of the high levels of K and Na was found to be the intensive farming around these springs [49]. It is clear that the springs have high concentrations of total and fecal coliform bacteria, which indicates the presence of pollution hotspots in that area, such as the cesspits or due to sewage water flowing near the springs (Table 4) [49]

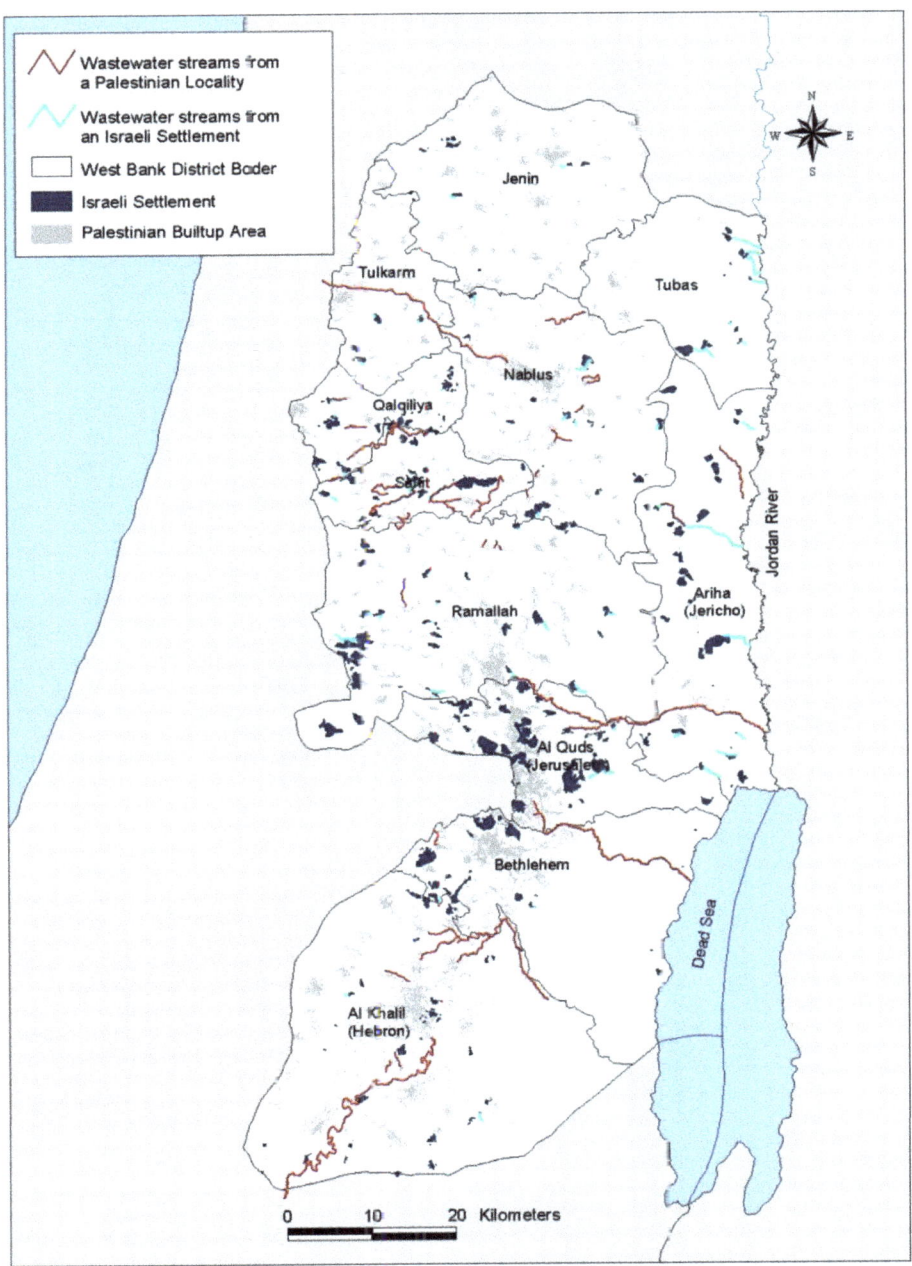

**Figure 2.** Wastewater streams from Palestinian location and Israeli settlements (ARIJ, 2008).

Based on the per capita wastewater generation in the West Bank, the total volume of wastewater generated for the year 2015 was estimated to be approximately 66 MCM/year [69]. Wastewater treatment plants are mainly present and used in urban centers, where approximately 60% of the population are using the public wastewater network, while the remainder use cesspits and septic tanks to dispose of wastewater [23,70]. The percentage of the population served by cesspits and the sewage network in the northern West Bank is 54.5% and 45.5%, respectively [71]. Table 5 shows the governorates, population, percentage of the population served by cesspits and the sewage network and the amount of water supply and wastewater generated in northern West Bank. Wastewater seeps through the soil and rocks to reach the groundwater, which is considered as the main source of drinking water in the West Bank, and thus experiences considerable contamination and pollution [46].

**Table 4.** Overview of the results for analysis of 50 springs in the West Bank sampled for major cations and anions and comparison with the limits of WHO and PSI standards.

|  | Ca | Mg | Na | K | Cl | $NO_3$ | $SO_4$ | $HCO_3$ |
|---|---|---|---|---|---|---|---|---|
| Mean (mg/L) | 91.64 | 39.25 | 53.60 | 26.03 | 40 | 5 | 27.7 | 200.8 |
| Max (mg/L) | 132.3 | 69.68 | 122.9 | 170.7 | 57.5 | 12.5 | 40.5 | 241 |
| Min (mg/L) | 26.48 | 15.56 | 11.12 | 0.204 | 28.5 | 1.9 | 15.8 | 156 |
| Median (mg/L) | 79.62 | 25.7 | 19.04 | 0.624 | 39.15 | 4.2 | 27.8 | 203 |
| WHO Standard (mg/L) | NA | NA | NA | NA | 250 | 50 | 250 | NA |
| PSI Standard (mg/L) | 100 | 100 | 200 | 10 | 250 | 50 | 200 | NA |

**Table 5.** Governorate, area, population, wastewater disposal systems, generation of solid waste, quantity of water supply, and quantity of wastewater generated in the northern West Bank [23,70,71].

| No | Governorate | Area (km$^2$) | Population | % Population Served by Cesspits | % Population Served by a Sewage Network | Quantity of Water Supply (Million m$^3$) | Quantity of Wastewater Generated (Million m$^3$) | Generation of Solid Waste Ton/Year |
|---|---|---|---|---|---|---|---|---|
| 1 | Salfit | 204 | 80,000 | 90% 73,889 | 10% | 5.4 | 3.51 | 29,200 |
| 2 | Qalqiliya | 166 | 120,000 | 55% 66,919 | 45% | 8.3 | 5.4 | 43,800 |
| 3 | Tulkarm | 246 | 200,000 | 57% 113,347 | 43% | 10.1 | 6.57 | 73,000 |
| 4 | Jenin | 583 | 335,000 | 60% 203,351 | 40% | 8.2 | 5.33 | 122,275 |
| 5 | Tubas | 402 | 65,000 | 65% 42,844 | 35% | 4 | 2.6 | 23,725 |
| 6 | Nablus | 605 | 410,000 | 52% 216,115 | 48% | 15.7 | 10.21 | 149,650 |
|  | Total | 2206 | 1,210,000 |  |  | 51.7 | 33.7 | 441,650 |

One of the main causes of groundwater contamination in the West Bank is the effluent (outflow) from septic tanks and cesspits. Approximately 28% of residential homes rely on cesspits and 10% on septic tanks [23]. The large number and widespread use of these systems means they are a serious source of pollution. Sewage systems can contaminate groundwater with bacteria, viruses, nitrates, detergents, oils, and chemicals [39,72]. Most of the communities in the rural areas of the West Bank suffer from a lack of adequate sewage systems for the disposal of sewage. Rural systems in Palestine are limited to cesspits and septic tanks [73]. On the other hand, the US Environmental Protection Agency has specific and mandatory legislation governing the operation of septic tanks and cesspits in the US, whereby it assists in lowering groundwater contamination levels [74]. In a recent study on the quality of spring water, 127 samples were collected from 300 springs distributed across the West Bank, Palestine. Most of the physical and chemical characteristics for water from springs were within the acceptable standard limits, with the exception of turbidity, chloride, and nitrates. Regarding biological contamination limits, 97% of the samples were classified as possessing no risk, and only 2% were classified as possessing a simple risk and thus require chlorination treatment [43]. The development of sewage networks, wastewater treatment plants, and reduction in cesspits in rural areas significantly reduces nitrate concentrations and biological contamination in spring water.

The West Bank's soils are exposed to a wide range of human activities, including agricultural and industrial operations, which have a negative impact on arable land fertility [41]. Fertilizer and pesticide overuse is one of the most serious problems affecting the land in the West Bank, where farmers are forced to use increasing amounts of fertilizers and pesticides to boost the productivity of agricultural land due to the enormous rise in population and the limited agricultural area. In the West Bank, the annual rate of agricultural fertilizer use has reached up to 30,000 tons of chemical fertilizers and manures [75]. Likewise, annual rates of pesticide use in agricultural activities has reached up to 502.7 tons, a considerable amount that is internationally banned for health reasons [75]. Agricultural fertilizers and pesticides have caused high levels of nitrate and potassium in the groundwater [45].

West Bank, Palestine, is facing the problem of solid waste for several reasons: (i) increasing population, (ii) lack of materials and resources needed for solid waste management, and finally (iii) weak technical expertise [46,76]. Both Israeli settlements and Palestinian communities use non-engineered solid waste dumping sites in the West Bank [11]. This waste is mostly industrial or domestic and poses a high risk to the environment and to both surface and groundwater. In 2017, it was estimated that an individual generates approximately 1.9 kg of solid waste per day in Israeli settlements in the West Bank [77]. In 2019, Palestinians produced roughly 4333 tons of solid waste per day, totaling around 1.58 million tons for the year or around 0.9 kg per capita per day. However, 441,650 ton/year was produced north of the West Bank (Figure 3) [78,79].

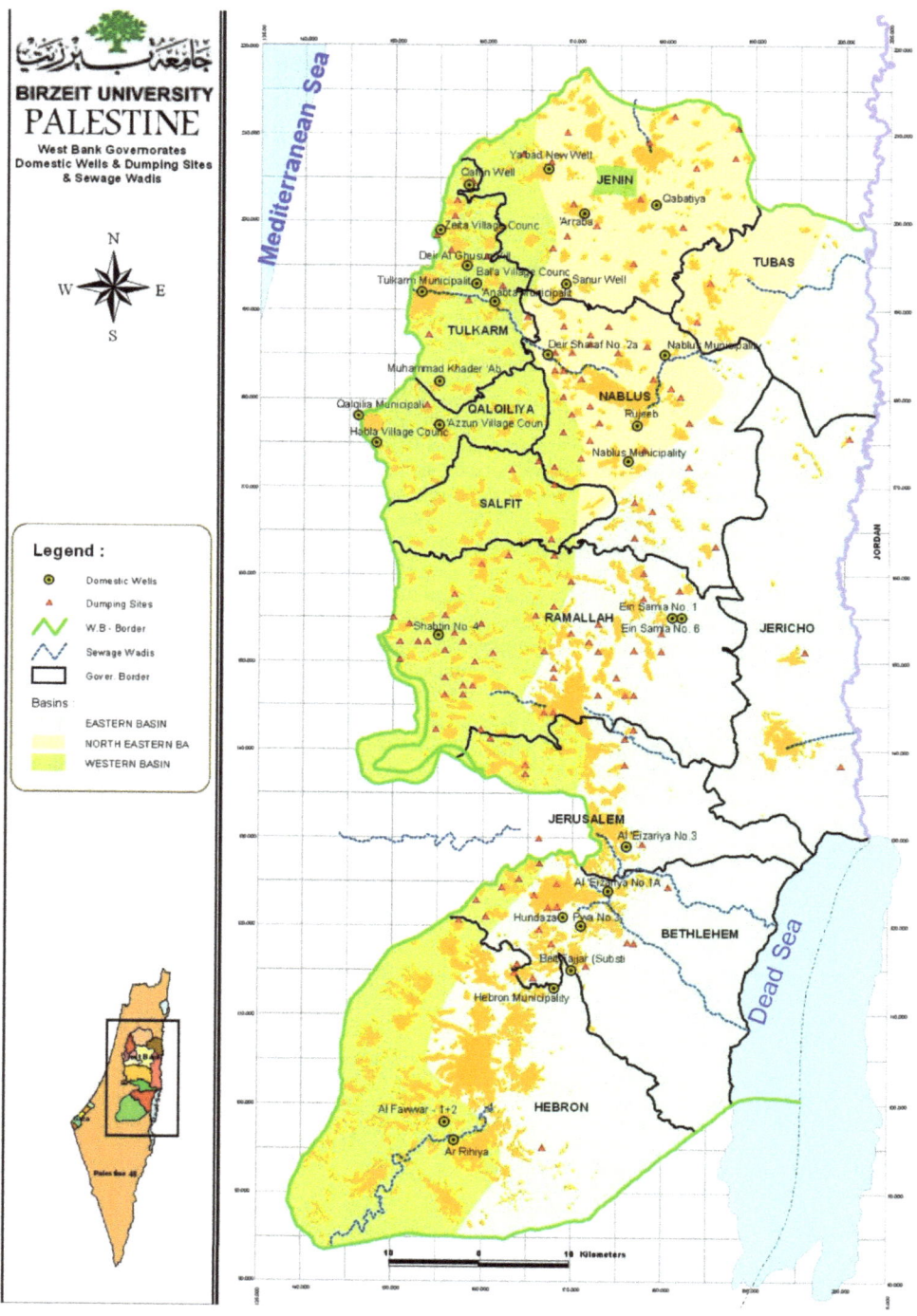

**Figure 3.** The non-engineered solid waste dumping sites in the West Bank—Birzeit University, Palestine [27].

## 4. Health Risk

The contamination of groundwater either from anthropogenic or natural sources poses risks to human health and results in the occurrence of waterborne diseases in humans, such as diarrhea and vomiting [80]. Biological contamination is considered the main cause of death worldwide, especially in poor and developing countries [18,81]. FC and TC results have shown that 1.3% of groundwater samples are contaminated in the northern governorates of the West Bank, Palestine [2]. In the study of water quality in Wadi al-Qlt area, the results showed that 47% of the samples are contaminated with FC bacteria, which indicates the leakage of pollutants in the area feeding the springs [35]. However, in 2019, a study of the quality of drinking water from springs in Palestine found that only 3% of the samples were contaminated with FC bacteria [43].

Chemical pollutants in groundwater have been a major concern due to health risks. Water contaminated with nitrates is not suitable for domestic use, since it causes diseases and health problems, such as shortness of breath, methemoglobinemia or (blue baby) syndrome, an increase in starchy deposits, and hemorrhaging at the spleen [82]. In the study of Daghara et el. (2019), 21% of West Bank groundwater samples were found to have a nitrate concentration above the permissible limit [43]. On the other hand, the high concentrations of sodium and potassium in drinking water may cause high blood pressure in humans [83]. Pollutants, such as heavy metals, affect waterbodies due to their strong toxicity even at low concentrations. For some minerals, such as Ca, Mg, K, and Na, their presence in normal proportions is important for sustaining life, but extensive exposure to heavy metals can cause poisoning with serious health effects [84] (Table 6).

Table 6. Study on diseases associated with chemical and biological contamination of drinking water around the world.

| Chemical and Biological Contaminants | Disease | Country | Source | Remarks |
|---|---|---|---|---|
| Fluoride | - Dental and skeletal fluorosis<br>- The results of non-carcinogenic health risk indicate health risk was higher in infants and children as compared to the adults | Global scale and India | [85,86] | Fluorosis still represents a serious and widespread health problem, particularly in rural communities, which depend on untreated water supplies |
| Nitrate (NO3) | - Methemoglobinemia or "blue baby" syndrome, birth defects, thyroid disease, and colon cancer<br>- Health risk was higher in infants and children compared to adults | United States, Europe and Global scale and India | [86–88] | Fertilizer and cesspits are the main source of NO3 in groundwater |
| Sodium (Na) and potassium (K) | May cause high blood pressure in humans | Coastal areas in Southeast Asia (coast of Bangladesh) | [83] | Drinking water salinity that contains high concentrations of sodium may affect pregnant women, which increases risk of hypertension and associated diseases |
| Arsenic | Cancer of the skin, lung, bladder, and probably liver | Global scale | [88] | Arsenic is responsible for a range of adverse effects, including hyperkeratosis and peripheral vascular disease |

Table 6. *Cont.*

| Chemical and Biological Contaminants | Disease | Country | Source | Remarks |
|---|---|---|---|---|
| Lead, cadmium and chromium | - Cognitive and developmental impairment, and hypertension<br>- Kidney disfunction and bone toxicity<br>- Cancer | Global scale | [89] | Chemical contamination in drinking water is a global issue affecting more than one billion people, placing them at risk of adverse health impacts and water scarcity |
| Zinc | Stomach cramps, nausea, and vomiting may occur | Global scale | [90] | Ingesting high levels of zinc for several months may cause anemia, damage the pancreas, and decrease levels of high-density lipoprotein (HDL) cholesterol |
| Fecal coliform (FC) | The occurrence of waterborne diseases to humans, such as diarrhea and vomiting | African countries and Ethiopia | [17,18] | Biological contamination is considered the main cause of death worldwide, especially in poor and developing countries |

## 5. Contamination of Drinking Water

The water analysis of wells and springs conducted by the PWA in 2016 in the West Bank has shown that 15% of the water sources used for drinking contain nitrate concentrations higher than the permissible levels (50 mg/L) according to the Palestinian standard. In addition, the microbial tests showed concentrations around or below 19% [21]. Poor drinking water quality results in many waterborne diseases. Understanding the factors that affect drinking water quality is very important and also essential for informing decisions aimed at protecting drinking water sources. The quality of drinking water is usually affected by the quality of the source. In rural areas, drinking water is usually pumped directly from wells and rivers without adequate treatment and, therefore, the quality of the source water plays a critical role in determining the quality of the drinking water [91,92].

Heavy metals, such as Mn, Fe, Co, Ni, Cu, Zn, Se, and Cr, are essential for the growth of organisms, while Pb, Cd, Hg, and As are not only biologically nonessential, but definitely toxic [93–95]. After entering the water, metals may precipitate, be adsorbed onto the solid surface, remain soluble or suspended in water, or be taken up by fauna. A very important biological property of metals is their tendency to accumulate [84]. Common water contaminants iron and manganese are not health hazards but can cause offensive taste, appearance, and staining [96]. The groundwater in the West Bank is potable, except for some cases where the water is not suitable for drinking due to excessive salinity, high nitrate concentration, and bacterial contamination. Importantly, the levels of heavy metals in the water, including Cr, Cu, Fe, Mn, Pb, Cd, and As, are well below the limits advised for human consumption [27].

## 6. Drinking Water Management

Under Palestinian law, the PWA is considered primarily responsible for managing the water sector in the West Bank and Gaza. The ministry of health (MoH) participates in monitoring water quality, and the Ministry of Local Government (MLG) participates through local authorities and Joint Service Councils (JSC) in providing the population with water and sanitation services [97]. There are different local government institutional entities that provide water services in the West Bank, within which around 17% of the population is served by independent utility firms that are formally established under their own law and are accountable to the board of directors of the local units they own. However, the services are provided to the rest of the West Bank residents and families by service providers under the auspices of the Ministry of Local Government. Large cities have municipal water departments that provide water and/or sewage services (76 in the West Bank). A number of smaller municipalities and villages joined together to form the Joint Service Boards (JSBs), which provide water and/or wastewater services to these localities. Likewise, another 162 village councils provide water and sanitation services directly to their residence. Currently, there are moves to begin with aggregation smaller service providers and to encourage service providers to strengthen transparency, accountability, and financial independence [42].

There has been a rapid increase in the Palestinian population in the West Bank and a decline in water security in recent years, with increasing demand for water and the dwindling of water resources, where the demand is already outstripping the supply. The situation is constantly deteriorating [42]. The internal renewable water resources are exploited to such a large extent that the quality of groundwater in the Gaza Strip is qualified as undrinkable because of seawater intrusion and wastewater discharging areas and solid waste dumping sites [98,99]. The Palestinians' access to additional water sources has become very difficult due to the political situation [100]. The residents of the Gaza Strip depend on desalination as a non-traditional water resource to cover the massive shortage of fresh water, and it also contributes to addressing global water scarcity issues [101,102]. Ensuring water security is a priority Water security requires adequate and well-managed water resources, including risk management and water resources that provide sustainable, efficient, and equitable services to improve water security in the Palestinian water sector [42].

A well-managed drinking water system must be managed from source to end-users (i.e., from wells and springs to the drinking taps). Drinking water is more likely to be safe if all steps of the process (e.g., extraction, treatment, and distribution) are working as they should [103] This requires proper management of the water sources during the planning, construction, installation, operation, and maintenance of the entire system (i.e., management of the catchment as a whole unit). A good understanding of these processes facilitates the early identification of potential vulnerabilities. Inadequate wastewater management in urban and rural areas means that drinking water may be at risk, and this will negatively affect public health [15,104].

Commitment to a future with sustainable management of water resources is a matter that requires integrated management and planning of water resources, which should involve all stakeholders [105]. The commitment of stakeholders is important because of their impact on water management through their joint efforts [105]. Addressing water scarcity, protecting ecosystems, preserving human health, and raising the level of economic development are among the most important factors to handle in the process of evaluating the integrated management of water resources [106]. The sound management of water resources is extremely important and has a significant impact on water quality and the health of end-users [65]. The process of monitoring groundwater quality that starts at the source and ends at the water networks and includes disinfection activities and supervision of water distribution to beneficiaries with fairness, equity, transparency, and governance of the water sector through integrated and sustainable management is important [107].

In the last decade, the expansion and diversity of water providers (e.g., private companies, municipalities, village councils, joint services councils, and others) underpin the importance of stakeholder involvement as a necessary tool to regulate and organize the sector. Working to establish a unified entity for water service providers could help to reduce financial, administrative, and technical burdens [108].

Access to an adequate amount of water is an important goal that countries aspire to achieve because of its importance to human development. Much of the world's population lacks access to water as a result of the great pressure on water resources and their pollution. Population increase, economic growth, and pollution have led to considerably increased competition and conflicts over fresh water [107]. There is currently a greater need to manage the available water resources rather than searching for new ones that may not be available at all. In light of climate change, drought, and complex political conditions, it has become demanding to implement an integrated and sustainable approach for the management of water resources [109]. A sound approach for the management of water resources should include governance, involvement of stakeholders, and the protection of ecological systems (e.g., controlling of water pollution, wastewater treatment, and the recycling and treatment of solid waste), the organization of agricultural sectors, and rationalized use of agricultural pesticides and fertilizers [109].

Human activities in the catchment area of water sources can lead to water pollution and negatively impact public health. Water catchment protection is therefore also important in securing clean and safe drinking water. The prevention of pollution is essential, and regular sanitary inspections to detect any sources of contamination will help in securing good quality drinking water year round [110].

As the human population increases, there is an increase in pollution and catchment destruction, inadequate sewage collection and treatment, and increase in the use of fertilizers to grow more food, which together result in increased water contamination. Catchment management is playing an increasingly important role in reducing the levels of potential contaminants in raw waters. An efficiently managed scheme will help to reduce pollution from agriculture and also help to control urban and chemical pollution from sites within a catchment. Due to the complex interactions between the natural environment and human action, which determine the quantity and quality of water resources, knowledge of water resources and (possible) pollution is often very low [111]. Determining the sources of pollution first and then linking their impact on water sources through water tests and studying the nature of waterbodies, with the help of GIS, will directly help in the good management of drinking water sources. Sufficient groundwater quality for future drinking water supply cannot be ensured unless the appropriate and effective environmental management of the pollution sources is implemented to ensure the sustainable availability of this fresh water sources also in the future [27].

## 7. Conclusions

The main reasons for water scarcity in the West Bank are the unique hydrological situation and diverse political and natural conditions, where the main natural conditions include scarcity and uneven distribution of rainfall due to extreme topographic variations within the region and the hydrogeological location of the West Bank. As for the political conditions, the Israeli government has virtually complete control over all West Bank. Although the PA has made it a priority to obtain safe and adequate drinking water, it is still struggling to achieve this goal, which is considered difficult in the light of the uncontrolled of PA over the groundwater sources. Furthermore, groundwater contamination from untreated wastewater, cesspits, solid waste dumping sites, and fertilizers has increased this problem. It is crucial to solve the problem of cesspits in Palestinian rural areas and replace them with sewage networks. Furthermore, solving the problem of untreated wastewater from Israeli settlements is important for reducing groundwater contamination.

The West Bank's groundwater is nevertheless drinkable, with the exception of a few instances where it is unfit for consumption due to high nitrate concentrations, excessive salinity, and bacterial contamination. It is important to note that the concentrations of heavy metals in the water, including Cr, Cu, Fe, Mn, Pb, Cd, and As, are significantly lower than the levels recommended for human consumption. Increasing the Palestinian share of drinking water is important, especially in light of the large growth in the population and the increase in water demand. A considerable group of West Bank residents is still struggling to obtain the minimum amount of drinking water stipulated by the World Health Organization, especially during the dry seasons. The quality of the groundwater and the health of end users are impacted by effective management of the available water resources. For this end, it is crucial to monitor the quality of groundwater from its source to its final destination. This process also involves disinfection procedures and the supervision of water distribution to beneficiaries with fairness, equity, transparency, and water sector governance. Sufficient groundwater quality for future drinking water supplies cannot be ensured unless the appropriate and effective environmental management of pollution sources is implemented to ensure the sustainable availability of this fresh water source also in the future. Future directions in research should look for new and creative methods for managing and monitoring groundwater, such as the use of smart technology that would enhance the protection of these sources from contamination and reduce health risks and thus maintain their sustainability. Detailed information on the quality of groundwater should be provided, and the data and information should be made publicly available, which will encourage scientific research aimed at providing more creative solutions for water shortage in the entire region.

In the West Bank, the number of Israeli settlements has exceeded over two hundred. According to the Israeli Beit Salem Foundation for Human Rights, the highest number of Israeli settlers was registered in 2020, with more than 900,000 living in the West Bank, where they consume vast amounts of groundwater and produce large quantities of contamination (Figure 4). On the other hand, the Palestinian communities need more water but generate different types of contamination. To date, in the West Bank, there has been a lack of studies that focus on developing suitable groundwater management solutions in terms of water quality and vulnerable zones considering the pollution sources and the degree of health risk. Therefore, the problem of access to safe drinking water persists. The recently announced Sustainable Development Goals (SDGs) also highlight the importance of universal and equitable access to safe and affordable drinking water, which was established as one of the 17 global goals (SDG 6) to be achieved by 2030. Unfortunately, these goals may not be achievable under these complex political and natural conditions. Finally, conducting more research in groundwater field is important and may contribute to protecting it from pollution and maintaining its sustainability.

**Figure 4.** Israeli settlements in West Bank, Palestine [112].

**Author Contributions:** A.Z.: conceptualization, methodology, visualization, investigation, writing—original draft. L.A.: supervision, validation, writing—review and editing. All authors have read and agreed to the published version of the manuscript.

**Funding:** This research was supported by the research projects GUP-2022-065 and XX-2022-008.

**Institutional Review Board Statement:** Not applicable.

**Informed Consent Statement:** Not applicable.

**Data Availability Statement:** Not applicable.

**Conflicts of Interest:** The authors declare that they have no known competing financial interests or personal relationships that could have appeared to influence the work reported in this paper.

## References

1. Shingne, M.C.; Gasteyer, P.S. *Water Justice as Social Policy: Tackling the Global Challenges to Water and Sanitation Access*; Bristol University Press: Bristol, UK, 2022; pp. 53–61.
2. Hejaz, B.; Al-Khatib, I.A.; Mahmoud, N. Domestic Groundwater Quality in the Northern Governorates of the West Bank, Palestine. *J. Environ. Public Health* 2020, *2020*, 6894805. [CrossRef] [PubMed]
3. Trottier, J. Palestinian Water Management–Policies and Pitfalls. 2019. Available online: https://hal.archives-ouvertes.fr/hal-02272810/file/water2019_final8thOct5%281%29%281%29.pdf (accessed on 19 April 2021).
4. PWA. Status Report of Water Resources in the Occupied State of Palestine-2012, no. October. 2013, p. 22. Available online: http://www.pwa.ps/userfiles/file/1/WRSTATUSReport-finaldraft2014-04-01.pdf (accessed on 11 September 2021).
5. Yadav, A.K. Physicochemical Studies on Assessment of Ground Water Quality of Kota District. Ph.D. Thesis, University of Kota, Kota, India, 2016; p. 208.
6. Mofor, N.A.; Njoyim, E.B.T.; Mvondo-Zé, A.D. Quality Assessment of Some Springs in the Awing Community, Northwest Cameroon, and Their Health Implications. *J. Chem.* 2017, *2017*, 3546163. [CrossRef]
7. WHO. *Guidelines for Drinking-water Quality*; WHO: Geneva, Switzerland, 2012.
8. Palestine Standards Institute (PSI). *The Second Working Draft of the Amended Drinking Water Standard, Ramallah*; Palestine Standards Institute: Ramallah, Palestine, 2004.
9. Jebreen, H.; Banning, A.; Wohnlich, S.; Niedermayr, A.; Ghanem, M.; Wisotzky, F. The Influence of Karst Aquifer Mineralogy and Geochemistry on Groundwater Characteristics: West Bank, Palestine. *Water* 2018, *10*, 1829. [CrossRef]
10. Anayah, F.M.; Almasri, M.N. Trends and occurrences of nitrate in the groundwater of the West Bank, Palestine. *Appl. Geogr.* 2009, *29*, 588–601. [CrossRef]
11. Aliewi, A.; Al-Khatib, I.A. Hazard and risk assessment of pollution on the groundwater resources and residents' health of Salfit District, Palestine. *J. Hydrol. Reg. Stud.* 2015, *4*, 472–486. [CrossRef]
12. Judeh, T.; Haddad, M.; Özerol, G. Assessment of water governance in the West Bank, Palestine. *Int. J. Glob. Environ. Issues* 2017, *16*, 119. [CrossRef]
13. Friday, O.; Ben-gurion, D. Oslo Accords (Declaration of Principles on Interim Self-Government Arrangements) (13 September 1993), no. May 1948. 2012. Available online: https://israeled.org/resources/documents/oslo-accords/ (accessed on 13 September 2022).
14. Kate, S.; Kumbhar, S.; Jamale, P. Water quality analysis of Urun-Islampur City, Maharashtra, India. *Appl. Water Sci.* 2020, *10*, 95. [CrossRef]
15. Cotruvo, J.A. 2017 WHO guidelines for drinking water quality: First addendum to the fourth edition. *J. Am. Water Work. Assoc.* 2017, *109*, 44–51. [CrossRef]
16. Zeidan, B.A. Groundwater Degradation and Remediation in the Nile Delta Aquifer. In *The Nile Delta*; Springer: Cham, Switzerland, 2017; pp. 159–232. [CrossRef]
17. Edessa, N.; Geritu, N.; Mulugeta, K.; Negera, E.; Nuro, G.; Kebede, M. Microbiological assessment of drinking water with reference to diarrheagenic bacterial pathogens in Shashemane Rural District, Ethiopia. *Afr. J. Microbiol. Res.* 2017, *11*, 254–263. [CrossRef]
18. Wen, X.; Chen, F.; Lin, Y.; Zhu, H.; Yuan, F.; Kuang, D.; Jia, Z.; Yuan, Z. Microbial Indicators and Their Use for Monitoring Drinking Water Quality—A Review. *Sustainability* 2020, *12*, 2249. [CrossRef]
19. EMCC. Environmental and Social Impact Assessment (ESIA) & Environmental and Social Management Plan (ESMP) For Gaza Water Supply and Sewage Systems Improvement Project (WSSSIP) Phase 1 and Additional Financing (AF). 2014, pp. 1–174. Available online: https://documents1.worldbank.org/curated/en/379371468143393852/text/E46460V10MNA0A00Box385335B00PUBLIC0.txt (accessed on 13 September 2022).
20. Amous, B.; Mahmoud, N.; Van Der Steen, P.; Lens, P.N.L. Septage composition and pollution fluxes from cesspits in Palestine. *J. Water Sanit. Hyg. Dev.* 2020, *10*, 905–915. [CrossRef]
21. PWA. Groundwater in West Bank, Palestine. 2018. Available online: http://www.pwa.ps/ar_page.aspx?id=J0h6J5a2717254815aJ0h6J5 (accessed on 5 January 2022).
22. Lazarou, E. *Water in the Israeli-Palestinian Conflict*; European Parliamentary Research Service: Brussels, Belgium, 2016; p. 8.

23. PCBS. Palestine in Figures2020. Palestinian Central Bureau of Statistics, Ramallah. 2021; pp. 1–105. Available online: http://www.pcbs.gov.ps (accessed on 13 September 2022).
24. Israel Military Order No. 92 Concerning Powers for the Purpose of the Water Provisions. 1967. Available online: http://www.geocities.ws/savepalestinenow/israelmilitaryorders/fulltext/mo0092.htm (accessed on 1 October 2022).
25. General Assembly; Security Council. *UNITED NATIONS as General Assembly Security Council*; UN: New York, NY, USA, 1997.
26. PCBS. *Sustainable Development Goals Statistical Report*; PCBS: Ramallah, Palestine, 2020.
27. Mahmoud, N.; Zayed, O.; Petrusevski, B. Groundwater Quality of Drinking Water Wells in the West Bank, Palestine. *Water* **2022**, *14*, 377. [CrossRef]
28. Isaac, J.; Qumsieh, V.; Owewi, M. *Assessing the Pollution of the West Bank Water Resources, no. 02*; ARIJ: Bethlehem, Palestine, 1995; p. 14.
29. Isaac, J.; Sabbah, W. The Intensifying Water Crisis in Palestine. *Appl. Res. Inst.–Jerus.* **1997**, *2*, 1–10.
30. PWA. *Summary of Palestinian Hydrologic Data 2000. Volume 1: West Bank*; PWA: Ramallah, Palestine, 2000; Volume 1.
31. Froukh, L.J. Transboundary Groundwater Resources of the West Bank. *Water Resour. Manag.* **2003**, *17*, 175–182. [CrossRef]
32. Ghanem, M. *Qualitative Water Demand Management for Rural Communities in the West Bank*; CIHEAM-IAMB: Valenzano, Italy, 2005; Volume 65, pp. 385–390.
33. Stephan, R.M. Legal Framework of Groundwater Management in the Middle East (Israel, Jordan, Lebanon, Syria and the Palestinian Territories). In *Water Resources in the Middle East*; Springer: Berlin/Heidelberg, Germany, 2007; Volume 2, pp. 293–299. [CrossRef]
34. Juaidi, M.S. Gis-Based Modeling of Groundwater Recharge for the West Bank, DSpace. 2008, pp. 1–129. Available online: https://repository.najah.edu/handle/20.500.11888/7561?show=full (accessed on 13 September 2022).
35. Daghrah, G.A. Water Quality Study of Wadi Al Qilt-West Bank-Palestine. *Asian J. Earth Sci.* **2009**, *2*, 28–38. [CrossRef]
36. Ghanem, M.; Samhan, S.; Carlier, E.; Ali, W. Groundwater Pollution Due to Pesticides and Heavy Metals in North West Bank. *J. Environ. Prot.* **2011**, *2*, 429–434. [CrossRef]
37. Mimi, Z.A.; Mahmoud, N.; Abu Madi, M. Modified DRASTIC assessment for intrinsic vulnerability mapping of karst aquifers: a case study. *Environ. Earth Sci.* **2011**, *66*, 447–456. [CrossRef]
38. Shreim, D.A. Environmental Assessment and Economic Valuation of Wastewater Generated from Israeli Settlements in the West Bank Ph.D. Thesis, Faculty of Graduate Studies, An-Najah National University, Nablus, Palestine, 2012; pp. 1–115.
39. Borst. B.; Mahmoud, N.J.; Van Der Steen, N.P.; Lens, P.N.L. A case study of urban water balancing in the partly sewered city of Nablus-East (Palestine) to study wastewater pollution loads and groundwater pollution. *Urban Water J.* **2013**, *10*, 434–446. [CrossRef]
40. Malassa, H.; Hadidoun, M.; Al-Khatib, M.; Al-Rimawi, F.; Al-Qutob, M. Assessment of Groundwater Pollution with Heavy Metals in North West Bank/Palestine by ICP-MS. *J. Environ. Prot.* **2014**, *5*, 54–59. [CrossRef]
41. Ghanem, M.G. Pollution Aspects Interconnections to Socio-economical impact of Natuf Springs—Palestine. *J. Geogr. Res.* **2021**, *4*, 1. [CrossRef]
42. World Bank Group. *Securing Water for Development in West Bank and Gaza*; World Bank: Washington, DC, USA, 2018. [CrossRef]
43. Daghara, A.; Al-Khatib, I.A.; Al-Jabari, M. Quality of Drinking Water from Springs in Palestine: West Bank as a Case Study. *J. Environ. Public Health* **2019**, *2019*, 8631732. [CrossRef] [PubMed]
44. Rudolph, M. Working Paper Water Governance under Occupation: A Contemporary Analysis of the Water Insecurities of Palestinians in the Jordan Valley, West Bank. *Inst. Soc. Stud.* **2020**, *655*, 1–71. Available online: https://www.iss.nl/en/news/water-governance-under-occupation-contemporary-analysis-water-insecurities-palestinians-jordan (accessed on 13 September 2022).
45. Almasri, M.N.; Judeh, T.G.; Shadeed, S.M. Identification of the Nitrogen Sources in the Eocene Aquifer Area (Palestine). *Water* **2020**, *12*, 1121. [CrossRef]
46. Daajna, D.H.M.A. Water Pollution Problems in the West Bank. *Maghreb J. Hist. Soc. Stud.* **2020**, *12*, 109–134.
47. Ahmad, W.; Ghanem, M. Effect of wastewater on the spring water quality of Sarida Catchment—West Bank. *Arab. J. Basic Appl. Sci.* **2021**, *28*, 292–299. [CrossRef]
48. Judeh, T.; Bian, H.; Shahrour, I. GIS-Based Spatiotemporal Mapping of Groundwater Potability and Palatability Indices in Arid and Semi-Arid Areas. *Water* **2021**, *13*, 1323. [CrossRef]
49. Ghanem, M.; Ahmad, W.; Keilani, Y.; Sawaftah, F.; Schelter, L.; Schuettrumpf, H. Spring water quality in the central West Bank, Palestine. *J. Asian Earth Sci. X* **2021**, *5*, 100052. [CrossRef]
50. Thaher, R.A.; Mahmoud, N.; Al-Khatib, I.A.; Hung, Y.-T. Cesspits as Onsite Sanitation Facilities in the Non-Sewered Palestinian Rural Areas: Users' Satisfaction, Needs and Perception. *Water* **2022**, *14*, 849. [CrossRef]
51. Government of Canada. Groundwater Contamination. 2017. Available online: https://www.canada.ca/en/environment-climate-change/services/water-overview/pollution-causes-effects/groundwater-contamination.html (accessed on 1 August 2022).
52. Elumalai, V.; Nethononda, V.G.; Manivannan, V.; Rajmohan, N.; Li, P.; Elango, L. Groundwater quality assessment and application of multivariate statistical analysis in Luvuvhu catchment, Limpopo, South Africa. *J. Afr. Earth Sci.* **2020**, *171*, 103967. [CrossRef]
53. Mititelu-Ionuș, O.; Simulescu, D.; Popescu, S.M. Environmental assessment of agricultural activities and groundwater nitrate pollution susceptibility: a regional case study (Southwestern Romania). *Environ. Monit. Assess.* **2019**, *191*, 501. [CrossRef] [PubMed]
54. Ravindra, K.; Thind, P.S.; Mor, S.; Singh, T.; Mor, S. Evaluation of groundwater contamination in Chandigarh: Source identification and health risk assessment. *Environ. Pollut.* **2019**, *255*, 113062. [CrossRef] [PubMed]

55. Ibe, F.C.; Opara, A.I.; Amaobi, C.E.; Ibe, B.O. Environmental risk assessment of the intake of contaminants in aquifers in the vicinity of a reclaimed waste dumpsite in Owerri municipal, Southeastern Nigeria. *Appl. Water Sci.* **2021**, *11*, 24. [CrossRef]
56. Troudi, N.; Hamzaoui-Azaza, F.; Tzoraki, O.; Melki, F.; Zammouri, M. Assessment of groundwater quality for drinking purpose with special emphasis on salinity and nitrate contamination in the shallow aquifer of Guenniche (Northern Tunisia). *Environ. Monit. Assess.* **2020**, *192*, 641. [CrossRef] [PubMed]
57. Zakhem, B.A.; Hafez, R. Hydrochemical, isotopic and statistical characteristics of groundwater nitrate pollution in Damascus Oasis (Syria). *Environ. Earth Sci.* **2015**, *74*, 2781–2797. [CrossRef]
58. Chitsazan, M.; Tabari, M.M.R.; Eilbeigi, M. Analysis of temporal and spatial variations in groundwater nitrate and development of its pollution plume: a case study in Karaj aquifer. *Environ. Earth Sci.* **2017**, *76*, 391. [CrossRef]
59. Khan, S.N.; Yasmeen, T.; Riaz, M.; Arif, M.S.; Rizwan, M.; Ali, S.; Tariq, A.; Jessen, S. Spatio-temporal variations of shallow and deep well groundwater nitrate concentrations along the Indus River floodplain aquifer in Pakistan. *Environ. Pollut.* **2019**, *253*, 384–392. [CrossRef] [PubMed]
60. Chotpantarat, S.; Parkchai, T.; Wisitthammasri, W. Multivariate Statistical Analysis of Hydrochemical Data and Stable Isotopes of Groundwater Contaminated with Nitrate at Huay Sai Royal Development Study Center and Adjacent Areas in Phetchaburi Province, Thailand. *Water* **2020**, *12*, 1127. [CrossRef]
61. Wegahita, N.K.; Ma, L.; Liu, J.; Huang, T.; Luo, Q.; Qian, J. Spatial Assessment of Groundwater Quality and Health Risk of Nitrogen Pollution for Shallow Groundwater Aquifer around Fuyang City, China. *Water* **2020**, *12*, 3341. [CrossRef]
62. Hirata, R.; Cagnon, F.; Bernice, A.; Maldaner, C.H.; Galvão, P.; Marques, C.; Terada, R.; Varnier, C.; Ryan, M.C.; Bertolo, R. Nitrate Contamination in Brazilian Urban Aquifers: A Tenacious Problem. *Water* **2020**, *12*, 2709. [CrossRef]
63. Feng, W.; Wang, C.; Lei, X.; Wang, H.; Zhang, X. Distribution of Nitrate Content in Groundwater and Evaluation of Potential Health Risks: A Case Study of Rural Areas in Northern China. *Int. J. Environ. Res. Public Health* **2020**, *17*, 9390. [CrossRef]
64. Naser, S.; Ghanem, M.G. Environmental and Socio- Economic Impact of Wastewater in Wadi- Qana Drainage Basin- Salfeet- Palestine. *J. Geogr. Res.* **2018**, *1*, 1–6. [CrossRef]
65. Hicham, G.; Mustapha, A.; Mourad, B.; Abdelmajid, M.; Ali, S.; Yassine, E.Y.; Mohamed, C.; Ghizlane, A.; Zahid, M. Assessment of the physico-chemical and bacteriological quality of groundwater in the Kert Plain, northeastern Morocco. *Int. J. Energy Water Resour.* **2021**, *6*, 133–147. [CrossRef]
66. Vaidya, S.R.; Labh, S.N. Determination of Physico-Chemical Parameters and Water Quality Index (WQI) for drinking water available in Kathmandu Valley, Nepal: A review Int. *J. Fish. Aquat. Stud.* **2017**, *5*, 188–190. Available online: www.fisheriesjournal.com (accessed on 12 December 2021).
67. PCBS. *Number of Israeli Settlements in the West*; PCBS: Ramallah, Palestine, 2021.
68. Arwenyo, B.; Wasswa, J.; Nyeko, M.; Kasozi, G.N. The impact of septic systems density and nearness to spring water points, on water quality. *Afr. J. Environ. Sci. Technol.* **2017**, *11*, 11–18. [CrossRef]
69. Salem, H.S.; Yihdego, Y.; Muhammed, H.H. The status of freshwater and reused treated wastewater for agricultural irrigation in the Occupied Palestinian Territories. *J. Water Health* **2020**, *19*, 120–158. [CrossRef]
70. Dare, A.E.; Mohtar, R.H.; Javfert, T.C.; Shomar, B.; Engel, B.; Boukchina, R.; Rabi, A. Opportunities and Challenges for Treated Wastewater Reuse In The West Bank, Tunisia, and Qatar. *Trans. ASABE* **2017**, *60*, 1563–1574. [CrossRef]
71. PCBS. *Palestinians at the End of 2020*; PCBS: Ramallah, Palestine, 2020; pp. 1–77.
72. Adegoke, A.A.; Stenstrom, T.-A. Part Four. Management of Risk from Excreta and Wastewater. *Constr. Wetl.* **2017**, 1–20.
73. Zimmo, O.; Petta, G. *Prospects of Efficient Wastewater Management and Water Reuse in Palestine*; Water Studies Institute, Birzeit University: Birzeit, Palestine, 2005; Available online: http://www.pseau.org/outils/ouvrages/enea_meda_water_iws_inwent_prospects_of_efficient_wastewater_management_and_water_reuse_in_palestine_2005.pdf (accessed on 14 December 2021).
74. U.S.E.P. Agency. *Report to Congress on The Prevalence Throughout the U.S. of Low- and Moderate-Income Households Without Access to a Treatment Works and The Use by States of Assistance under Section 603(c) (12) of the Federal Water Pollution Control Act*; Environmental Protection Agency: Chicago, IL, USA, 2021; Volume 603, pp. 1–51.
75. PCBS. The Palestinian Central Bureau of Statistics (PCBS) issues a press release on World Environment Day. The Palestinian environment to where? *Palest. Cent. Bur. Stat.* **2010**, *2009*, 1–4.
76. Radfard, M.; Yunesian, M.; Nabizadeh, R.; Biglari, H.; Nazmara, S.; Hadi, M.; Yousefi, N.; Yousefi, M.; Abbasnia, A.; Mahvi, A.H. Drinking water quality and arsenic health risk assessment in Sistan and Baluchestan, Southeastern Province, Iran. *Hum. Ecol. Risk Assessment: Int. J.* **2018**, *25*, 949–965. [CrossRef]
77. Daskal, S.; Ayalon, O.; Shechter, M. The state of municipal solid waste management in Israel. *Waste Manag. Res. J. Sustain. Circ. Econ.* **2018**, *36*, 527–534. [CrossRef] [PubMed]
78. Atallah, N. Palestine: Solid Waste Management Under Occupation. 2020. Available online: https://ps.boell.org/en/2020/10/07/palestine-solid-waste-management-under-occupation (accessed on 10 August 2021).
79. Thöni, V.; Matar, S.K.I. Solid Waste Management in The Occupied Palestinian Territory. In Overview Report; Palestinian. 2019. Available online: https://docslib.org/doc/6676400/solid-waste-management-in-the-occupied-palestinian-territory-west-bank-including-east-jerusalem-gaza-strip (accessed on 9 September 2022).
80. Kanyangarara, M.; Allen, S.; Jiwani, S.S.; Fuente, D. Access to water, sanitation and hygiene services in health facilities in sub-Saharan Africa 2013–2018: Results of health facility surveys and implications for COVID-19 transmission. *BMC Health Serv. Res.* **2021**, *21*, 601. [CrossRef] [PubMed]

81. Hasan, K.; Shahriar, A.; Jim, K.U. Water pollution in Bangladesh and its impact on public health. *Heliyon* **2019**, *5*, e02145. [CrossRef] [PubMed]
82. Camargo, J.A.; Alonso, Á. Ecological and toxicological effects of inorganic nitrogen pollution in aquatic ecosystems: A global assessment. *Environ. Int.* **2006**, *32*, 831–849. [CrossRef] [PubMed]
83. Scheelbeek, P.F.; Khan, A.E.; Mojumder, S.; Elliott, P.; Vineis, P. Drinking Water Sodium and Elevated Blood Pressure of Healthy Pregnant Women in Salinity-Affected Coastal AreasNovelty and Significance. *Hypertension* **2016**, *68*, 464–470. [CrossRef] [PubMed]
84. Li, C.; Zhou, K.; Qin, W.; Tian, C.; Qi, M.; Yan, X.; Han, W. A Review on Heavy Metals Contamination in Soil: Effects, Sources, and Remediation Techniques. *Soil Sediment Contam. Int. J.* **2019**, *28*, 380–394. [CrossRef]
85. Kimambo, V.; Bhattacharya, P.; Mtalo, F.; Mtamba, J.; Ahmad, A. Fluoride occurrence in groundwater systems at global scale and status of defluoridation–State of the art. *Groundw. Sustain. Dev.* **2018**, *9*, 100223. [CrossRef]
86. Adimalla, N.; Qian, H.; Nandan, M. Groundwater chemistry integrating the pollution index of groundwater and evaluation of potential human health risk: A case study from hard rock terrain of south India. *Ecotoxicol. Environ. Saf.* **2020**, *206*, 111217. [CrossRef]
87. Ward, M.H.; Jones, R.R.; Brender, J.D.; De Kok, T.M.; Weyer, P.J.; Nolan, B.T.; Villanueva, C.M.; Van Breda, S.G. Drinking Water Nitrate and Human Health: An Updated Review. *Int. J. Environ. Res. Public Health* **2018**, *15*, 1557. [CrossRef]
88. IARC. Some drinking-water disinfectants and contaminants, including arsenic. *IARC Monogr. Eval. Carcinog. Risks Hum.* **2004**, *84*, 1–477.
89. Amrose, S.E.; Cherukumilli, K.; Wright, N.C. Chemical Contamination of Drinking Water in Resource-Constrained Settings: Global Prevalence and Piloted Mitigation Strategies. *Annu. Rev. Environ. Resour.* **2020**, *45*, 195–226. [CrossRef]
90. Atsdr. Public Health Statement Zinc. no. CAS#: 7440-66-6. 2005; pp. 1–7. Available online: https://www.atsdr.cdc.gov/ToxProfiles/tp60-c1-b.pdf (accessed on 5 August 2022).
91. Tal-Spiro, O. Israeli-Palestinian Cooperation on Water Issues. *Knesset Res. Inf. Cent.* **2011**, 1–17.
92. Li, P.; Wu, J. Drinking Water Quality and Public Health. *Expo. Health* **2019**, *11*, 73–79. [CrossRef]
93. Obasi, P.N.; Akudinobi, B.B. Potential health risk and levels of heavy metals in water resources of lead–zinc mining communities of Abakaliki, southeast Nigeria. *Appl. Water Sci.* **2020**, *10*, 184. [CrossRef]
94. Harvard Medical School. Precious Metals and Other Important Minerals for Health. 2021. Available online: https://www.health.harvard.edu/staying-healthy/precious-metals-and-other-important-minerals-for-health (accessed on 1 May 2022).
95. Ferner, D.J. Toxicity, Heavy Metals. *Art. J. Anal. Chem* **2001**, *2*, 1.
96. Podgorski, J.; Araya, D.; Berg, M. Geogenic manganese and iron in groundwater of Southeast Asia and Bangladesh—Machine learning spatial prediction modeling and comparison with arsenic. *Sci. Total Environ.* **2022**, *833*, 155131. [CrossRef]
97. The President of the State of Palestine. *Decree No. (14) for the Year 2014 Relating to the Water Law, Chapter One-Definitions & General Provisions Article (1) Definitions*. no. 1664; Palestinian Water Authority: Ramallah, Palestine, 2014; pp. 1–26.
98. El Baba, M.; Kayastha, P.; Huysmans, M.; De Smedt, F. Evaluation of the Groundwater Quality Using the Water Quality Index and Geostatistical Analysis in the Dier al-Balah Governorate, Gaza Strip, Palestine. *Water* **2020**, *12*, 262. [CrossRef]
99. Shomar, B.; Abu Fakher, S.; Yahya, A. Assessment of Groundwater Quality in the Gaza Strip, Palestine Using GIS Mapping. *J. Water Resour. Prot.* **2010**, *2*, 93–104. [CrossRef]
100. Council, S. *The Allocation of Water Resources in the Occupied Palestinian Territory, including East Jerusalem*; United Nations: New York, NY, USA, 2011; Volume 32, pp. 1–24.
101. Shatat, M.; Arakelyan, K.; Shatat, O.; Forster, T.; Mushtaha, A.; Riffat, S. Low Volume Water Desalination in the Gaza Strip—Al Salam Small Scale RO Water Desalination Plant Case Study. *Future Cities Environ.* **2018**, *4*, 1–8. [CrossRef]
102. Abualtayef, M.T.; Salha, M.; Qahman, K. Study of the Readiness for Receiving Desalinated Seawater—Gaza City Case Study. *J. Eng. Res. Technol.* **2022**, *9*, 1–5. [CrossRef]
103. *The Water Safety Plan (WSP) Approach of WHO WSP steps UBA's Activities for Safe Management of Drinking-Water Supplies*; WHO: Geneva, Switzerland, 2019; pp. 2013–2015. Available online: https://www.umweltbundesamt.de/en/publikationen/das-water-safety-plan-wsp-konzept-fuer-gebaeude (accessed on 13 September 2022).
104. WHO. *Water Safety Plan: A Field Guide to Improving*; WHO: Geneva, Switzerland, 2014.
105. Jordaan, P.; Brand, M. Integrated Water Resources Management Plans. 2009, 125, pp. 154–161. Available online: www.gwpforum.org (accessed on 13 September 2022).
106. Baldwin, C.; Hamstead, M. Integrated Water Resource Planning. *Integr. Water Resour. Plan.* **2014**. [CrossRef]
107. Bain, R.; Johnston, R.; Slaymaker, T. Drinking water quality and the SDGs. *npj Clean Water* **2020**, *3*, 37. [CrossRef]
108. Syafiuddin, A.; Boopathy, R.; Hadibarata, T. Challenges and Solutions for Sustainable Groundwater Usage: Pollution Control and Integrated Management. *Curr. Pollut. Rep.* **2020**, *6*, 310–327. [CrossRef]
109. Fitch, P.; Brodaric, B.; Stenson, M.; Booth, N. Integrated Groundwater Data Management. In *Integrated Groundwater Management*; Springer: Cham, Switzerland, 2016; pp. 667–692.
110. Waarde, V.D.J.M.; Tebong, H.; Ischer, M. Water Catchment Protection Handbook. Helvetas. pp. 1–32. Available online: www.helvetascameroon.org (accessed on 13 September 2022).

111. Kollarits, S.; Kuschnig, G.; Veselic, M.; Pavicic, A.; Soccorso, C.; Aurighi, M. Decision-support systems for groundwater protection: Innovative tools for resource management. *Environ. Earth Sci.* **2006**, *49*, 840–848. [CrossRef]
112. Williams, J.; Zarracina, J. The Growth of Israeli Settlements, Explained in 5 Charts. VOX. 2016. Available online: https://www.vox.com/world/2016/12/30/14088842/israeli-settlements-explained-in-5-charts (accessed on 13 September 2022).

Article

# Variation in Spectral Characteristics of Dissolved Organic Matter and Its Relationship with Phytoplankton of Eutrophic Shallow Lakes in Spring and Summer

Yimeng Zhang [1,†], Fang Yang [2,3,†], Haiqing Liao [2,*], Shugang Hu [1], Huibin Yu [2], Peng Yuan [2], Bin Li [2,*] and Bing Cui [2]

1. College of Safety and Environmental Engineering, Shandong University of Science and Technology, Qingdao 266590, China
2. Chinese Research Academy of Environmental Science, Beijing 100012, China
3. Inner Mongolia Enterprise Key Laboratory of Damaged Environment Appraisal, Evaluation and Restoration, Hohhot 010010, China
* Correspondence: liaohq@craes.org.cn (H.L.); libin0318@yeah.net (B.L.)
† These authors contributed equally to this work.

**Citation:** Zhang, Y.; Yang, F.; Liao, H.; Hu, S.; Yu, H.; Yuan, P.; Li, B.; Cui, B. Variation in Spectral Characteristics of Dissolved Organic Matter and Its Relationship with Phytoplankton of Eutrophic Shallow Lakes in Spring and Summer. *Water* **2022**, *14*, 2999. https://doi.org/10.3390/w14192999

Academic Editors: George Arhonditsis and Jun Yang

Received: 26 July 2022
Accepted: 21 September 2022
Published: 23 September 2022

**Publisher's Note:** MDPI stays neutral with regard to jurisdictional claims in published maps and institutional affiliations.

**Copyright:** © 2022 by the authors. Licensee MDPI, Basel, Switzerland. This article is an open access article distributed under the terms and conditions of the Creative Commons Attribution (CC BY) license (https://creativecommons.org/licenses/by/4.0/).

**Abstract:** The compositional characteristics of dissolved organic matter (DOM) have important implications for lake water quality and aquatic ecology. Seasonal changes of dissolved organic matter (DOM) as well as phytoplankton abundance and composition in Shahu Lake from April to July were characterized by three-dimensional fluorescence spectroscopy (3DEEMs) combined with parallel factor (PARAFAC) analysis. The relationship between the response of components of the DOM and phytoplankton abundance were explored via Pearson correlation and redundancy analysis (RDA) in the overlying water. The results showed that the DOM was composed mainly of tryptophan-like (C2+C4), fulvic-acid-like (C3), humic-acid-like (C1), and tyrosine-like (C5) compounds that accounted for 44.47%, 20.18%, 20.04%, and 15.31%, respectively, of the DOM. The DOM was derived from both endogenous and terrestrial sources. With seasonal changes, endogenous DOM produced by phytoplankton growth and metabolism gradually increased. In spring and summer (April–July), Chl-a concentrations were significantly correlated with C3 ($p < 0.01$) and C5 ($p < 0.05$). The concentration of protein-like fractions (C2+C4, C5) were correlated with *Cyanobacteria* abundance, and the concentrations of humic-like component content (C1, C3) were correlated with the abundance of *Xanthophyta*, *Chlorophyta*, and *Cryptophytes*. Overall, phytoplankton density and Chl-a content increased by 125% and 197%, respectively, and the abundance of C3 and C5 in the DOM increased by 7.7% and 22.15% in parallel. Thus, seasonal phytoplankton growth had an important influence on the composition of the DOM.

**Keywords:** dissolved organic matter (DOM); three-dimensional excitation–emission matrix spectroscopy; parallel factor (PARAFAC) analysis; seasonal characteristics; Shahu Lake

## 1. Introduction

Lake eutrophication is one of the most prominent problems in the global aquatic environment [1] and seriously affects water quality by triggering the overproduction of phytoplankton and increasing the risk of algal toxin pollution [2]. This not only endangers the survival of aquatic organisms in lakes but also cause serious social problems [3]. The endogenous dissolved organic matter (DOM) is released from the phytoplankton production and the metabolism process, and it also plays an important role in the process of material and energy cycle in lakes [4]. The sources of DOM include exogenous and endogenous sources: exogenous DOM mainly comes from land, the atmosphere, rivers, and groundwater; endogenous DOM mainly originates from the extinction and degradation of aquatic organisms, including microorganisms, macrophytes, and phytoplankton [5,6].

DOM is an important carbon source for phytoplankton growth as well as a major product of phytoplankton metabolites [7,8]. Daggett et al. [9] found that the increases in DOM was synchronized with the increases in phytoplankton biomass. Regardless of the nutrition limitation patterns, the increase in DOM also stimulated the reproduction of *Chlorophyta*, *Bacillariophyta*, and *Chrysophyta* [9]. Phytoplankton production and metabolism is an important participant in DOM migration and transformation. Rochelle et al. [10] found a significant positive correlation between DOM content and chlorophyll a (Chl-a) in summer. The composition and structure of DOM in aquatic ecosystems is closely related to the transport and transformation of nutrients as well as the water bloom in lakes [11]. Therefore, investigating the sources, composition, and structure of DOM and the relationship between DOM and phytoplankton can help in the analysis of the causes of lake pollution, which is important for lake pollution control and lake water quality improvement.

Three-dimensional excitation emission matrix spectroscopy (3DEEMs) is an important technology in the analysis of the compositions of DOM in lakes and rivers. Song et al. [12] used 3DEEMs to characterize the composition and distribution of DOM in Taihu Lake and analyzed the source information of DOM in different areas of the lake. They found that the DOM in Taihu Lake had the dual characteristics of internal and external sources, and DOM in most of the area of the lake was mainly composed by protein-like substances that are produced by algae [12]. Zhao et al. [13] analyzed the seasonal changes of DOM in the overlying water of Erhai Lake by using 3DEEMs, finding that the changes of DOM fluorescent components could indicate the eutrophication of the lake. Moreover, parallel factor analysis (PARAFAC) can be combined with 3DEEMs as semi-quantitative analysis of DOM fluorescence composition in order to avoid the influence of overlapping and interference of fluorescence peaks in 3DEEMs. Wang et al. [14] used the 3DEEMs-PARAFAC method to qualitatively and quantitatively characterize the information of DOM components of Ebinur Lake and developed a diagnostic model based on the fluorescence index set that is applicable to surface water salinity in arid regions. Zhang et al. [15] used PARAFAC to characterize the composition of DOM in the overlying water of the Chaobai River and determined the relationship between the compositions of different types of DOM, anthropogenic inputs, microbial activity, and phytoplankton. 3DEEMs-PARAFAC analysis is undoubtedly an important research tool in the field of lake organic matter cycling and eutrophication studies.

As a typical eutrophic shallow lake, Shahu Lake was chosen as the research object. Shahu Lake is located in an arid and semi-arid region in Ningxia, China. In mid to early June 2019, the proliferation of *Prymnesiaceae* produced algal toxins, which caused large-scale fish death, and then the water quality had a tendency to deteriorate [16]. After active remediation by the local government, algal toxin events did not occur in 2020, but water quality compliance of Shahu Lake remains unstable. However, there are a few studies that have been conducted on the causes of water quality changes of Shahu Lake, or the source as well as the data of continuous monitoring of the DOM in Shahu Lake. This resulted a disadvantage of precise management of Shahu Lake. Therefore, this study used 3DEEMs-PARAFAC technology to (1) continuously monitor the DOM in the overlying water of Shahu Lake; (2) conduct a study on the structural characteristics, spatial and temporal distribution rules, and sources of DOM in the overlying water of Shahu Lake from spring (April–June) to early summer (July) in 2020; (3) investigate the relationship between structural changes of DOM components and phytoplankton response; and (4) provide an important data reference and theoretical support for the precise management of the water environment and early warning of water ecology in Shahu Lake.

## 2. Materials and Methods

### 2.1. Study Area

Shahu Lake is a brackish water lake located in a semi-arid desert area in Shizuishan City, Ningxia Hui Autonomous Region of China. It has an average water depth of 2.2 m, a water area of about 8.2 km$^2$ [17], an average annual temperature of 9.74 °C, an average annual precipitation of 172.5 mm, and an average annual evaporation of 1755.1 mm [18]. The main source of the Shahu Lake is the Yellow River recharge [19]; the water outlet located on the west side of the lake and the two canals on the south side of the lake are interconnected irrigation canal. Due to the natural climate change and tourism development, the water quality of Shahu Lake is unstable, and thus the ecological and environmental problems of Shahu Lake are critical.

### 2.2. Sample Collection and Processing

The data was conducted from continued sampling and monitoring of water quality in the area of the lake from April to July in 2020. Sampling points were set up in the area of the lake according to the distribution of the lake drainage, the location of the inlet, and the area of the lake (Figure 1). The water samples were collected into the pretreated polyethylene wide-mouth bottles, stored at 4 °C, and brought back to the laboratory for testing. The standard analytical methods for water quality indicators are shown in Table 1. The water samples for spectroscopy were filtered using glass fiber filters (Millipore, 0.45 μm fiber Ø) and then scanned for three-dimensional fluorescence spectroscopy. All monitoring work was completed within one day after sampling.

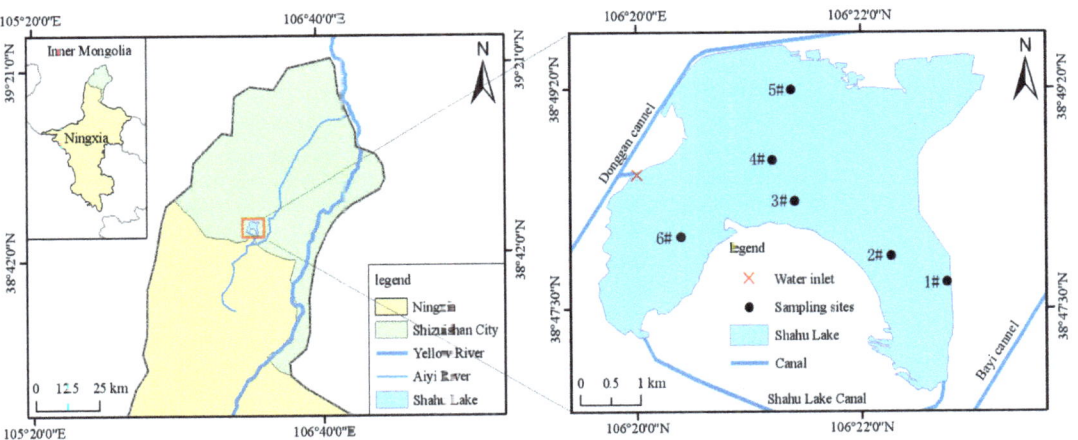

**Figure 1.** Sampling sites of overlying water in Shahu Lake.

**Table 1.** Analytical methods for selected parameters of water samples (GB3838-2002) [20].

| Parameter | Methods | Units |
|---|---|---|
| COD$_{Cr}$ | Potassium dichromate method | mg/L |
| TN | Ultraviolet spectrophotometry | mg/L |
| TP | Ammonium molybdate method | mg/L |
| Chl-a | Spectrophotometry (HJ 897-2017) [21] | μg/L |

### 2.3. 3DEEM Measurement and Indices Calculation

The samples to be tested were measured using a Hitachi Fluorescence Spectrophotometer (F-7000) at 25 °C to determine the three-dimensional fluorescence spectroscopy data. The instrument excitation light source was a 150 W xenon lamp, and Mill-Q ultrapure water was set as a blank control. The emission wavelength ($E_m$) was 260–550 nm, the excitation wavelength ($E_x$) was 200–450 nm, the interval between $E_m$ and $E_x$ was 5 nm, and the scanning speed was fixed at 2400 nm/min.

The fluorescence index (FI), autochthonous index (BIX), and humification index (HIX) were calculated from the data of 3DEEMs to assess the source and properties of DOM. FI refers to the ratio of $E_m$ fluorescence intensity at 470 nm and 520 nm, when $E_x$ = 370 nm [22]. When FI < 1.4, the DOM is dominated by exogenous input; when FI > 1.9, the DOM is dominated by autogenous source; when $1.4 \leq FI \leq 1.9$, the DOM has dual properties of internal and external sources [14,22]. BIX is the ratio of the fluorescence intensity at the emission wavelengths of 380 and 430 nm, when $E_x$ = 310 nm [22]. When 0.8 < BIX < 1.0, the DOM has a new autogenic feature; when BIX > 1.0, the DOM has a strong autogenous feature [23]. HIX is often used to indicate the degree of humification of DOM, being the ratio of the fluorescence area integral at the emission wavelength of 435~480 nm to the fluorescence area integral at 300~345 nm at $E_x$ = 255 nm [24]. The degree of humification is positively correlated with the value of HIX.

PARAFAC analysis of the 3D fluorescence spectral data was carried out using MATLAB 7.0 software. Rayleigh scattering and Raman scattering interferences in the spectral data were removed by interpolation to avoid the model from generating pseudo-peaks in specific regions [25]. The optimal number of components was then extracted by half-split validation and residual analysis, and the resulting maximum fluorescence intensity ($F_{max}$) could be used to represent the relative concentration of PARAFAC components [26].

*2.4. Data Analysis*

Sampling point layouts were mapped using ArcGIS 10.2, the redundancy analysis (RAD) used CANOCO 5, the Pearson analysis used SPSS 17.0, and data processing and mapping analysis used Excel 2019 and Origin 2021.

## 3. Results and Discussion

*3.1. DOM Spectral Fingerprint Analysis of Overlying Water in Shahu Lake*

Seven fluorescence peaks with different intensities appeared in the DOM 3D fluorescence spectra of the overlying water of the Shahu Lake (Figure 2), and the peak positions of the fluorescence spectra of the sampling points in different months were essentially similar. Therefore, the sampling point #4 with the most significant variation of the characteristic peak intensity was selected as the illustrated case. As shown in Figure 2, peaks T1 ($E_x/E_m$ = 225~235/340~345 nm) and T2 ($E_x/E_m$ = 270~280/340~370 nm) were related to tryptophan-like fluorescence in the ultraviolet and visible regions, respectively, associated with aromatic-protein-like substances produced by microbial decomposition and being easily degradable [27]. Peak B ($E_x/E_m$ = 265~285/290~315 nm) was concerned with tyrosine-like fluorescence substance in the UV region, mainly generated by microbial life activity [28]. Peaks A ($E_x/E_m$ = 240~270/375~445 nm) and C ($E_x/E_m$ = 300~370/400~500 nm) were associated with fulvic acid-like fluorescence peaks in the UV and visible regions, respectively, and were related to the degree of humification of the overlying water DOM [29]. Peaks F ($E_x/E_m$ = 260~300/470~510 nm) and H ($E_x/E_m$ = 350~380/475~510 nm) were linked to the fluorescence of humic-acid-like substances in the UV and visible regions, respectively, characterizing terrestrial humic substances, mainly from exogenous inputs such as domestic sewage and industrial and agricultural wastewater [30]. As seen in Figure 2, the fluorescence intensity of fulvic acid and tryptophan-like (peak A and T1) substances were significantly higher than the other five peaks, indicating a high content of fulvic acid and tryptophan-like substances in the DOM of the overlying water of the Shahu Lake.

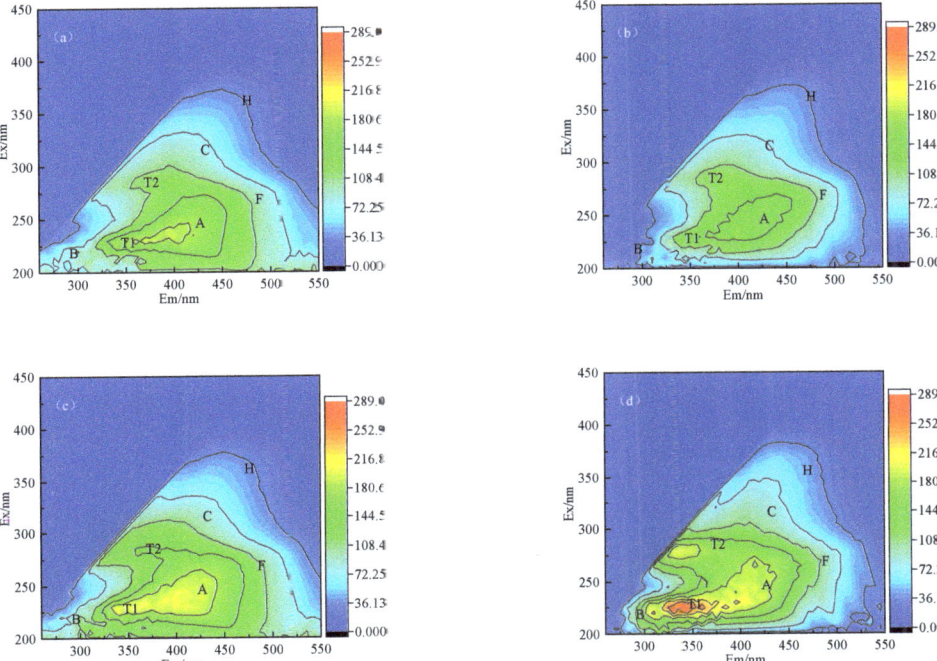

**Figure 2.** Three-dimensional excitation–emission matrix spectroscopy of the dissolved organic matter from the sampling point #4 of Shahu Lake in April (**a**), May (**b**), June (**c**), and July (**d**).

Fluorescence abundance showed an increasing trend at each sampling point between April and July. From spring to summer (April–July), the fluorescence abundance of tryptophan-like and tyrosine-like substances gradually increased, meaning that the amounts of protein-like substances increased. Zhang et al. [3] studied the changes of DOM molecular composition in June and November, which was during and after the algal bloom in Taihu Lake, finding that during the peak of the *Cyanobacterial* bloom in June, DOM was already combined with algal metabolites and showed the characteristics of biological mixture. They suggested that the temporal variation in the molecular characteristics of DOM in the Taihu Lake basin from summer to winter might be related to microbial metabolism [3]. Jia et al. [7] studied the relationship between DOM and phytoplankton response in Baiyang Lake in autumn (September–November) and found that the tryptophan-like contribution showed a clear downward trend from early to late autumn as phytoplankton abundance decreased. Therefore, it can be speculated that changes in DOM composition in overlying water of Shahu Lake might be related to phytoplankton and microbial activities.

Five major fractions were identified using PARAFAC for fraction extraction of DOM from Shahu Lake from April to July (Table 2, Figure 3). C1 ($E_x/E_m = 260/460$ nm) and C3 ($E_x/E_m = 245/400$ nm) were jointly classified as humic-like substances. C1 corresponds to the UV-zone humic-acid-like fluorescence peak of the F-peak, which generally represents terrestrially derived organic matter [31]. C3 is a microbially transformed authigenic humic-like component with a similar spectral pattern to the A-peak [32], which indicates the UV zone fulvic acid-like fluorescence peak and is probably derived from algal and microbial residue degradation [33]. C2 ($E_x/E_m = 230, 280/340$ nm) and C4 ($E_x/E_m = 300/365$ nm) are tryptophan-like substances that correspond to the peaks T1 and T2 of tryptophan-like fluorescence, and they are generally the metabolites of the microbial degradation [15,34,35]. This will be represented by (C2+C4) in the subsequent discussion. C5 ($E_x/E_m = 220/300$ nm) was identified as a tyrosine-like substance, corresponding to a tyrosine-like fluorescence peak in the UV region of the B peak, mainly an endogenous-protein-like substance produced by microorganisms and phytoplankton [28].

**Table 2.** Spectral characteristics of five fluorescent components in the overlying water of the surface of Shahu Lake.

| Component | $\lambda E_x/\lambda E_m$/nm | Characteristics | Traditional Peak Value |
|---|---|---|---|
| C1 | 260/460 | Terrestrial-humic-like, humic-acid-like | 260/488 [31] |
| C2 | 230, 280/340 | Tryptophan-like | 230/345 [15] 220–230, 280/320–336 [34] |
| C3 | 245/400 | Authigenic-humic-like, fulvic-acid-like | 235/397 [32] |
| C4 | 300/365 | Tryptophan-like | 295/370 [35] |
| C5 | 220/300 | Tyrosine-like | 220/305 [28] |

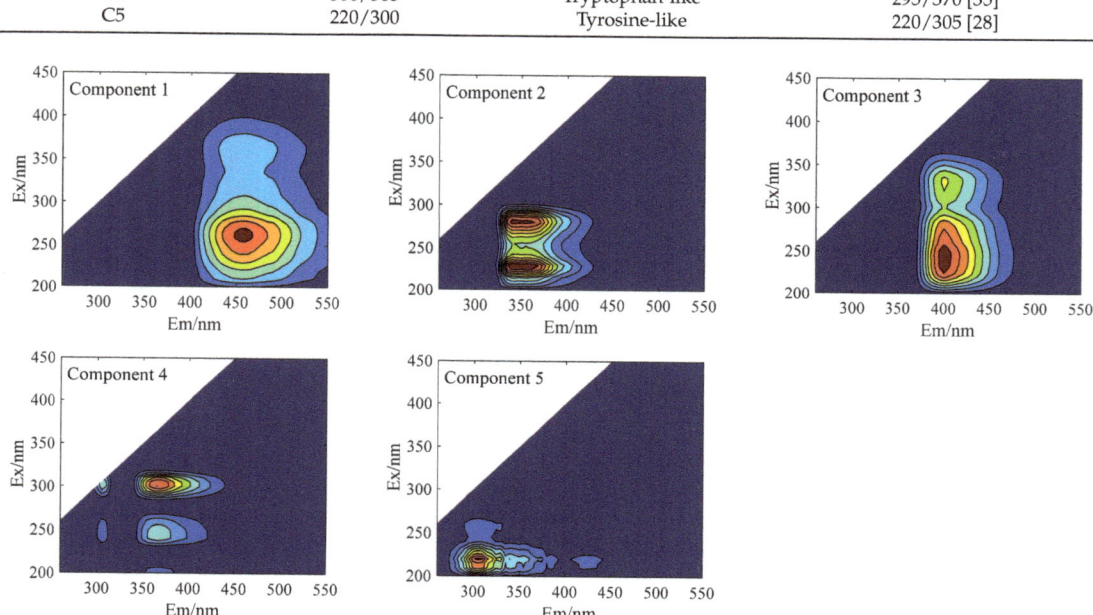

**Figure 3.** Five different components identified by the PARAFAC model.

### 3.2. Variation Characteristics of Phytoplankton in Shahu Lake from April to July

The phytoplankton species and density of Shahu Lake were counted between April and July of 2020. The data contained a total of 61 species from 8 phyla (Figure 4a). *Chlorophyta* had the most occurrences with 22 species, accounting for 36% of the total number of phytoplankton species present during the survey, followed by *Cyanobacteria* and *Bacillariophyta*, each accounting for 23% of the total number of phytoplankton species. Other algae such as *Euglenophyta*, *Pyrrophyta*, *Chrysophyta*, *Cryptophyta*, and *Xanthophyta* were relatively few in number, accounting for 8.2%, 3.3%, 3.3%, 1.6%, and 1.6%, respectively. The total amount of *Chlorophyta*, *Cyanobacteria*, and *Bacillariophyta* species accounted for 82.1% of the total number of phytoplankton species present during the research process. Therefore, *Chlorophyta*, *Cyanobacteria*, and *Bacillariophyta* phyla were represented by the greatest number of taxa in the Shahu Lake.

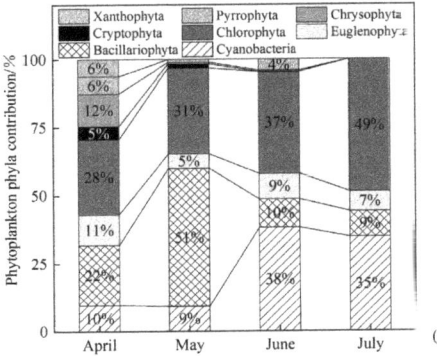 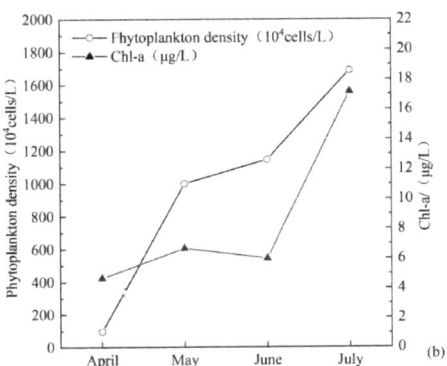

**Figure 4.** Monthly proportion of phytoplankton phyla (**a**) and monthly variation diagram of phytoplankton density and Chl-a (**b**).

With seasonal changes, phytoplankton began to grow and multiply in the water body (Figure 4b). The average of Chl-a content was 5.78 µg/L in spring (April–June) and became 17.17 µg/L in early summer (July), which was a percentage growth of 197%. The average phytoplankton densities from April to June were $9.5 \times 10^5$ cells/L, $1 \times 10^7$ cells/L, $1.15 \times 10^7$ cells/L, and $1.69 \times 10^7$ cells/L in that order, presenting a trend of monthly increase. The average phytoplankton density in spring (April–June) was $0.75 \times 10^7$ cells/L, and in early summer (July) was $1.69 \times 10^7$ cells/L, which was a percentage growth of 125%. The phytoplankton density in early summer was significantly higher than that in spring, and thus the growth and reproduction of the phytoplankton were affected by seasonal changes. As the seasons change, the increasement of temperatures and sunlight intensity will promote the reproduction of *Chlorophyta*, *Cyanobacteria*, and *Bacillariophyta* [9]. In temperate plain water systems, in addition to physical factors such as lake temperature, light, and hydrological effects, the composition of DOM also affects phytoplankton's growth and metabolism [36]. The DOM, as the most active organic matter in the lake water environment, provides available nutrients such as organic carbon, organic nitrogen, and organic phosphorus for phytoplankton growth and reproduction [37]. Moreover, with the increasement of phytoplankton biomass could enhance the metabolism, and the growth, metabolism, and decay of phytoplankton will promote the increase in DOM content in the aquatic environment [38].

### 3.3. Analysis of DOM Component Characteristics and Main Sources

Due to the different fluorescence quantum yields of different substances, the fluorescence intensity cannot directly characterize the concentration of the DOM component, but its relative content can be expressed in terms of $F_{max}$ [26]. The spatial distribution of total $F_{max}$ in Shahu Lake did not vary greatly, and was at each point ranging from 533.67 to 700.54 A.U. (Figure 5). Among them, sampling point #6 in the southwestern part of the lake had the lowest total $F_{max}$, while points #1 and #2 in the southeastern part of the lake were higher, which might have been due to the southeastern part of the lake being farther from the recharge outlet. The DOM of the overlying water of Shahu Lake was mainly composed of 59.78% of protein-like (tryptophan-like and tyrosine-like) and 40.12% of humic substances (including terrestrial and authigenic sources). The magnitude of each fluorescent fraction to the total $F_{max}$ was in the following order: (C2+C4) > C3 > C1 > C5, with the percentages of each fraction being 44.47%, 20.18%, 20.04%, and 15.31%, respectively. Zhang et al. [3] pointed out that the DOM, which consists mainly of protein-like and humic substances, was closely linked to phytoplankton community dynamics.

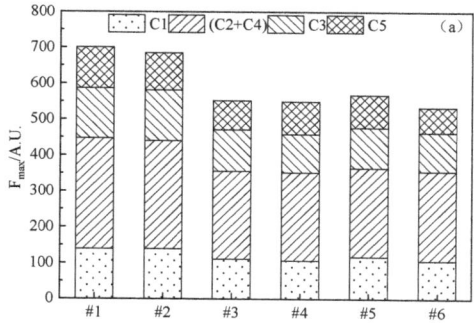 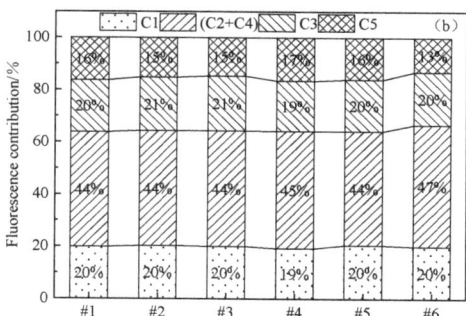

**Figure 5.** Total fluorescence abundance (**a**) and relative contribution ratio (**b**) of DOM components in the overlying water of Shahu Lake.

In order to study the variation of DOM content in the overlying water of Shahu Lake with time and its influencing factors, the variation of $F_{max}$ and spectral indices of each fluorescent component of DOM in spring and summer (April–July) was analyzed. There was little overall variation of C1 from spring to early summer (Figure 6), which was related to the continuous recharge of water in the Shahu Lake basin between April and July [19]. In spring (April–June), the abundance of (C2+C4) showed an upward trend, while the C3 showed a gentle change and the trend of C5 was raised in volatility. From spring to summer (June–July), the abundance of (C2+C4) decreased; the abundance of C3 and C5 increased obviously; and the fluorescence abundance increased by 7.7% and 22.15%, respectively, which was similar to the variation of phytoplankton density and Chl-a content. Moreover, research has shown that the humic-like content was controlled by hydrological processes, whereas protein-like fluorescent substances were more closely linked to biological processes [39]. Algae can release large amounts of tyrosine-like substances when they accumulate and die, which leads to the enhancement of the biogenic characteristics of DOM in water. Low-molecular-weight protein-like substances can both originate from phytoplankton metabolites and act as organic nutrients to promote phytoplankton growth and reproduction [40]. The above analysis shows that humic-like and protein-like material in DOM is strongly linked to changes in phytoplankton biomass over time in Shahu Lake.

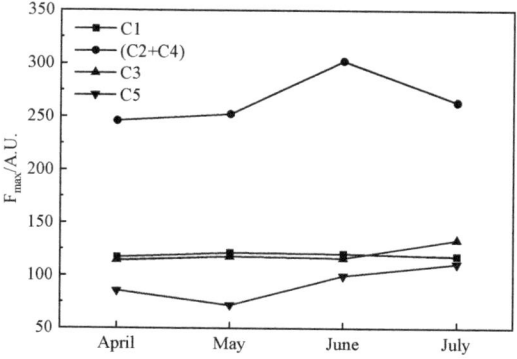

**Figure 6.** Variation components of DOM component content in spring (April to June) and summer (July).

The variation of spectral indices such as FI, BIX, and HIX in the area of the lake was analyzed over time. There was some spatial and temporal variation among the fluorescence index values in the area of the lake: the FI was ranging from 1.58 to 2.08 with a mean value of 1.68 ± 0.10 (1.4 < FI < 1.9) (Table 3), which indicated that the DOM of Shahu Lake has dual properties of internal and external sources [14]. The average of FI in spring

(April–June) was 1.66 ± 0.05, and in early summer was 1.77 ± 0.16. With seasonal changes, the DOM of Shahu Lake was gradually enhanced by endogenous influence. The BIX values ranged from 0.98 to 1.46, with a mean value of 1.15 ± 0.14, which indicated that the DOM in the overlying water of Shahu Lake has authigenic characteristics [23]. HIX values ranged from 1.68 to 2.94 with a mean value of 2.31 ± 0.36, indicating a low degree of overall humification of the DOM [24]. The average value of HIX was 2.35 ± 0.36 in spring and 2.19 ± 0.33 in early summer with a slight decrease in the humification index. This indicated an increase in the degree of endogenous influence on DOM. As the season changes from spring to summer, the temperature rises, and the above three parameters all indicated that the DOM in the overlying water of Shahu Lake was predominantly endogenous. The endogenous generation of DOM is closely related to biological activity. Phytoplankton growth and metabolism and decomposition of residues release DOM into the water body [4], influencing the molecular composition characteristics of DOM [41].

**Table 3.** Average FI, BIX, and HIX of Shahu Lake.

| Parameter | Spring (April–July) | Summer (July) | Shahu Lake |
|---|---|---|---|
| FI | 1.66 ± 0.05 | 1.77 ± 0.16 | 1.68 ± 0.10 |
| BIX | 1.20 ± 0.12 | 1.01 ± 0.02 | 1.15 ± 0.14 |
| HIX | 2.35 ± 0.36 | 2.19 ± 0.33 | 2.31 ± 0.36 |

*3.4. Influence Factors of DOM Composition in Overlying Water of Shahu Lake*

In order to determine the influencing factors of the DOM component in the overlying water of Shahu Lake, Pearson correlation analysis and RDA were performed on the four fluorescence components with the densities of TN, TP, $COD_{Cr}$, Chl-a, and phytoplankton. The length of the arrow in the RDA results ranking chart (Figure 7) indicates the importance of the factor: two arrows in the same direction indicate a positive correlation, a smaller angle between the arrows indicates a greater correlation, and a near right angle indicates a small correlation [30]. The eigenvalues of RDA−1 and RDA−2 in Figure 7a were 0.3009 and 0.1604, respectively, which explain 30.29% and 16.04% of the variance of data. With the same method, the eigenvalues of RDA−1 and RDA−2 in Figure 7b explain 24.25% and 4.71% of the variance of data, respectively.

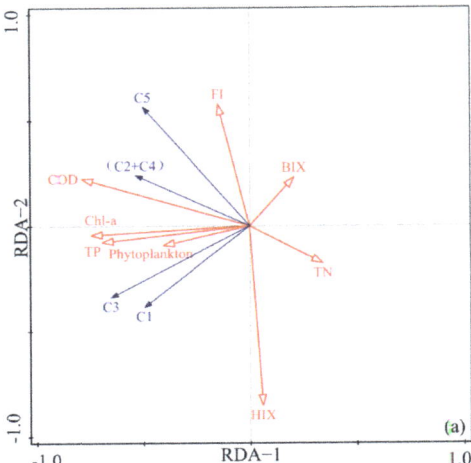
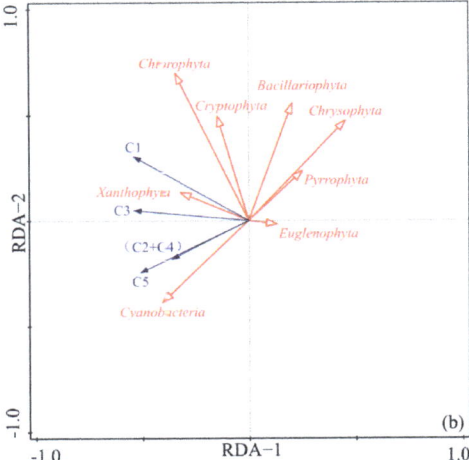

**Figure 7.** RDA results of DOM components, environmental factors (**a**), and phytoplankton phyla (**b**).

Pearson correlation analysis and RDA results showed that Chl-a was positively correlated with phytoplankton density ($p < 0.01$) (Table 4). Chl-a is not only an important component of algal cells but also a type contained in all phytoplankton phyla. Therefore, the Chl-a and phytoplankton density were generally well correlated, and both can reflect phytoplankton standing crop [42]. However, the correlations between Chl-a and phytoplankton density and other environmental factors showed slight differences, which were related to the changes in the dominant phytoplankton species, the water quality condition, and the hydrological characteristics in the lake [43]. Chl-a showed a highly significant positive correlation with $COD_{Cr}$ ($p < 0.01$), a significant negative correlation with TN ($p < 0.05$), and a statistically positive correlation with TP ($r^2 > 0.3$). COD can reflect the degree of water pollution. The degradation of algal residues will release organic matter and nutrient elements (such as N and P) and consume dissolved oxygen by microbial decomposition or photodegradation. The growth and reproduction of phytoplankton could be promoted by sufficient organic matter and nutrients, and excessive phytoplankton will consume more dissolved oxygen and result in water quality degradation [44].

**Table 4.** Correlation coefficients between DOM components and environmental factors. "*" indicates $p < 0.05$, "**" indicates $p < 0.01$.

| Parameter | Chl-a | Phytoplankton Density | TN | TP | $COD_{Cr}$ | C1 | (C2+C4) | C3 | C5 | FI | BIX | HIX |
|---|---|---|---|---|---|---|---|---|---|---|---|---|
| Chl-a | 1 | - | - | - | - | - | - | - | - | - | - | - |
| Phytoplankton density | 0.697 ** | 1 | - | - | - | - | - | - | - | - | - | - |
| TN | −0.411 * | −0.60 ** | 1 | - | - | - | - | - | - | - | - | - |
| TP | 0.381 | 0.297 | −0.294 | 1 | - | - | - | - | - | - | - | - |
| $COD_{Cr}$ | 0.559 ** | 0.427 * | −482 * | 0.591 ** | 1 | - | - | - | - | - | - | - |
| C1 | 0.277 | 0.14 | −0.064 | 0.463 * | 0.333 | 1 | - | - | - | - | - | - |
| (C2+C4) | 0.224 | 0.139 | −0.254 | 0.386 | 0.459 * | 0.627 ** | 1 | - | - | - | - | - |
| C3 | 0.645 ** | 0.395 | −0.183 | 0.381 | 0.410 * | 0.824 ** | 0.568 ** | 1 | - | - | - | - |
| C5 | 0.430 * | 0.168 | −0.247 | 0.292 | 0.532 ** | 0.344 | 0.503* | 0.336 | 1 | - | - | - |
| FI | 0.331 | 0.346 | −0.187 | −0.108 | −0.027 | −0.216 | 0.083 | −0.004 | 0.482 * | 1 | - | - |
| BIX | −0.580 ** | −0.261 | −0.117 | −0.037 | −0.124 | −0.183 | 0.374 | −0.364 | −0.238 | −0.185 | 1 | - |
| HIX | −0.068 | 0.134 | −0.149 | 0.087 | −0.337 | 0.308 | −0.183 | 0.222 | −0.545 ** | −0.368 | 0.009 | 1 |

Chl-a was significantly correlated with BIX ($p < 0.01$) and C3 ($p < 0.01$), and positively correlated with C5 ($p < 0.05$), indicating that phytoplankton biomass in Shahu Lake was related to the production of endogenous DOM. Previous research has shown that the DOM from in situ phytoplankton degradation contains approximately 25% humic-like fluorescence [45]. Mangal et al. [46] suggested that the protein-like material in the lake probably came from phytoplankton production and metabolism, as well as the decay and degradation of residues. Moreover, Chl-a and phytoplankton density were positively correlated with the DOM (C1, C2+C4, C3, C5) (Figure 7a), which indicated that the variation of DOM components in the overlying water of Shahu Lake were influenced by phytoplankton. The accumulation of phytoplankton and degradation of residues were the important source of contribution to DOM [8,10]. Meanwhile, (C2+C4) and C5 were correlated well with *Cyanobacteria*, and C1 and C3 were correlated positively with *Xanthophyta*, *Chlorophyta*, and *Cryptophyta* (Figure 7b). Low-molecular-weight DOMs, such as protein-like substances, can be absorbed directly by phytoplankton (such as *Cyanobacteria*); high-molecular-weight DOMs, such as humic substances, can be devoured by mixotrophic phytoplankton (*Chlorophyta* and *Cryptophyta*) or degraded by microbial decomposition [41,47]. The endogenous DOM in Shahu Lake may be an important "source" of nutrients for phytoplankton growth and an important "sink" for phytoplankton production of metabolic organic matter. The changes in phytoplankton biomass in the Shahu Lake water column are important drivers of changes in DOM composition characteristics. Therefore, it is important to pay attention to the changes in DOM composition in order to maintain the health of the water environment and improve water quality in Shahu Lake.

## 4. Conclusions

In this study, the main composition of DOM in the overlying water of Shahu Lake and its relationship with the seasonal changes of phytoplankton biomass were continuously monitored. According to 3DEEM and PARAFAC analysis, the DOM in the overlying water of Shahu Lake was mainly composed of tryptophan-like (C2+C4), fulvic-acid-like (C3), humic-acid-like (C1), and tyrosine-like (C5) substances, and their fluorescence contributions were 44.47%, 20.18%, 20.04%, and 15.31%, respectively. Protein-like (C2+C4, C5) material accounted for a higher proportion (59.78%) of the DOM, which was gradually enhanced with seasonal changes due to more endogenous influences (BIX > 1.0). The endogenous DOM was mainly derived from phytoplankton reproduction and decay in the area of the lake. Continuous water quality monitoring (April to July in 2020) showed that from spring to summer (April to July 2020), phytoplankton density increased from $0.75 \times 10^7$ cells/L to $1.69 \times 10^7$ cells/L, an increase of 125%, and Chl-a content increased from 5.78 to 17.17 μg/L, an increase of 197%, while the autotrophic humic fraction and the tyrosine-like fractions increased in abundance by 7.7 and 22.15 percent, respectively. Among them, the concentration of the Chl-a showed a significant positive correlation with C3 ($p < 0.01$) and showed a relatively positive correlation with C5 ($p < 0.05$). The protein-like components (C2+C4, C5) were closely correlated with *Cyanobacteria*, and the humic-like components (C1, C3) were well correlated with *Xanthophyta*, *Chlorophyta*, and *Cryptophyta*. This indicated that seasonal growth of phytoplankton was an important driving factor of the changes in DOM composition in the research period of Shahu Lake. This work provided a theoretical foundation and important data support for the analysis of the causes of seasonal water quality changes in eutrophic lakes.

**Author Contributions:** Y.Z. wrote the paper and analyzed the data. B.C. performed the experiments. F.Y., H.L. and B.L. polished this manuscript. All authors (Y.Z., F.Y., H.L., S.H., H.Y., P.Y., B.L. and B.C.) modified the manuscript. All authors have read and agreed to the published version of the manuscript.

**Funding:** This research was funded by National Natural Science Foundation of China (41907338) and National Key Research and Development Program of China (2019YFC0409205, 2019YFC0409202).

**Data Availability Statement:** Data will be made available on request.

**Conflicts of Interest:** The authors declare no conflict of interest.

## References

1. Ho, J.C.; Michalak, A.M.; Pahlevan, N. Widespread global increase in intense lake phytoplankton blooms since the 1980s. *Nature* **2019**, *574*, 667–670. [CrossRef] [PubMed]
2. Zhou, J.; Qin, B.Q.; Han, X.X.; Zhu, L. Turbulence increases the risk of microcystin exposure in a eutrophic lake (Lake Taihu) during cyanobacterial bloom periods. *Harmful Algae* **2016**, *55*, 213–220. [CrossRef] [PubMed]
3. Zhang, F.F.; Harir, M.; Moritz, F.; Zhang, J.; Witting, M.; Wu, Y.; Schmitt-Kopplin, P.; Fekete, A.; Gaspar, A.; Hertkorn, N. Molecular and structural characterization of dissolved organic matter during and post cyanobacterial bloom in Taihu by combination of NMR spectroscopy and FTICR mass spectrometry. *Water Res.* **2014**, *57*, 280–294. [CrossRef] [PubMed]
4. Zhou, Y.; Davidson, T.A.; Yao, X.; Zhang, Y.; Erik, J.; Garcia, D.; Wu, H.; Shi, K.; Qin, B. How autochthonous dissolved organic matter responds to eutrophication and climate warming: Evidence from a cross-continental data analysis and experiments. *Earth Sci. Rev.* **2018**, *185*, 928–937. [CrossRef]
5. Zhang, H.; Zheng, Y.; Wang, X.C.; Wang, Y.; Dzakpasu, M. Characterization and biogeochemical implications of dissolved organic matter in aquatic environments. *J. Environ. Manag.* **2021**, *294*, 113041. [CrossRef] [PubMed]
6. Derrien, M.; Brogi, S.R.; Goncalves-Araujo, R. Characterization of aquatic organic matter: Assessment, perspectives and research priorities. *Water Res.* **2019**, *163*, 114908. [CrossRef]
7. Jia, S.Q.; Yang, F.; Wang, X.H.; Bai, Y.W.; Ma, W.J.; Li, Q.H.; Liao, H.Q.; Chen, S.H. Relationship between DOM components and phytoplankton in Baiyangdian Lake in autumn. *Environ. Pollut. Control.* **2021**, *43*, 1101–1107. [CrossRef]
8. Zhang, Y.; Dijk, M.A.V.; Liu, M.L.; Zhu, G.W.; Qin, B.Q. The contribution of phytoplankton degradation to chromophoric dissolved organic matter (CDOM) in eutrophic shallow lakes: Field and experimental evidence. *Water Res.* **2009**, *43*, 4685–4697. [CrossRef] [PubMed]
9. Carmen, T.D.; Jasmine, E.S.; Lafrancois, B.M.; Kevir, S.S. Effects of increased concentrations of inorganic nitrogen and dissolved organic matter on phytoplankton in boreal lakes with differing nutrient limitation patterns. *Aquat. Sci.* **2015**, *77*, 511–521.

10. Rochelle-Newall, E.J.; Fisher, T.R. Production of chromophoric dissolved organic matter fluorescence in marine and estuarine environments: An investigation into the role of phytoplankton. *Mar. Chem.* **2002**, *77*, 7–21. [CrossRef]
11. Li, Y.P.; Zhang, L.; Wang, S.R.; Zhao, H.C.; Zhang, R. Composition, structural characteristics and indication of water quality of dissolved organic matter in Dongting Lake sediments. *Ecol. Eng.* **2016**, *97*, 370–380. [CrossRef]
12. Song, X.N.; Yu, T.; Zhang, Y.; Zhang, Y.; Yin, X.Y. Distribution characterization and source analysis of dissolved organic matters in Taihu Lake using three dimensional fluorescence excitation-emission matrix. *Acta Sci. Circumstantiae* **2010**, *30*, 2321–2331. [CrossRef]
13. Zhao, H.C.; Li, Y.P.; Wang, S.R.; Jiao, L.X.; Zhang, L. Fluorescence characteristics of DOM in overlying water of Erhai Lake and its indication of eutrophication. *Spectrosc. Spectr. Anal.* **2019**, *39*, 3888–3896.
14. Wang, X.P.; Zhang, F.; Yang, S.T.; Ayinuer, Y.; Chen, Y. Rapid diagnosis of surface water salt content (WSC) in Ebinur Lake watershad based on 3-D fluorescence technology. *Spectrosc. Spectr. Anal.* **2018**, *38*, 1468–1475.
15. Zhang, L.; Sun, Q.X.; You, Y.; Zhang, K.; Gao, C.D.; Peng, Y.Z. Compositional and structural characteristics of dissolved organic matter in overlying water of the Chaobai River and its environment significance. *Environ. Sci. Pollut. Res.* **2021**, *28*, 59673–59686. [CrossRef]
16. Cui, B. Study on Spectral Fingerprint Characteristics of Dissolved Organic Matter in Sand Lake and Its Response Mechanism to Water Quality and Algal Bloom. Master's Thesis, Shandong Normal University, Jinan, China, 2021.
17. Li, J.Y.; Zhang, Y.F.; Yang, Z.; Wang, M. Bacterial diversity in Shahu lake, northwest China is significantly affected by nutrient composition rather than location. *Ann. Microbiol.* **2017**, *67*, 469–478. [CrossRef]
18. Li, P.Y.; Feng, W.; Xue, C.Y.; Tian, R.; Wang, S.T. Spatiotemporal Variability of Contaminants in Lake Water and Their Risks to Human Health: A Case Study of the Shahu Lake Tourist Area, Northwest China. *Expo. Health* **2017**, *9*, 213–225. [CrossRef]
19. Chen, J.; Qian, H.; Gao, Y.Y.; Wang, H.K.; Zhang, M.S. Insights into hydrological and hydrochemical processes in response to water replenishment for lakes in arid regions. *J. Hydrol.* **2020**, *581*, 124386. [CrossRef]
20. MEP of PRC (Ministry of Environmental Protection of the People's Republic of China). *Environmental Quality Standards for Surface Water (GB3838-2002)*; China Environmental Science Press: Beijing, China, 2002; Volume 6, pp. 8–9.
21. MEP of PRC (Ministry of Environmental Protection of the People's Republic of China). *Water Quality—Determination of Chlorophyll A—Spectrophotometric Method (HJ 897-2017)*; China Environmental Science Press: Beijing, China, 2017.
22. Wilske, C.; Herzsprung, P.; Lechtenfeld, O.J.; Kamjunke, N.; Einax, J.W.; von Tümpling, W. New Insights into the Seasonal Variation of DOM Quality of a Humic-Rich Drinking-Water Reservoir—Coupling 2D-Fluorescence and FTICR MS Measurements. *Water* **2021**, *13*, 1703. [CrossRef]
23. Huguet, A.; Vacher, L.; Relexans, S.; Saubusse, S.; Froidefond, J.M.; Parlanti, E. Properties of fluorescent dissolved organic matter in the Gironde Estuary. *Org. Geochem.* **2009**, *40*, 706–719. [CrossRef]
24. Ohno, T. Fluorescence inner-filtering correction for determining the humification index of dissolved organic matter. *Environ. Sci. Technol.* **2002**, *36*, 742–746. [CrossRef] [PubMed]
25. Stedmon, C.; Bro, R. Characterizing Dissolved Organic Matter Fluorescence with Parallel Factor Analysis: A Tutorial. *Limnol. Oceanogr.* **2008**, *6*, 572–579. [CrossRef]
26. Zeng, Z.; Zheng, P.; Ding, A.Q.; Zhang, M.; Abbas, G.; Li, W. Source analysis of organic matter in swine wastewater after anaerobic digestion with EEM-PARAFAC. *Environ. Sci. Pollut. Res.* **2017**, *24*, 6770–6778. [CrossRef] [PubMed]
27. Yu, B.X.; Liu, D.P.; Wang, J.; Sun, Y.X. Insight into removals of PARAFAC components from dissolved and particulate organic matter in wastewater treatment process by two-dimensional correlation and structure equation modeling. *Environ. Sci. Eur.* **2021**, *33*, 118. [CrossRef]
28. Baker, A.; Inverarity, R. Protein-like fluorescence intensity as a possible tool for determining river water quality. *Hydrol. Process.* **2004**, *18*, 2927–2945. [CrossRef]
29. Baker, A.; Curry, M. Fluorescence of leachates from three contrasting landfills. *Water Res.* **2004**, *38*, 2605–2613. [CrossRef]
30. Lu, K.T.; Gao, H.J.; Yu, H.B.; Liu, D.P.; Zhu, N.M.; Wan, K.L. Insight into variations of DOM fractions in different latitudinal rural black-odor waterbodies of eastern China using fluorescence spectroscopy coupled with structure equation model. *Sci. Total Environ.* **2022**, *816*, 151531. [CrossRef]
31. Dainard, P.G.; Gueguen, C.; Yamamoto-Kawai, M.; Williams, W.J.; Hutchings, J.K. Interannual Variability in the Absorption and Fluorescence Characteristics of Dissolved Organic Matter in the Canada Basin Polar Mixed Waters. *J. Geophys. Res. Ocean.* **2019**, *124*, 5258–5269. [CrossRef]
32. Zhang, Y.L.; Zhang, E.L.; Yin, Y.; van Dijk, M.A.; Feng, L.Q.; Shi, Z.Q.; Liu, M.L.; Qin, B.Q. Characteristics and sources of chromophoric dissolved organic matter in lakes of the Yungui Plateau, China, differing in trophic state and altitude. *Limnol. Oceanogr.* **2010**, *55*, 2645–2659. [CrossRef]
33. Du, Y.X.; Lu, Y.H.; Roebuck, J.A.; Liu, D.; Chen, F.Z.; Zeng, Q.F.; Xiao, K.; He, H.; Liu, Z.W.; Zhang, Y.L.; et al. Direct versus indirect effects of human activities on dissolved organic matter in highly impacted lakes. *Sci. Total Environ.* **2021**, *752*, 141839. [CrossRef]
34. Liu, Q.; Jiang, Y.; Tian, Y.L.; Hou, Z.J.; He, K.J.; Fu, L.; Xu, H. Impact of land use on the DOM composition in different seasons in a subtropical river flowing through a region undergoing rapid urbanization. *J. Clean. Prod.* **2019**, *212*, 1224–1231. [CrossRef]
35. Ishii, S.K.L.; Boyer, T.H. Behavior of Reoccurring PARAFAC Components in Fluorescent Dissolved Organic Matter in Natural and Engineered Systems: A Critical Review. *Environ. Sci. Technol.* **2012**, *46*, 2006–2017. [CrossRef]

36. Mayora, G.; Devercelli, M.; Frau, D. Spatial variability of chromophoric dissolved organic matter in a large floodplain river: Control factors and relations with phytoplankton during a low water period. *Ecohydrology* **2016**, *9*, 487–497. [CrossRef]
37. Burpee, B.; Saros, J.E.; Northington, R.M.; Simon, K.S. Microbial nutrient limitation in Arctic lakes in a permafrost landscape of southwest Greenland. *Biogeosciences* **2016**, *13*, 365–374. [CrossRef]
38. Kissman, C.E.H.; Williamson, C.E.; Rose, K.C.; Saros, J.E. Nutrients associated with terrestrial dissolved organic matter drive changes in zooplankton: Phytoplankton biomass ratios in an alpine lake. *Freshw. Biol.* **2017**, *62*, 40–51. [CrossRef]
39. Fellman, J.B.; Hood, E.; Spencer, R.G.M. Fluorescence spectroscopy opens new windows into dissolved organic matter dynamics in freshwater ecosystems: A review. *Limnol. Oceanogr.* **2010**, *55*, 2452–2462. [CrossRef]
40. Liu, Q.; Tian, Y.L.; Liu, Y.; Yu, M.; Hou, Z.J.; He, K.J.; Xu, H.; Cui, B.S.; Jiang, Y. Relationship between dissolved organic matter and phytoplankton community dynamics in a human-impacted subtropical river. *J. Clean. Prod.* **2021**, *289*, 125144. [CrossRef]
41. Wei, M.J.; Gao, C.; Zhou, Y.J.; Duan, P.F.; Li, M. Variation in spectral characteristics of dissolved organic matter in inland rivers in various trophic states, and their relationship with phytoplankton. *Ecol. Indic.* **2019**, *104*, 321–332. [CrossRef]
42. Zhao, H.X.; Qiu, X.C.; Yang, Y.M.; Li, G.D. Evaluation and analysis on trophic level in Shahu Lake. *Hubei Agric. Sci.* **2010**, *49*, 2414–2417.
43. Yu, H.Y.; Zhou, B.; Hu, Z.Y.; Ma, Y.; Chao, A.M. Study on correlation between chlorophyll a and algal density of biological monitoring. *Environ. Monit. China* **2009**, *25*, 40–43. [CrossRef]
44. Kim, D.; Lim, J.-H.; Chun, Y.; Nayna, O.K.; Begum, M.S.; Park, J.-H. Phytoplankton nutrient use and CO2 dynamics responding to long-term changes in riverine N and P availability. *Water Res.* **2021**, *203*, 117510. [CrossRef] [PubMed]
45. Kramer, G.D.; Herndl, G.J. Photo- and bioreactivity of chromophoric dissolved organic matter produced by marine bacterioplankton. *Aquat. Microb. Ecol.* **2004**, *36*, 239–246. [CrossRef]
46. Mangal, V.; Stock, N.L.; Guéguen, C. Molecular characterization of phytoplankton dissolved organic matter (DOM) and sulfur components using high resolution Orbitrap mass spectrometry. *Anal. Bioanal. Chem.* **2016**, *408*, 1891–1900. [CrossRef] [PubMed]
47. Hounshell, A.G.; Peierls, B.L.; Osburn, C.L.; Paerl, H.W. Stimulation of Phytoplankton Production by Anthropogenic Dissolved Organic Nitrogen in a Coastal Plain Estuary. *Environ. Sci. Technol.* **2017**, *51*, 13104–13112. [CrossRef]

Article

# Simultaneous Removal of $COD_{Mn}$ and Ammonium from Water by Potassium Ferrate-Enhanced Iron-Manganese Co-Oxide Film

Yingming Guo [1,*], Ben Ma [1], Shengchen Yuan [1], Yuhong Zhang [1], Jing Yang [1], Ruifeng Zhang [1] and Longlong Liu [2]

1. School of Urban Planning and Municipal Engineering, Xi'an Polytechnic University, Xi'an 710048, China
2. Shaanxi LangMingRun Environmental Protection Technology Co., Xi'an 710061, China
* Correspondence: guoyingming@xpu.edu.cn

**Abstract:** Iron-manganese co-oxide film ($MeO_x$) has a high removal efficiency for ammonium ($NH_4^+$) and manganese ($Mn^{2+}$) in our previous studies, but it cannot effectively remove $COD_{Mn}$ from water. In this study, the catalytic oxidation ability of $MeO_x$ was enhanced by dosage with potassium ferrate ($K_2FeO_4$) to achieve the simultaneous removal of $COD_{Mn}$ and $NH_4^+$ from water in a pilot-scale experimental system. By adding 1.0 mg/L $K_2FeO_4$ to enhance the activity of $MeO_x$, the removal efficiencies of $COD_{Mn}$ (20.0 mg/L) and $NH_4^+$ (1.1 mg/L) were 92.5 ± 1.5% and 60.9 ± 1.4%, respectively, and the pollutants were consistently and efficiently removed for more than 90 days. The effects of the filtration rate, temperature and pH on the removal of $COD_{Mn}$ were also explored, and excessive filtration rate (over 11 m/h), lower temperature (below 9.2 °C) and pH (below 6.20) caused a significant decrease in the removal efficiency of $COD_{Mn}$. The removal of $COD_{Mn}$ was analyzed at different temperatures, which proved that the kinetics of $COD_{Mn}$ oxidation was pseudo-first order. The mature sands ($MeO_x$) from column IV were taken at different times for microscopic characterization. Scanning electron microscope (SEM) showed that some substances were formed on the surface of $MeO_x$ and the ratio of C and O elements increased significantly, and the ratio of Mn and Fe elements decreased significantly on the surface of $MeO_x$ by electron energy dispersive spectrometer (EDS). However, the elemental composition of $MeO_x$ would gradually recover to the initial state after the dosage of $Mn^{2+}$. According to X-ray photoelectron spectroscopy (XPS) analysis, the substance attached to the surface of $MeO_x$ was [(-(CH$_2$)$_4$O-)$_n$], which fell off the surface of $MeO_x$ after adding $Mn^{2+}$. Finally, the mechanism of $K_2FeO_4$-enhanced $MeO_x$ for $COD_{Mn}$ removal was proposed by the analysis of the oxidation process.

**Keywords:** $COD_{Mn}$ and $NH_4^+$; potassium ferrate; $MeO_x$; catalytic oxidation

**Citation:** Guo, Y.; Ma, B.; Yuan, S.; Zhang, Y.; Yang, J.; Zhang, R.; Liu, L. Simultaneous Removal of $COD_{Mn}$ and Ammonium from Water by Potassium Ferrate-Enhanced Iron-Manganese Co-Oxide Film. *Water* **2022**, *14*, 2651. https://doi.org/10.3390/w14172651

Academic Editors: Weiying Feng, Fang Yang and Jing Liu

Received: 30 May 2022
Accepted: 25 August 2022
Published: 28 August 2022

**Publisher's Note:** MDPI stays neutral with regard to jurisdictional claims in published maps and institutional affiliations.

**Copyright:** © 2022 by the authors. Licensee MDPI, Basel, Switzerland. This article is an open access article distributed under the terms and conditions of the Creative Commons Attribution (CC BY) license (https://creativecommons.org/licenses/by/4.0/).

## 1. Introduction

$COD_{Mn}$ and ammonium ($NH_4^+$) are the main indicators for water quality evaluation of drinking water sources in China [1]. $COD_{Mn}$ is a comprehensive index for determining the relative content of organic matter, and it is a key water pollutant index controlled by China. The excessive intake of organic matter into the human body may cause chronic poisoning and reproductive and genetic issues [2–5]. $NH_4^+$ is the main component of essential nutrients for aquatic plants and animals, but a high concentration of $NH_4^+$ can lead to eutrophication in surface water [6,7] and produce toxic disinfection byproducts in water plants [8,9]. In China, the maximum levels of $COD_{Mn}$ and $NH_4^+$ in drinking water cannot exceed 3.0 and 0.5 mg/L, respectively.

The general methods for removing $COD_{Mn}$ and $NH_4^+$ in the drinking water treatment process include an adsorption method, membrane separation technology and a biofiltration process. Activated alumina was used to adsorb $COD_{Mn}$ in water, and the removal efficiency of 4.3 mg/L $COD_{Mn}$ could reach 79.07% by reducing the hardness and chloride ions in water [10]. Green iron oxide nanoparticles synthesized on zeolite were used to remove

10 mg/L $NH_4^+$ and $PO_4^{3-}$, and the removal efficiency of $NH_4^+$ was about 56.57% [11]. The adsorption method has a high removal efficiency and simple operation, but it is difficult to guarantee the quality of the effluent after adsorption saturation, and the adsorption material needs to be replaced and regenerated regularly. The removal efficiency of 3.73 mg/L $COD_{Mn}$ could be 45.38% by an ultrafiltration-nanofiltration (UF-NF) double-membrane separation technology [12]. Guo et al. [13] combined continuous sand filtration (CSF) and ultrafiltration (UF) to treat raw water; the removal efficiencies of $NH_4^+$ and $COD_{Mn}$ exceeded 70% and 30%, respectively. Although the membrane separation technology has a good removal effect for $NH_4^+$ and $COD_{Mn}$, its operation and maintenance costs are expensive, and the membrane is easily fouled. The simultaneous removal of $NH_4^+$ and $COD_{Mn}$ could be achieved using an aerated bioactive filter with suspended filter media, and the removal efficiencies of $NH_4^+$ and $COD_{Mn}$ were 88.11% and 57.49%, respectively [14]. The influence of the filter material thickness on the zeolite-ceramic aerated biological filter was studied, and the removal efficiencies of $COD_{Mn}$ and $NH_4^+$ reached 38.62% and 93.02%, respectively [15]. The biological treatment process is less expensive to operate, but it has a long start-up period and is easily affected by low temperature [16].

In a previous study, the iron-manganese co-oxide film ($MeO_x$) with catalytic oxidation activity could be formed on the surface of the quartz sand filter material in a pilot-scale filtration system. $MeO_x$ could be used to efficiently remove $NH_4^+$, iron ($Fe^{2+}$) and manganese ($Mn^{2+}$) from groundwater and surface water sources [17,18]. However, the removal effect of $COD_{Mn}$ was very poor by $MeO_x$. As an emerging green water treatment agent, potassium ferrate ($K_2FeO_4$) has the advantages of strong oxidation and no secondary pollution, but a high dosage concentration of $K_2FeO_4$ was required when it was used to remove $COD_{Mn}$ from water [19]. Khoi et al. [20] explored the application of ferrate as the oxidant in river water purification, and the removal efficiency of $COD_{Mn}$ could reach 86.2% by adding 20 mg/L of ferrate.

In this study, the mature quartz sands with $MeO_x$ were used as the filter material in a pilot-scale filtration experimental system, and a small dose of $K_2FeO_4$ was used to enhance the catalytic oxidation activity of $MeO_x$ so that $NH_4^+$ and $COD_{Mn}$ could be removed simultaneously. The strengthening effect of $K_2FeO_4$ on $MeO_x$, the optimal dosage of $K_2FeO_4$ and the effects of different filtration rates, pH and water temperature (T) on the $COD_{Mn}$ removal process was mainly studied. Finally, some microscopic characterization techniques were used to explore the changes in the $MeO_x$ in these experiments, and the mechanism of the $COD_{Mn}$ removal process was determined.

## 2. Materials and Methods
### 2.1. Raw Water Quality and the Pilot-Scale System

The raw water was a drinking water source in Xi'an, China. As shown in Table 1, the $COD_{Mn}$ concentration and $NH_4^+$ concentration were significantly lower than the surface water quality standards, so they cannot be directly used for the experimental research. The $COD_{Mn}$ concentration and $NH_4^+$ concentration in the influent could be increased by adding glucose and ammonium chloride, respectively.

Table 1. Raw water quality.

| Index | Unit | Value | Surface Water Quality Standard Class III (GBT3838-2002) |
|---|---|---|---|
| Ammonium | mg·L$^{-1}$ | 0–0.2 | ≤1.0 |
| COD$_{Mn}$ | mg·L$^{-1}$ | 0.87–2.10 | ≤6.0 |
| Nitrate | mg·L$^{-1}$ | 3.8–4.3 | ≤10.0 |
| Manganese | mg·L$^{-1}$ | 0–0.05 | ≤0.1 |
| pH | - | 7.5–8.0 | 6.0~9.0 |
| Iron | mg·L$^{-1}$ | 0.051–0.062 | ≤0.3 |
| Temperature | °C | 14.9–26.5 | - |
| Dissolved oxygen (DO) | mg·L$^{-1}$ | 8.0–9.5 | ≥5.0 |

As can be seen in Figure 1, the pilot-scale filter system includes four identical filter columns (inner diameter = 0.1 m, height = 3.0 m), the dosing system, the water distribution system and the backwashing system. Using potassium permanganate to continuously oxidize manganese and ferrous ions from raw water have been used to form the MeO$_x$ on the surface of virgin quartz sand quickly [17,18]. There was a 30 cm support layer (70–150 mm pebbles) at the bottom of the filter column. There are seven sampling ports on one side of the filter column. Eight dosing pumps were used for dosing different chemicals, and the filtration rate was controlled by the valve. The backwashing system includes air washing and water washing, and a flow meter was set to adjust the washing intensity. When the water level reached about 2.5 m above the bed layer, or the effluent water quality deteriorated, the pilot-scale column was backwashed, and the operation method of backwashing the filter column was as in a previous study [18]. The same batch of filter media was replaced after each experiment was completed.

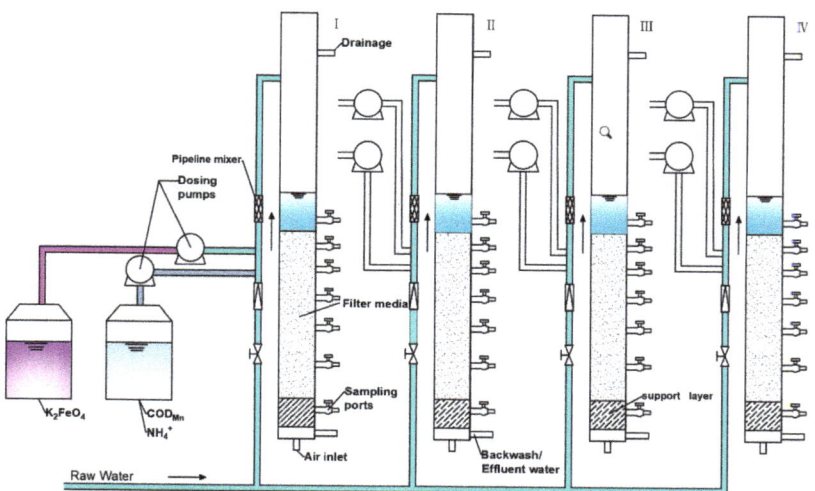

Figure 1. Schematic diagram of the pilot filter system.

## 2.2. Pollutant Removal Experiments

### 2.2.1. K$_2$FeO$_4$-Enhanced Filtration to Remove COD$_{Mn}$

Columns I, II and III were used for this experiment. The filter material was virgin quartz sands in column I, and in columns II and III, the filter material was mature sands with MeO$_x$. The K$_2$FeO$_4$ solution (0.1 mg/L, prepared from potassium ferrate) and glucose solution (20.0 mg/L COD$_{Mn}$) were dosed into the static mixer by the dosing pump. K$_2$FeO$_4$

and a glucose solution were added to columns I and II, and only the glucose solution was added to column III, as shown in Table 2. The filtration rate was 7 m/h in this experiment, and all columns were run continuously for 10 days.

Table 2. The operating conditions.

| Column | Filter Material | COD$_{Mn}$ | K$_2$FeO$_4$ |
|---|---|---|---|
| I | virgin quartz sands | 20.0 mg/L | 0.1 mg/L |
| II | mature sands | 20.0 mg/L | 0.1 mg/L |
| III | mature sands | 20.0 mg/L | 0 |

2.2.2. Simultaneous Removal of COD$_{Mn}$ and NH$_4^+$

The COD$_{Mn}$ concentration and NH$_4^+$ concentration in the influent were $20.0 \pm 0.6$ and $1.1 \pm 0.1$ mg/L, respectively, and different initial concentrations of K$_2$FeO$_4$ (about 0.1, 0.5, 1.0 and 2.0 mg/L) were added into the influent, which was used to determine the optimal dosage of K$_2$FeO$_4$. Each experimental condition was examined in triplicate. After the optimal dosage of K$_2$FeO$_4$ was determined, the experiment for the simultaneous removal of COD$_{Mn}$ and NH$_4^+$ was performed in column IV, and the experiment was run for 90 days with daily sampling. The K$_2$FeO$_4$ (0.1 mg/L) was added for the entire 90 days, and $1.0 \pm 0.1$ mg/L Mn$^{2+}$ was continuously added into the influent after day 47.

2.3. Influential Factors on the Removal of COD$_{Mn}$

Columns I, II and III were used to explore the experiment for influential factors on the removal of COD$_{Mn}$, and the COD$_{Mn}$ concentration and K$_2$FeO$_4$ concentration in the influent were $20.0 \pm 0.6$ and $0.10 \pm 0.03$ mg/L, respectively. Each condition was run for 48 h, and all samples were taken and measured the change in the COD$_{Mn}$ concentration along the filter column. The effect of K$_2$FeO$_4$ on the enhancement of MeO$_x$ to remove COD$_{Mn}$ was explored under different filtration rates, pH and T.

The filtration rate (6–11 m/h) was controlled by the flow meters in the filter column During the experiment, the water temperature was $20.0 \pm 0.5$ °C, and the pH was $8.0 \pm 0.2$ in the influent.

Hydrochloric acid (36% ($w/w$)) was used to adjust the pH value (in the range of 6.20–8.04) of the influent. The water temperature was $20.0 \pm 0.5$ °C, and the filtration rate was 7 m/h during the experiment.

The different initial temperatures (6.0–22.0 °C) of the influent were controlled by adding some ice cubes to the original water bucket. The filtration rate was maintained at 7 m/h, and the pH was $8.0 \pm 0.2$ in the influent.

2.4. Analytic Methods and Characterization Methods

The experimental reagents are glucose, potassium ferrate, sodium oxalate, potassium permanganate, ammonium chloride, mercury iodide, potassium sodium tartrate, potassium iodide, potassium periodate, potassium pyrophosphate, sodium acetate, sodium hydroxide and hydrochloric acid (36% ($w/w$)). All the above chemicals are of analytical grade. The hydrochloric acid (36% ($w/w$)) was purchased from Merck Ltd. (Beijing, China), and the rest of the chemicals were purchased from Shanghai Macklin Biochemical Co., Ltd (Shanghai, China).

The concentration of NH$_4^+$ was determined using Nessler reagent spectrophotometry, Mn$^{2+}$ concentration was monitored by potassium periodate oxidation spectrophotometry, and the COD$_{Mn}$ concentration was measured by the acid method according to the water and wastewater detection and analysis method [21]. The temperature, pH and DO were detected using a portable instrument (HACH, HQ30d, Loveland, CO, USA).

The microtopography of MeO$_x$ was characterized by scanning electron microscope (SEM) (FEI Quanta 600F, Portland, OR, USA), and the elemental composition was determined by energy-dispersive X-ray spectroscopy (EDS) (INCA Energy 350, Oxford, UK).

The binding energy of C, O and Mn were analyzed using X-ray photoelectron spectroscopy (XPS) (Thermo Scientific K-Alpha, Waltham, MA, USA), and the XPS spectra were analyzed and peak fitted by bundled software (Avantage 5.9921, Thermo Scientific, Waltham, MA, USA).

## 3. Results

### 3.1. The Removal of $COD_{Mn}$ and $NH_4^+$

#### 3.1.1. $K_2FeO_4$-Enhanced Filtration to Remove $COD_{Mn}$

The process of $K_2FeO_4$-enhanced $MeO_x$ for the removal of $COD_{Mn}$ was explored. From Figure 2 and Table 2, when 0.1 mg/L $K_2FeO_4$ was added to columns I and III, the removal efficiency of $COD_{Mn}$ in water was only $5.0 \pm 0.3\%$ by the virgin quartz sands, while the removal efficiency of $COD_{Mn}$ could reach $92.5 \pm 1.5\%$ by the mature sands ($MeO_x$). When the filter media was the same batch of mature sand in columns II and III, the removal efficiency of $COD_{Mn}$ was only $10.0 \pm 0.3\%$ without adding $K_2FeO_4$. To sum up, the presence of $K_2FeO_4$ enhanced the catalytic oxidation activity of $MeO_x$, so the removal efficiency of $COD_{Mn}$ was significantly improved.

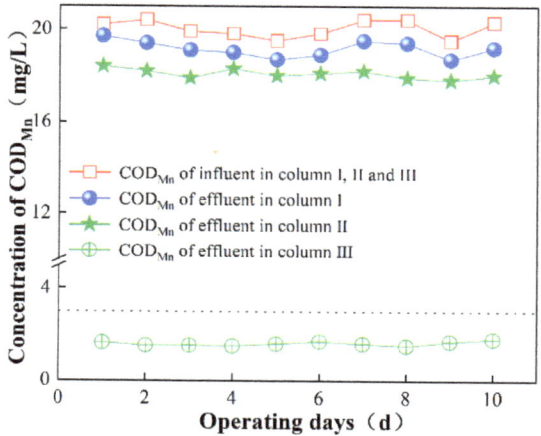

**Figure 2.** $K_2FeO_4$ enhanced the $MeO_x$ for the removal process of $COD_{Mn}$.

#### 3.1.2. Simultaneous Removal of $COD_{Mn}$ and $NH_4^+$

The optimal dosage of $K_2FeO_4$ was determined, and the results are shown in Figure 3a,b. As shown in Figure 3a,b, the $COD_{Mn}$ concentration and $NH_4^+$ concentration in the effluent gradually increased with the gradual decrease in the dosage of $K_2FeO_4$. However, only 1.0 mg/L $K_2FeO_4$ was added to the influent, and the concentration of pollutants in the effluent could meet the standard, so the optimal dosage of $K_2FeO_4$ was determined to be 1.0 mg/L.

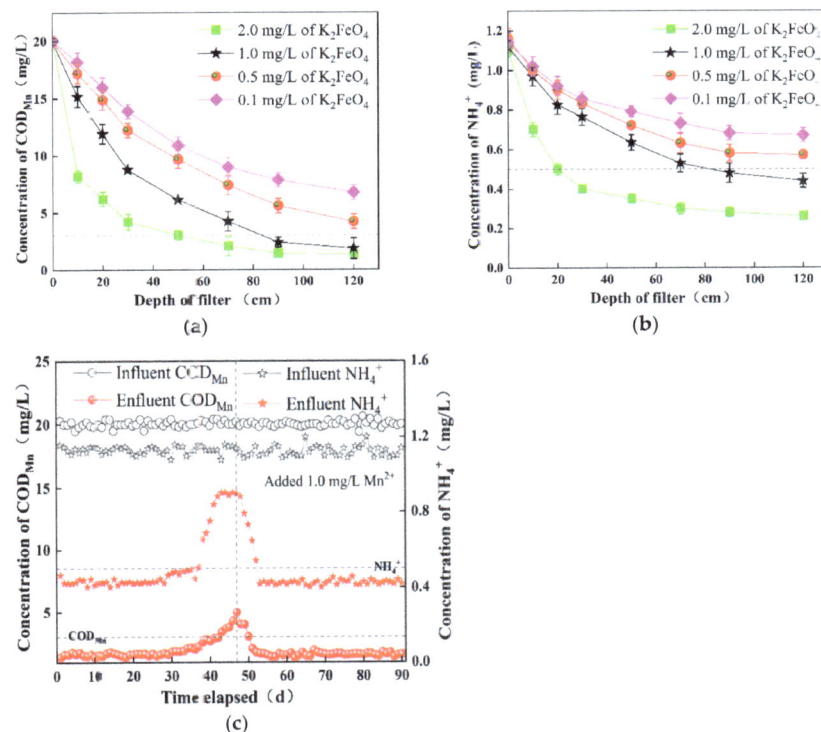

**Figure 3.** Effect of $K_2FeO_4$ concentration on the removal of (**a**) $COD_{Mn}$ and (**b**) $NH_4^+$; (**c**) The removal of $COD_{Mn}$ and $NH_4^+$ over the continuous operational period in the pilot-scale filter system.

Column IV was continuously operated for more than 90 days, and the concentrations of $COD_{Mn}$ and $NH_4^-$ in the influent and effluent are shown in Figure 3c. During the initial 30 days, the removal efficiency of $COD_{Mn}$ and $NH_4^+$ remained stable. The concentration of the pollutants in the effluent began to gradually increase when the pilot-scale system was run for 37 days. There was no $Mn^{2+}$ in the influent for a long time, and the $MeO_x$ on the surface of the filter media could not be renewed, so the activity of the MeOx gradually decreased [22]. $Mn^{2+}$ was continuously added into the influent on the 47th day, and the concentration of pollutants in the effluent gradually decreased and returned to the same level after 5 days. The recovery of the oxide film activity could be achieved by the continuous addition of $Mn^{2+}$ in the influent.

## 3.2. Influential Factors on the Removal of $COD_{Mn}$

### 3.2.1. Effect of Filtration Rate

The effect of the filtration rate on the removal of $COD_{Mn}$ is shown in Figure 4. When the filtration rate was 6.0 m/h, the $COD_{Mn}$ concentration reached the effluent standard at the 20-cm-deep filter layer. When the filtration rate increased from 6.0 to 11.0 m/h, the $COD_{Mn}$ concentration in the effluent also increased gradually. However, the removal efficiency of $COD_{Mn}$ was more than 80% even if the filtration rate reached 11.0 m/h, so the effect of the filtration rate on the removal of $COD_{Mn}$ was not obvious when the filtration rate was 6.0–11.0 m/h.

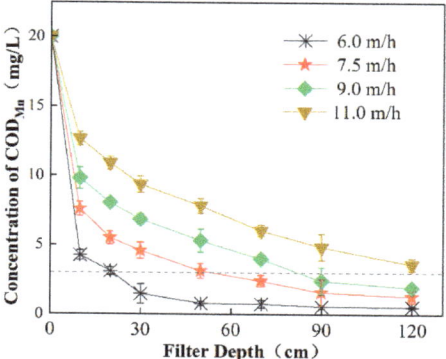

**Figure 4.** The effect of the filtration rate on the removal of $COD_{Mn}$.

3.2.2. Effect of pH

The effect of pH on the removal of $COD_{Mn}$ is shown in Figure 5. From Figure 5, when the pH value of the influent was 6.2, the removal efficiency of $COD_{Mn}$ was only 64.0 ± 3.2%. The removal efficiency of $COD_{Mn}$ increased with the increase in pH value. Considering the different reduction products of $K_2FeO_4$ at different pH [23], $Fe^{3+}$ exists in the dissolved state under acidic conditions.

$$FeO_4^{2-} + 8H^+ + 3e^- \rightarrow Fe^{3+} + 4 H_2O \quad (1)$$

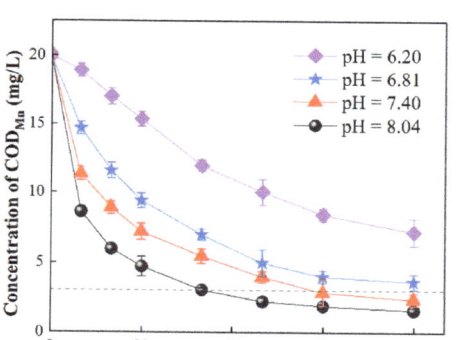

**Figure 5.** The effect of pH on the removal of $COD_{Mn}$.

Under neutral and alkaline conditions, $Fe^{3+}$ exists in the form of $Fe(OH)_3$ precipitation. Neutral condition:

$$FeO_4^{2-} + 4H^+ + 3e^- \rightarrow Fe(OH)_3 + OH \quad (2)$$

Alkaline condition:

$$FeO_4^{2-} + 4H_2O + 3e^- \rightarrow Fe(OH)_3 + 5OH \quad (3)$$

The $Fe(OH)_3$ colloid had an adsorption effect on $COD_{Mn}$ in water under alkaline conditions, which could further improve the removal efficiency of $COD_{Mn}$, so the removal efficiency of $COD_{Mn}$ was lower under acidic conditions than under alkaline conditions.

### 3.2.3. Effect of Temperature

The water temperature had a significant influence on the removal of $COD_{Mn}$, and the removal efficiency of $COD_{Mn}$ decreased with the decrease in temperature, as shown in Figure 6a. The removal efficiency of $COD_{Mn}$ was only 53.92 ± 0.82% when the temperature was reduced to 6.0 °C. Since the activity of $MeO_x$ was affected by the low temperature [18], the removal efficiency of $COD_{Mn}$ was significantly reduced.

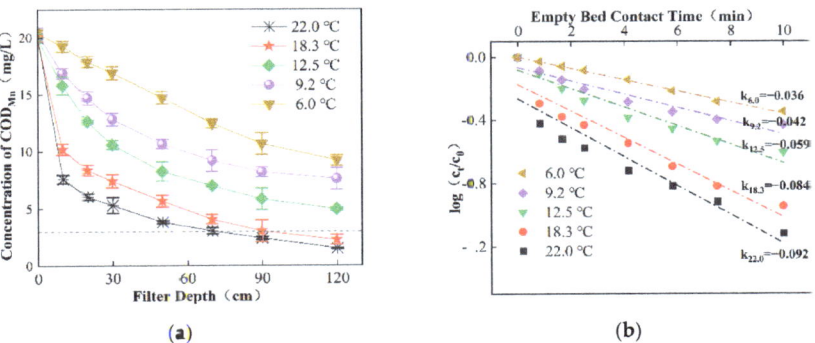

**Figure 6.** (**a**) The concentration changes of $COD_{Mn}$ along with the filter depth, (**b**) linear regression analysis of $COD_{Mn}$ depletion with the EBCT at different temperatures.

The oxidation kinetics of $COD_{Mn}$ at different temperatures are shown in Figure 6b. By maintaining the concentration of DO and pH in the influent constant, the $COD_{Mn}$ consumption rate was assumed to be pseudo-first order: $-d[COD_{Mn}]/dt = k [COD_{Mn}]$, where $k$ is the rate constant (min$^{-1}$) [24]. The plot of $\log\{[COD_{Mn}]_t/[COD_{Mn}]_0\}$ versus empty bed contact time (EBCT) were linear at all temperatures (6.0–22.0 °C), confirming that the kinetics of $COD_{Mn}$ oxidation was pseudo-first order.

### 3.3. Surface Property Variation of $MeO_x$

#### 3.3.1. The Morphology of the $MeO_x$

The filter media at different stages (the 1st, 47th and 90th day) were taken in column IV for the microscopic characterization analysis. As shown in Figure 7a,b, $MeO_x$ on the surface of quartz sand was smooth and dense, and the pore structure was relatively developed on the 1st day. From Figure 7c,d, the experimental system was continuously operated until the 47th day, part of the structure of the $MeO_x$ was broken, and the pore structure was blocked by some substances. It was due to the oxidation of organic matter by $K_2FeO_4$ to form the substances, which were attached to the surface of $MeO_x$. After adding $Mn^{2+}$ into the influent (Figure 7e,f), the surface structure and pore structure of $MeO_x$ were gradually recovered to smooth and dense, and the dosage of $Mn^{2+}$ was oxidized to form the manganese oxides, which could be used to restore the activity of the $MeO_x$.

**Figure 7.** The morphology of the oxide film on the 1st, 47th and 90th day: (**a**) 1st day filter × 100, (**b**) 1st day filter × 10,000, (**c**) 47th day filter × 100, (**d**) 47th day filter × 10,000, (**e**) 90th day filter × 100, (**f**) 90th day filter × 10,000.

### 3.3.2. Characterization of EDS

The EDS analysis results are shown in Figure S1. At the beginning of the experiment (the 1st day), the content of Mn was significantly higher than other elements on the $MeO_x$ surface. Due to the continuous dosage of $COD_{Mn}$ into the influent, the content of Mn reduced, and the proportion of C and O increased significantly on the surface of $MeO_x$. The main reason was that the organic matter was oxidized and covered on the surface of $MeO_x$. After the addition of $Mn^{2+}$, the proportions of C, Mn and O on the surface of $MeO_x$ were restored to the original state.

### 3.3.3. XPS of the Oxide Film

The XPS analysis was performed on the binding energies of C1s, O1s and Mn 3/2p, and the results are shown in Figure 8. By analyzing the binding energy of C1s, the organic matter was oxidized by $K_2FeO_4$ to form $[(-(CH_2)_4O-)_n]$ [25] on the surface of $MeO_x$; this substance is more likely caused by the addition of glucose. In addition, the Si-C content increased due to $MeO_x$ exfoliation on the oxide film surface. From Figure 8b, the Mn (2p3/2) mainly exists in the form of manganese oxides, mainly including $Mn_2O_3$ [26], MnO [27] and $Mn_3O_4$ [28]. From the binding energy of O1s, the compound form of O was gradually changed from manganese oxide to $[(-(CH_2)_4O-)_n]$ and a small amount of MnO [29] with the dosage of $COD_{Mn}$. The activity of $MeO_x$ was recovered after adding $Mn^{2+}$, and the compound form of O on the surface of $MeO_x$ is mainly C=O and part of $Mn_2O_3$, which was the intermediate product of $COD_{Mn}$ after oxidation.

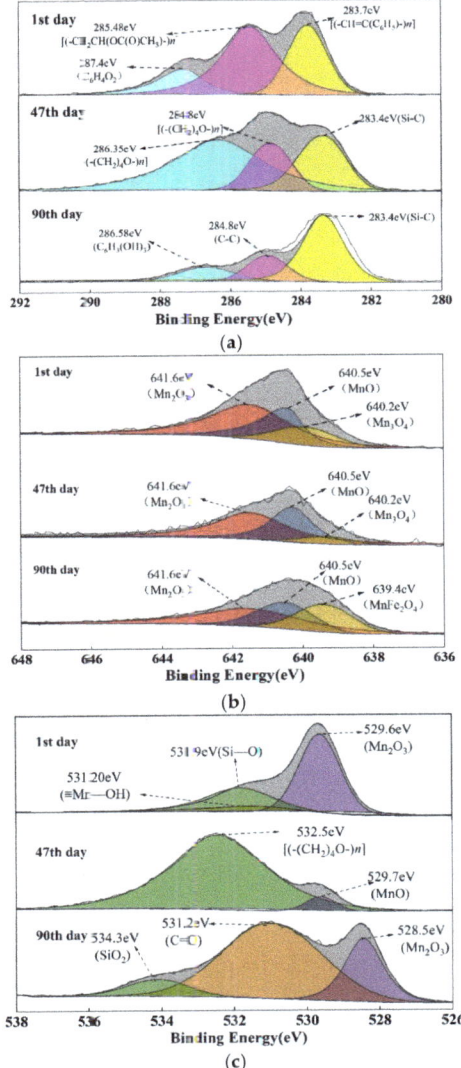

**Figure 8.** XPS energy spectra of (a) C1s, (b) Mn2p3/2 and (c) O1s with different experimental stages

### 3.4. Proposed Mechanism for $COD_{Mn}$ Removal

In previous studies, the removal mechanism of $NH_4^+$ and $Mn^{2+}$ was inferred. $NH_4^+$ could be catalytically oxidized by $MeO_x$ to $NO_3^-$ and $H^+$ [19]. $Mn^{2+}$ could be adsorbed by the surface of $MeO_x$, and a new active oxide film and some loose oxides would be generated after a series of reactions [30].

As shown in Figure 9, a schematic presentation of the removal mechanism of $COD_{Mn}$ by $K_2FeO_4$ enhanced filtration was proposed. The enhanced filtration process of $K_2FeO_4$ could be presented as three main steps: (1) Adsorption of $FeO_4^{2-}$ onto the surface of $MeO_x$; (2) organic matter (glucose molecules) were adsorbed to the surface of $[MeO_x] \cdot FeO_4^{2-}$, and the reaction occurs to generate $[(-(CH_2)_4O-)_n]$ and $[MeO_x] \cdot FeO_4^{2-}$, and $[MeO_x] \cdot FeO_4^{2-}$ was reduced to $[MeO_x] \cdot Fe^{3+}$; (3) $[MeO_x] \cdot Fe^{3+}$ was still oxidized and finally reduced to $[MeO_x] \cdot Fe(OH)_3$, and $Fe(OH)_3$ was released from $MeO_x$ after backwashing.

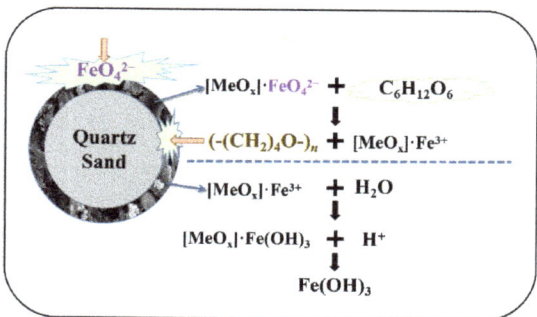

**Figure 9.** Mechanism of $K_2FeO_4$-enhanced $MeO_x$ removal of $COD_{Mn}$.

### 4. Conclusions

By adding 0.1 mg/L $K_2FeO_4$ into the influent, the removal efficiency of 20.0 mg/L $COD_{Mn}$ reached 92.5 ± 1.5% by $MeO_x$. The filtration rate of the influent was lower than 11 m/h, which had little effect on the removal of $COD_{Mn}$. The removal efficiency of $COD_{Mn}$ increased as the pH value increased from 6.20 to 8.04. Too low of a temperature (about 6.0 °C) would affect the activity of $MeO_x$, and the removal efficiency of $COD_{Mn}$ would drop to 53.92 ± 0.82%. The kinetics of $COD_{Mn}$ oxidation was pseudo-first order. The optimal dosage of $K_2FeO_4$ for the simultaneous removal of 20.0 mg/L $COD_{Mn}$ and 1.1 mg/L $NH_4^+$ was determined to be 1.0 mg/L. After the simultaneous removal of $COD_{Mn}$ and $NH_4^+$ after about 30 days, the removal efficiency of the pollutants gradually decreased. From SEM characterization, the surface of $MeO_x$ was blocked by some substances. EDS analysis found that the proportion of C and O on the surface of $MeO_x$ increased significantly, while the proportion of Mn decreased by 45.93 ± 0.64%. The surface of $MeO_x$ was found to be covered with $[(-(CH_2)_4O-)_n]$ using XPS analysis. After $Mn^{2+}$ was continuously added to the influent, the catalytic activity of $MeO_x$ was recovered after 5 days, and the efficient removal of $COD_{Mn}$ remained stable until the 90th day of continuous operation.

**Supplementary Materials:** The following supporting information can be downloaded at: https://www.mdpi.com/article/10.3390/w14172651/s1, Figure S1: The elemental composition of the filter film at different experimental stages.

**Author Contributions:** Conceptualization, Y.G. and B.M.; methodology, S.Y. and Y.Z.; formal analysis, J.Y. and L.L.; writing—original draft preparation, B.M. and R.Z.; project administration, Y.G.; funding acquisition, J.Y. and Y.G. All authors have read and agreed to the published version of the manuscript.

**Funding:** This work was supported by the Scientific Research Program Funded by Shaanxi Provincial Education Department (21JK0650), the Natural Science Basic Research Program of Shaanxi (2021JQ-688), the Scientific Research Project of Shaanxi province of China (2021GY-147) and the Graduate Scientific Innovation Fund for Xi'an Polytechnic University, China (chx2022030).

**Institutional Review Board Statement:** Not applicable.

**Informed Consent Statement:** Not applicable.

**Data Availability Statement:** Not applicable.

**Conflicts of Interest:** The authors declare no conflict of interest.

# References

1. You, Q.; Fang, N.; Liu, L.; Yang, W.; Zhang, L.; Wang, Y. Effects of land use, topography, climate and socio-economic factors on geographical variation pattern of inland surface water quality in China. *PLoS ONE* **2019**, *14*, e0217840.
2. Yu, Y.; Zhang, C.; Ding, W.; Zhang, Z.; Wang, G.G.X. Determining the performance for an integrated process of COD removal and $CO_2$ capture. *J. Clean. Prod.* **2020**, *275*, 122845.
3. Dos Santos, N.O.; Teixeira, L.A.; Zhou, Q.; Burke, G.; Campos, L.C. Fenton pre-oxidation of natural organic matter in drinking water treatment through the application of iron rails. *Environ. Technol.* **2021**, *43*, 2590–2603. [CrossRef] [PubMed]
4. Sillanpää, M.; Ncibi, M.C.; Matilainen, A. Advanced oxidation processes for the removal of natural organic matter from drinking water sources: A comprehensive review. *J. Environ. Manag.* **2018**, *208*, 56–76. [CrossRef] [PubMed]
5. Sillanpää, M.; Ncibi, M.C.; Matilainen, A.; Vepsäläinen, M. Removal of natural organic matter in drinking water treatment by coagulation: A comprehensive review. *Chemosphere* **2018**, *190*, 54–71.
6. Yadu, A.; Sahariah, B.P.; Anandkumar, J. Influence of COD/ammonia ratio on simultaneous removal of $NH_4^+$-N and COD in surface water using moving bed batch reactor. *J. Water Process Eng.* **2018**, *22*, 66–72. [CrossRef]
7. Zhang, R.; Qi, F.; Liu, C.; Zhang, Y.; Wang, Y.; Song, Z.; Kumirska, J.; Sun, D. Cyanobacteria derived taste and odor characteristics in various lakes in China: Songhua Lake, Chaohu Lake and Taihu Lake. *Ecotoxicol. Environ. Saf.* **2019**, *181*, 499–507. [CrossRef]
8. Tabassum, S. A combined treatment method of novel Mass Bio System and ion exchange for the removal of ammonia nitrogen from micro-polluted water bodies. *Chem. Eng. J.* **2019**, *378*, 122217. [CrossRef]
9. Dos Santos, P.R.; Daniel, L.A. A review: Organic matter and ammonia removal by biological activated carbon filtration for water and wastewater treatment. *Int. J. Environ. Sci. Technol.* **2020**, *17*, 591–606. [CrossRef]
10. Szatyłowicz, E.; Skoczko, I. Studies on the efficiency of groundwater treatment process with adsorption on activated alumina. *J. Ecol. Eng.* **2018**, *18*, 211–218. [CrossRef]
11. Xu, Q.; Li, W.; Ma, L.; Cao, D.; Owens, G.; Chen, Z. Simultaneous removal of ammonia and phosphate using green synthesized iron oxide nanoparticles dispersed on to zeolite. *Sci. Total Environ.* **2020**, *703*, 135002. [CrossRef] [PubMed]
12. Yuan, C. Experimental Study on UF-NF filtration purification of pipe drinking water. *J. Phys. Conf. Ser.* **2019**, *1176*, 062021. [CrossRef]
13. Guo, Y.; Bai, L.; Tang, X.; Huang, Q.; Xie, B.; Wang, T.; Wang, J.; Li, G.; Liang, H. Coupling continuous sand filtration to ultrafiltration for drinking water treatment: Improved performance and membrane fouling control. *J. Membr. Sci.* **2018**, *567*, 18–27. [CrossRef]
14. Xu, X.; Wang, J.; Li, J.; Wang, Z.; Lin, Z. Attapulgite suspension filter material for biological aerated filter to remove $COD_{Mn}$ and ammonia nitrogen in micro-polluted drinking water source. *Environ. Prot. Eng.* **2020**, *46*, 21–40.
15. Liu, J.; Xie, S.; Cheng, C.; Lou, J.; Li, S. Effect on bed material heights to the performance of ZCBAF in the treatment of micro-polluted raw water. *Appl. Mech. Mater.* **2012**, *209–211*, 2053–2057. [CrossRef]
16. Terry, L.G.; Summers, R.S. Biodegradable organic matter and rapid-rate biofilter performance: A review. *Water Res.* **2018**, *128*, 234–245. [CrossRef]
17. Cheng, Y.; Zhang, S.; Huang, T.; Li, Y. Arsenite removal from groundwater by iron–manganese oxides filter media: Behavior and mechanism. *Water Environ. Res.* **2019**, *91*, 536–545. [CrossRef]
18. Guo, Y.; Huang, T.; Wen, G.; Cao, X. The simultaneous removal of ammonium and manganese from groundwater by iron-manganese co-oxide filter film: The role of chemical catalytic oxidation for ammonium removal. *Chem. Eng. J.* **2017**, *308*, 322–329. [CrossRef]
19. Sharma, V.K.; Zboril, R.; Varma, R.S. Ferrates: Greener oxidants with multimodal action in water treatment technologies. *Acc. Chem. Res.* **2015**, *48*, 182–191. [CrossRef]
20. Tran, T.K.; Nguyen, D.H.C.; Hoang, G.P.; Nguyen, T.T.; Nguyen, N.H. Application of ferrate as coagulant and oxidant alternative for purifying Saigon river water. *VN J. Sci. Earth Environ. Sci.* **2020**, *36*, 1–7.
21. State Environmental Protection Administration. *Water and Wastewater Monitoring and Analysis Methods*, 4th ed.; China Environmental Science Press: Beijing, China, 2002; pp. 224–226.
22. Guo, Y.; Ma, B.; Huang, J.; Wang, J.; Zhang, R. The simultaneous removal of bisphenol A, manganese and ammonium from groundwater by $MeO_x$: The role of chemical catalytic oxidation for bisphenol A. *Water Supply* **2022**, *22*, 2106–2116. [CrossRef]
23. Zhang, J.; Wang, D.; Zhang, H. Oxidative degradation of emerging organic contaminants in aqueous solution by high valent manganese and iron. *Prog. Chem.* **2021**, *33*, 1201–1211.
24. Cheng, Y.; Zhang, S.; Huang, T.; Hu, F.; Gao, M.; Niu, X. Effect of alkalinity on catalytic activity of iron-manganese co-oxide in removing ammonium and manganese: Performance and mechanism. *Int. J. Environ. Res. Public Health* **2020**, *17*, 784. [CrossRef] [PubMed]

25. López, G.P.; Castner, D.G.; Ratner, B.D. XPS O1s binding energies for polymers containing hydroxyl, ether, ketone and ester groups. *Surf. Interface Anal.* **1991**, *17*, 267–272. [CrossRef]
26. Hercule, B.R. Surface spectroscopic characterization of Mn/Al$_2$O$_3$ catalysts. *J. Chem. Phys.* **1984**, *88*, 4922–4929.
27. Oku, M.; Hirokawa, K. X-ray photoelectron spectroscopy of Co$_3$O$_4$, Fe$_3$O$_4$, Mn$_3$O$_4$, and related compounds. *J. Electron Spectrosc. Relat. Phenom.* **1976**, *8*, 475–481. [CrossRef]
28. Audi, A.A.; Sherwood, P.M.A. Valence-band X-ray photoelectron spectroscopic studies of manganese and its oxides interpreted by cluster and band structure calculations. *Surf. Interface Anal.* **2002**, *33*, 274–282. [CrossRef]
29. Oku, M.; Hirokawa, K.; Ikeda, S. X-ray photoelectron spectroscopy of manganese–oxygen systems. *J. Electron Spectrosc. Relat. Phenom.* **1975**, *7*, 465–473. [CrossRef]
30. Guo, Y.; Zhang, J.; Chen, X.; Yang, J.; Huang, J.; Huang, T. Kinetics and mechanism of Mn$^{2+}$ removal from groundwater using iron-manganese co-oxide filter film. *Water Supply* **2019**, *19*, 1711–1717. [CrossRef]

MDPI
St. Alban-Anlage 66
4052 Basel
Switzerland
www.mdpi.com

*Water* Editorial Office
E-mail: water@mdpi.com
www.mdpi.com/journal/water

Disclaimer/Publisher's Note: The statements, opinions and data contained in all publications are solely those of the individual author(s) and contributor(s) and not of MDPI and/or the editor(s). MDPI and/or the editor(s) disclaim responsibility for any injury to people or property resulting from any ideas, methods, instructions or products referred to in the content.

www.ingramcontent.com/pod-product-compliance
Lightning Source LLC
LaVergne TN
LVHW070732100526
838202LV00013B/1214